Aspergillus

BIOTECHNOLOGY HANDBOOKS

Series Editors: Tony Atkinson and Roger F. Sherwood
PHLS Centre for Applied Microbiology and Research
Division of Biotechnology
Salisbury, Wiltshire, England

Volume 1 *PENICILLIUM* AND *ACREMONIUM*
Edited by John F. Peberdy

Volume 2 *BACILLUS*
Edited by Colin R. Harwood

Volume 3 CLOSTRIDIA
Edited by Nigel P. Minton and David J. Clarke

Volume 4 *SACCHAROMYCES*
Edited by Michael F. Tuite and Stephen G. Oliver

Volume 5 METHANE AND METHANOL UTILIZERS
Edited by J. Colin Murrell and Howard Dalton

Volume 6 PHOTOSYNTHETIC PROKARYOTES
Edited by Nicholas H. Mann and Noel G. Carr

Volume 7 *ASPERGILLUS*
Edited by J. E. Smith

Aspergillus

Edited by

J. E. Smith

University of Strathclyde
Glasgow, Scotland

Plenum Press • New York and London

Library of Congress Cataloging-in-Publication Data

Aspergillus / edited by J.E. Smith.
p. cm. -- (Biotechnology handbooks ; v. 7)
Includes bibliographical references and index.
ISBN 0-306-44545-X
1. Aspergillus--Biotechnology. 2. Fungi--Biotechnology.
I. Smith, John E. II. Series.
TP248.27.F86A873 1993
660'.62--dc20 93-39484
CIP

ISBN 0-306-44545-X

A Division of Plenum Publishing Corporation
233 Spring Street, New York, N.Y. 10013

Printed in the United States of America

Contributors

J. G. Anderson • Department of Bioscience and Biotechnology, University of Strathclyde, Glasgow G1 1XW, Scotland

M. Bensoussan • ENSBANA, Départment de Microbiologie-Biotechnologie, Université de Bourgogne, 21000 Dijon, France

W. M. Fogarty • Department of Industrial Microbiology, University College, Dublin 4, Ireland

P. Gervais • ENSBANA Laboratoire de Génie des Procédés Alimentaires et Biotechnologiques, Université de Bourgogne, 21000 Dijon, France

L. M. Harvey • Department of Bioscience and Biotechnology, University of Strathclyde, Glasgow G1 1XW, Scotland

J. R. Kinghorn • Plant Science Laboratory, School of Biological and Medical Science, University of St. Andrews, Fife KY16 9TH, Scotland

Z. Kozakiewicz • International Mycological Institute, Bakeham Lane, Egham, Surrey TW20 9TY, England

C. W. Lewis • Department of Bioscience and Biotechnology, University of Strathclyde, Glasgow G1 1XW, Scotland

B. McNeil • Department of Bioscience and Biotechnology, University of Strathclyde, Glasgow G1 1XW, Scotland

Robert A. Samson • Centraalbureau voor Schimmelcultures, 3740 AG Baarn, The Netherlands

D. Smith • International Mycological Institute, Bakeham Lane, Egham, Surrey TW20 9TY, England

J. E. Smith • Department of Bioscience and Biotechnology, University of Strathclyde, Glasgow G1 1XW, Scotland

S. E. Unkles • Plant Science Laboratory, School of Biological and Medical Science, University of St. Andrews, Fife KY16 9TH, Scotland

Shigeomi Ushijima • Research and Development Division, Kikkoman Corporation, 399 Noda, Noda City, Chiba Prefecture 278, Japan

Preface

The genus *Aspergillus* has a worldwide distribution and is one of the most common of all groups of fungi. They are possibly the greatest contaminants of natural and man-made organic products, and a few species can cause infections in man and animals. The aspergilli are also one of the most important mycotoxin-producing groups of fungi when growing as contaminants of cereals, oil seeds, and other foods. Not all aspergilli are viewed as troublesome contaminants, however, as several species have had their metabolic capabilities harnessed for commercial use.

The aspergilli have long been associated in the Far East with the koji stage of several food fermentations, particularly soy sauce and miso, and subsequently as a source of useful enzymes. The ability of these fungi to produce several organic acids, especially citric acid, has created major industrial complexes worldwide. Traditional methods of strain development have been extensively studied with the industrial strains, while more recently, recombinant DNA technology has been applied to the aspergilli with emphasis on heterologous protein production.

In compiling this book, I have been fortunate to have the full enthusiastic involvement of the authors, and to them I extend my very grateful thanks for mostly being on time and for producing such readable and authoritative chapters.

Collectively, we hope that our efforts will strengthen the scientific understanding of this intriguing group of filamentous fungi and further their use in the field of biotechnology.

John Smith

Glasgow, Scotland

Contents

Chapter 5

Chapter 6

Taxonomy—Current Concepts of *Aspergillus* Systematics 1

ROBERT A. SAMSON

1. INTRODUCTION

Species of *Aspergillus* belong to the first fungal organisms that were cultivated on artifical media and studied for their biochemical properties, and they are one of the most common fungi in man's environment. Since ancient times, *Aspergillus* species have been used in fermentation of food in Japan and other Asian countries, and the early discovery of their ability to produce organic acids was made at the turn of the century. It is therefore not surprising that the aspergilli used or encountered in biotechnology play a significant role.

The first description of *Aspergillus* dates from 1727, when P. A. Micheli provided a meager diagnosis and simple illustrations, but a clear depiction of the characteristics of the genus. After this first description, many taxa were described, and Thom and Church (1926) were the first to revise the genus *Aspergillus* with the acceptance of 69 species divided into 11 groups based on colony morphology and characteristics of the aspergillum. Later, Thom and Raper (1945) comprised 77 species with 10 varieties placed in 14 groups. In their monographic treatment, Raper and Fennell (1965) accepted 132 species subdivided into 18 groups. The generic and species concept was well circumscribed, and this monograph is still accepted. Since 1965, many new taxa have been described, and the validity of these species has been reviewed by Samson (1979; 1992). The genus now contains more than 200 species.

Two International Workshops on *Penicillium* and *Aspergillus* (Samson and Pitt, 1985, 1990) had a great impact on the current taxonomic

ROBERT A. SAMSON • Centraalbureau voor Schimmelcultures, 3740 AG Baarn, The Netherlands.

Aspergillus, edited by J. E. Smith. Plenum Press, New York, 1994.

schemes of *Aspergillus*. The contributions have also revealed the significance for the systematics of this important genus of a multidisciplinary approach including biochemistry, molecular biology, and serology. Recently, several molecular studies on problematic but important aspergilli have been performed, and the potential taxonomic value is apparent. An accurate, fast detection of species and isolates might be possible in the near future. This will be particularly required for *Aspergillus* strains that have undergone various modifications for specific biotechnogical applications. Often, these strains will not produce typical morphological structures any more, and recognition will be possible only at the molecular level. At this time, however, the results of molecular studies need careful evaluation with a good comparison of morphological characters.

2. CRITERIA FOR CLASSIFICATION AND IDENTIFICATION

2.1. Morphology

Morphological structures are still important characters for classification. *Aspergillus* species typically produce the aspergillum, a conidiophore with an aseptate stipe terminating in a vesicle, on which the conidiogenous cells (phialides and metulae) are borne. These phialides produce conidia in long, dry chains with different pigmentation and ornamentation. These structures can be readily seen with a light microscope equipped with an oil immersion lens. In addition to the typical structure of the aspergillum, sclerotia and fruiting bodies with Hülle cells that can be typical for the species can also be produced. Also used for species delimitation, however, are colony color and diameter, characteristics of the vegetative mycelium, exudate (droplets on colony surface), reverse, and soluble pigment (colors in the agar around the colony).

Scanning electron microscopy (SEM) has been applied in the systematics of *Aspergillus;* in particular, the ornamentation of thick-walled ascospores allows examination by SEM in the unfixed, air-dried state. Conidia and other structures are less resistant to shrinkage, and artifacts may develop during fixation and examination of conidia by SEM.

Most aspergilli can be readily identified when cultivated on Czapek and malt extract agars. Samson and Pitt (1985) recommended Czapek yeast agar (CYA) and malt extract agar (MEA) for most anamorphic species. For teleomorphic species, cornmeal agar, oatmeal agar, and CBS malt agar were recommended. For xerophilic species, Czapek agar or CYA with 20% sucrose or malt yeast extract 40% sucrose agar should be used. Cultures should be incubated for 5–7 days at 25°C for anamorphic species and 10–14 days for teleomorphic species.

Some specialized media are used for the isolation and identification of potentially aflatoxigenic aspergilli. Bothast and Fennell (1974) and Pitt *et al.* (1983) proposed Aspergillus Differential medium and Aspergillus Flavus and Parasiticus Agar, respectively, which make detection of *A. flavus* and *A. parasiticus* possible within 3 days. Another medium containing the antibiotic bleomycin distinguishes *A. parasiticus* from *A. sojae* (Klich and Mullaney, 1989). On this medium, growth of both species is somewhat reduced, but *A. sojae* produced very restricted colonies, while *A. parasiticus* colonies were at least 3 mm in diameter within 6 days. These selective media are especially appropriate for fast detection of mycotoxinogenic isolates in food commodities.

Practical laboratory guides for identification of common aspergilli that are based primarily on morphology and colony characters were published by Pitt and Hocking (1985), Samson and Van Reenen-Hoekstra (1988), Klich and Pitt (1988a), and Tzean *et al.* (1990).

2.2. Other Methods for Classification and Identification

Besides the convential characters of the colony and morphological structure, several new approaches have been investigated (Samson and Frisvad, 1991a,b; Samson, 1992).

The use of enzyme profiles has been studied for a limited number of *Aspergillus* isolates by a number of investigators (Nealson and Garber, 1967; Nasuno, 1971, 1972a,b, 1974; Kurzeja and Garber, 1973; Cruickshank and Pitt, 1990; Sugiyama and Yamatoya, 1990; Yamatoya *et al.*, 1990). In some cases, taxonomic relationships could be elucidated, but as yet this technique has not provided a reliable system for identification. However, profiles of secondary metabolites including mycotoxins have been shown to be useful in *Aspergillus* systematics. Samson *et al.* (1990) clearly distinguished nine taxa within the genus *Neosartorya* by ascospore morphology and the profile of secondary metabolites. Frisvad and Samson (1990) compared the morphological characteristics of many isolates of *A. fumigatus* and related taxa with their secondary metabolite profiles and were able to confirm the existing species delimitation. Klich and Pitt (1988b) investigated the relationship between species and secondary metabolite production in Section *Flavi* and found a good correlation between the production of aflatoxins $B_1 + B_2$, $G_1 + G_2$, and cyclopiazonic acid and the morphological characteristics of the species.

The distribution of ubiquinone systems, often in combination with electrophoretic comparison of enzymes, has been used by several Japanese researchers (Kuraishi *et al.*, 1990; Matsuda *et al.*, 1992). These chemotaxonomic approaches indicate relationships between groups, but also clarify the relatedness among important species. Matsuda *et al.* (1992) found, for example, that clinical and nonclinical isolates of *A. fumigatus* showed homogeneity, and that glucose-6-phosphate dehydrogenase and glutamate dehydrogenase were key enzymes for differentiating isolates of this pathogenic species.

Several molecular studies on *Aspergillus* species have been carried out. Kozlowski and Stepien (1982) used restriction-fragment-length polymorphisms (RFLPs) to determine the relationships among several species of the genus *Aspergillus.* Using plasmids containing cloned fragments of *A. nidulans* mitochondrial DNA (mtDNA) as probes, they found this method to be a useful aid in resolving certain taxonomic disputes in the genus *Aspergillus.* However, they also found that *A. oryzae* and *A. tamarii* were identical and that *A. awamori* and *A. niger* were not closely related to one another. Several molecular studies have been conducted on members of Section *Flavi,* which are difficult to distinguish morphologically. Kurtzman *et al.* (1986) studied relative DNA reassociation values and were unable to distinguish among the closely related members of this group and subsequently proposed to place the species in subspecific categories. Their taxonomic conclusions have received much criticism because their methods have been questioned, and the proposed names are unlikely to be accepted by researchers in the applied field. In contrast, RFLPs created with the restriction enzyme *Sma*I using total DNA consistently differentiated the morphologically similar species *A. oryzae* and *A. flavus* (Klich and Mullaney, 1987). Gomi *et al.* (1989) were also able to distinguish four species of Section *Flavi* using *Sma*I digests of total DNA. Moody and Tylen (1990a,b), using mtDNA restriction fragments, found a clear separation of the three taxa *A. flavus, A. parasiticus,* and *A. nomius* (Kurtzman *et al.,* 1987). However, when they used RFLPs, a limited correlation with geographic location of the isolates was observed. Kusters-Van Someren *et al.* (1990, 1991) used pectin lyase production and pectin lyase probes to screen restriction enzyme DNA digests and ribosomal banding patterns of chromosomal digests to study the black aspergilli. Their findings are discussed in Section 4.6.3.

Moderately repetitive sequences have recently been isolated by Girardin *et al.* (1993) from chromosomal DNA of *A. fumigatus,* and the authors demonstrated that they could be used as probes in Southern blot hybridization protocols to assess genetic relatedness of *A. fumigatus.* The hybridization patterns, with 15–20 bands of varying intensities, are stable within a single strain of *A. fumigatus* over many generations and vary between independently isolated strains. Moreover, moderately repetitive probes from *A. fumigatus* have also proven to be species-specific and do not hybridize with other *Aspergillus* species, and these probes could be ideal tools for fingerprinting monospore isolates of *A. fumigatus*.

For the fast detection of aspergilli, sensitive enzyme immunoassay systems and production of monoclonal antibodies that identify unique epitopes are being developed. Practical serological methods for the detection of heat-stable extracellular polysaccharides, an enzyme-linked immunosorbent assay (ELISA) method, and a latex agglutination method are available (Notermans and Heuvelman, 1985; Notermans *et al.*, 1986, 1988; Kamphuis *et al.*, 1989; Kamphuis and Noterman, 1992; Cousin *et al.*, 1989). A latex agglutination test for the detection of aspergilli in the serum of patients with invasive aspergillosis has now been modified for the detection of food-borne aspergilli and penicillia (Stynen *et al.*, 1992). As yet, the immunoassays are not specific for species or strains.

3. NOMENCLATURE

Pitt and Samson (1990a) have pointed out that the taxonomy of *Aspergillus* has suffered for several decades by the incorrect application of the rules of the International Code of Botanical Nomenclature (ICBN) (Greuter, 1988). Problems were caused by neglecting priority of old names and the failure to specify dried types for accepted or newly described species they accepted. Typification of *Aspergillus* names has now been solved because the majority of the *Aspergillus* names used by Raper and Fennell (1965) were typified by Samson and Gams (1985) and Kozakiewicz (1989).

Raper and Fennell (1965) subdivided *Aspergillus* into "groups," an infrageneric classification without nomenclatural status under the ICBN. To meet the requirement of the ICBN, another important change of nomenclature was made in which the group names were replaced by names of subgenera and sections by Gams *et al.* (1985).

3.1. Anamorph–Teleomorph Names

Many *Aspergillus* species have an ascomycete teleomorph (Table I), but the names used were not in accordance with the ICBN because the generic names in *Aspergillus* were retained for both teleomorph and anamorph. Samson and Gams (1985) and Kozakiewicz (1989) carried out the necessary nomenclatural changes (see also Pitt, 1989).

Table I. Ascomycetous Genera with an *Aspergillus* Anamorph

Eurotium Link:Fr., anamorphs in Section *Aspergillus*
Dichlaena Mont. & Durieu, anamorphs in Section *Aspergillus*
Neosartorya Malloch & Cain., anamorphs in Section *Fumigati*
Warcupiella Subram., anamorphs in Subgenus *Ornati*
Sclerocleista Subram., anamorphs in Subgenus *Ornati*
Hemicarpenteles Sarbhoy & Elphick, anamorphs in Subgenus *Ornati*
Emericella Berk. & Br., anamorphs in Section *Nidulantes*
Fennellia Wiley & Simmons, anamorphs in Section *Flavipedes*
Petromyces Malloch & Cain, anamorphs in Section *Circumdati*
Chaetosartorya Subram., anamorphs in Section *Cremei*

3.2. Conservation and Protection of Names

In her revision of the black aspergilli, Al-Musallam (1980) found that two older species, *A. phoenicis* (Corda) Thom 1840 and *A. ficuum* (Reichardt) Hennings 1867, accepted by Thom and Raper (1945) and Raper and Fennell (1965), were synonymous with *A. niger*. Consequently, following the rules of priority of names, the epithet *A. niger* cannot be used, a proscription that probably will not be accepted by workers in the field. Therefore, a strong case exists for conserving the name *A. niger* because it is the source of commercial production of citric and other organic acids around the world, and is clearly of major economic importance. Conservation of *A. niger* was proposed by Frisvad *et al.* (1990) and Kozakiewicz *et al.* (1992). Another example is the case for conservation of the name *A. nidulans*, a species used in a wide range of significant genetic studies. *Aspergillus nidulans* is threatened by the legitimate name for the anamorph, *A. nidulellus* Samson and Gams (1985), a name unlikely to be accepted by geneticists and biochemists.

Recently, a new approach for protecting names has been developed. Pitt and Samson (1993), on behalf of the International Commission on *Penicillium* and *Aspergillus* (ICPA) (a commission of the International Union of Microbiological Societies), proposed a draft list of species names in current use in the family Trichocomaceae (Fungi, Eurotiales). This compilation was generated *de novo* from discussions among the authors and other ICPA members. In order to protect *Aspergillus* names in current use from being threatened or displaced by names that are no longer in use and to eliminate uncertainties regarding their application, spelling, gender, and date and place or valid publication, published lists of names can be approved by an International Botanical Congress. The list of names of 205 aspergilli and 69 teleomorphs with an *Aspergillus* anamorph in current use was proposed to the XVth International

Botanical Congress in order to grant protection to listed names in current use.

4. CURRENT TAXONOMIC SCHEME

The scheme outlined in this section summarizes the subgenera and sections of the genus and the accepted species. This scheme and the species accepted differ from the "group" concept recently proposed by Kozakiewicz (1989). For a more detailed discussion and account, the reader is referred to Samson (1992).

4.1. Subgenus *Aspergillus*

4.1.1. Section *Aspergillus* (= *A. glaucus* Group Thom & Church)

This section consists of the xerophilic species with *Eurotium* teleomorphs, which are common as contaminants on food and pharmaceutical products. Strains of *Eurotium* have a limited use in the fermentation of dried fish (katsuobushi) in Japan. All species grow optimally on low water activity media (e.g., MEA or Czapek agar with additional sucrose), and most species form both anamorph and teleomorph under these conditions.

The taxonomy proposed by Pitt (1985) is based on ascospore morphology and on cultural and colony characteristics of cultures grown on Czapek yeast extract agar with 20% sucrose (CY20S). The treatment of Kozakiewicz (1989) emphasized conidial ornamentation as seen by SEM and reached different taxonomic conclusions (see also Samson, 1992).

Accepted Species

Eurotium amstelodami Mangin (anamorph: *A. vitis* Novobranova)
Eurotium appendiculatum Blaser (anamorph: *A. appendiculatus* Blaser)
Eurotium athecium (Raper & Fennell) v. Arx (anamorph: *A. atheciellus* Samson & Gams)
Eurotium carnoyi Malloch & Cain (anamorph: *A. neocarnoyi* Kozakiewicz)
Eurotium chevalieri Mangin (anamorph: *A. equitis* Samson & Gams)
Eurotium cristatum (Raper & Fennell) Malloch & Cain (anamorph: *A. cristatellus* Kozakiewicz)
Eurotium echinulatum Delacr. (anamorph: *A. brunneus* Delacr.)
Eurotium glabrum Blaser (anamorph: *A. glaber* Blaser)
Eurotium halophilicum Christensen *et al.* (anamorph: *A. halophilicus* Christensen *et al.*)

Eurotium herbariorum (Wiggers: Fr.) Link (anamorph: *A. glaucus* Link)
Eurotium intermedium Blaser (anamorph: *A. intermedius* Blaser)
Eurotium leucocarpum Hadlok & Stolk (anamorph: *A. leucocarpus* Hadlok & Stolk)
Eurotium medium Meissner (anamorph: *A. medius* Thom & Church)
Eurotium niveoglaucum (C. Thom & Raper) Malloch & Cain (anamorph: *A. glauconiveus* Samson & Gams)
Eurotium pseudoglaucum (Blochwitz) Malloch & Cain (anamorph: *A. glaucoaffinis* Samson & W. Gams)
Eurotium repens De Bary (anamorph: *A. repens* de Bary)
Eurotium rubrum König *et al.* (anamorph: *A. rubrobrunneus* Samson & Gams)
Eurotium tonophilum Ohtsuki (anamorph: *A. tonophilus* Ohtsuki)
Eurotium xerophilum Samson & Mouchacca (anamorph: *A. xerophilus* Samson & Mouchacca)

4.1.2. Section *Restricti* (= *A. restrictus* Group Raper & Fennell)

Pitt and Samson (1990b) revised the Section *Restricti* and accepted three species, *A. caesiellus, A. penicillioides,* and *A. restrictus,* which are characterized by their xerophilic growth and therefore found on substrates with a low water activity. Species belonging to this section have as yet no biotechnological significance.

4.2. Subgenus *Fumigati* Gams *et al.*

4.2.1. Section *Fumigati* (= *A. fumigatus* Group)

The taxonomy of the members of this section is important because these aspergilli play a significant role as human pathogens and as contaminants in processes at elevated temperatures. Frisvad and Samson (1990) found that isolates of *A. fumigatus, A. brevipes, A. viridinutans, A. duricaulis,* and *A. unilateralis* produced unique profiles of secondary metabolites, different from those in *A. fumigatus,* indicating that the taxa in this section are well defined.

Accepted Species

ANAMORPHIC SPECIES

A. brevipes Smith
A. duricaulis Raper & Fennell
A. fumigatus Fresenius
A. unilateralis Thrower

A. viridinutans Ducker & Thrower

TELEOMORPHIC SPECIES

Neosartorya aurata (Warcup) Malloch & Cain (anamorph: *A. igneus* Kozakiewicz)

Neosartorya aureola (Fennell & Raper) Malloch & Cain (anamorph: *A. aureoluteus* Samson & Gams)

Neosartorya fennelliae Kwon-Chung & Kim (anamorph: *A. fennelliae* Kwon-Chung & Kim)

Neosartorya fischeri Wehmer (anamorph: *A. fischeranus* Samson & Gams)

Neosartorya glabra (Fennell & Raper) Kozakiewicz (anamorph: *A. neoglaber* Kozakiewicz)

Neosartorya hiratsukae Udagawa *et al.* (anamorph: *A. hiratsukae* Udagawa *et al.*)

Neosartorya pseudofischeri Petersen [anamorph: *A. thermomutatus* (Paden) Petersen]

Neosartorya quadricincta (Vuill) Malloch & Cain (anamorph: *A. quadricingens* Kozakiewicz)

Neosartorya spathulata Takada & Udagawa (anamorph: *A. spathulatus* Takada & Udagawa)

Neosartorya spinosa (Raper & Fennell) Kozakiewicz (anamorph: *A. spinosus* Kozakiewicz)

Neosartorya stramenia (E. H. Novak & Raper) Malloch & Cain (anamorph: *A. paleaceus* Samson & Gams)

4.2.2. Section *Cervini* Gams *et al.* (= *A. cervinus* Group Raper & Fennell)

Species of this section are not common and are mostly isolated from soil. The taxa belonging to this section are: *A. cervinus, A. kanagawanesis, A. parvulus,* and *A. nutans*. *Aspergillus bisporus* has some similarity with this section, but its placement in this section should be reconsidered.

4.3. Subgenus *Ornati* Gams *et al.* (= *A. ornatus* Group Raper & Fennell)

This section is characterized by three teleomorphic genera, which indicates that the section is heterogeneous. This has been substantiated by both isoenzyme and ubiquinone data (Sugiyama and Yamatoya, 1990). *Aspergillus apicalis* B. S. Mehrotra and Basu showed a low level of relationship to any other species in its ubiquinone system, and its placement in Section *Ornati* is questionable.

Accepted Species

ANAMORPHIC SPECIES

A. apicalis B. S. Mehrotra & Basu
A. brunneo-uniseriatus Singh & Bakshi
A. raperi Stolk

TELEOMORPHIC SPECIES

Hemicarpenteles acanthosporus Udagawa & Takada (anamorph: *A. acanthosporus* Udagawa & Takada)
Hemicarpenteles paradoxus Sarbhoy & Elphick (anamorph: *A. paradoxus* Fennell & Raper)
Sclerocleista thaxteri Subramanian [anamorph: *A. citrisporus* (von Höhnel) Raper *et al.*]
Sclerocleista ornata (Raper *et al.*) Subramanian (anamorph: *A. ornatulus* Samson & Gams)
Warcupiella spinulosa (Warcup) Subramanian (anamorph: *A. warcupii* Samson & Gams)

4.4. Subgenus *Clavati* Gams *et al.*

4.4.1. Section *Clavati* (= *A. clavatus* Group Thom & Church)

The section includes four morphologically similar species, *A. giganteus, A. clavatus, A. clavatonanica,* and *A. longivesica. Aspergillus clavatus* is an important mycotoxin producer [patulin, cytochalasin E, and tryptoquivalins (Frisvad and Samson, 1991a,b)] and can play an important role as a comtaminant during food fermentations.

4.5. Subgenus *Nidulantes*

4.5.1. Section *Nidulantes* (= *A. nidulans* Group Thom & Church)

The Section *Nidulantes* contains many species with the teleomorphic genus *Emericella.* Except for *A. nidulans,* which has been extensively used in genetic studies, the other species often found in soil have no biotechnological application. The various *Emericella* species are characterized primarily by the shape and ornamentation of the ascospores. Two species of *Emericella, E. similis* and *E. desertorum,* have all the characters of the ascomycete genus, but they do not produce an *Aspergillus* anamorph. The genus *Fennellia* is reminiscent of *Emericella,* but differs by the ascoma initials, hyaline to pale yellow ascospores, bright yellow ascomata, and anamorphs belonging to Section *Flavipedes.*

Accepted Species

Anamorphic Species

A. aeneus Sappa
A. aureolatus Muntanola-Cvetkoviç & Bata
A. crustosus Raper & Fennell
A. eburneo-cremeus Sappa
A. multicolor Sappa
A. recurvatus Raper & Fennell
A. subsessilis Raper & Fennell

Teleomorphic Species

Emericella acristata (Fennell & Raper) Horie (anamorph: *A. nidulans* var. *acristatus* Fennell & Raper)
Emericella astellata (Fennell & Raper) Horie (anamorph: *A. variecolor* var. *astellatus* Fennell & Raper)
Emericella aurantio-brunnea (Atkins *et al.*) Malloch & Cain [anamorph: *A. aurantiobrunneus* (Atkins *et al.*) Raper & Fennell]
Emericella bicolor Christensen & States (anamorph: *A. bicolor* Christensen & States)
Emericella corrugata Udagawa & Horie (anamorph: *A. corrugata* Udagawa & Horie)
Emericella dentata (Sandhu & Sandhu) Horie (anamorph: *A. nidulans* var. *dentatus* Sandhu & Sandhu)
Emericella desertorum Samson & Mouchacca (no known *Aspergillus* anamorph)
Emericella echinulata (Fennell & Raper) Horie (anamorph: *A. nidulans* var. *echinulatus* Fennell & Raper)
Emericella falconensis Horie *et al.* (anamorph: *A. falconensis* Horie *et al.*)
Emericella foeniculicola Udagawa (anamorph: *A. foeniculicola* Udagawa)
Emericella foveolata Horie (anamorph: *A. foveolata* Horie)
Emericella fruticulosa (Raper & Fennell) Malloch & Cain (anamorph: *A. fruticans* Samson & Gams)
Emericella heterothallica (Kwon *et al.*) Raper & Fennell (anamorph: *A. compatibilis* Samson & Gams)
Emericella navahoensis Christensen & States (anamorph: *A. navahoensis* Christensen & States)
Emericella nidulans (Eidam) Vuill. [anamorph *A. nidulans* (Eidam) Wint.]
Emericella nidulans (Eidam) Vuill. var. *lata* (Thom & Raper) Subramanian (anamorph: *A. nidulans* var. *latus* Thom & Raper)
Emericella parvathecia (Raper & Fennell) Malloch & Cain (anamorph: *A. microthecius* Samson & Gams)

Emericella purpurea Samson & Mouchacca (anamorph: *A. purpurea* Samson & Mouchacca)
Emericella rugulosa (C. Thom & Raper) C. R. Benjamin (anamorph: *A. rugulovalvus* Samson & Gams)
Emericella similis Horie (no known *Aspergillus* anamorph)
Emericella spectabilis Christensen (anamorph: *A. spectabilis* Christensen)
Emericella striata (Rai *et al.*) Malloch & Cain (anamorph: *A. striatulus* Samson & Gams)
Emericella sublata Horie (anamorph: *A. sublata* Horie)
Emericella quadrilineata (C. Thom & Raper) C. R. Benjamin (anamorph: *A. tetrazonus* Samson & Gams)
Emericella undulata Kong & Qi (anamorph: *A. undulata* Kong & Qi)
Emericella unguis Malloch & Cain [anamorph: *A. unguis* (Emile-Weil & Gaudin) Thom & Raper]
Emericella variecolor Berkeley & Broome (anamorph: *A. stellifer* Samson & Gams)
Emericella violacea (Fennell & Raper) Malloch & Cain (anamorph: *A. violaceobrunneus* Samson & Gams)

4.5.2. Section *Versicolores* Gams *et al.* (= *A. versicolor* Group Thom & Church)

Accepted Species

A. alahabadii Mehrotra & Agnihotri
A. ambiguus Sappa
A. amylovorus Panasenko ex Samson
A. arenarius Raper & Fennell
A. asperescens Stolk
A. caespitosus Raper & Thom
A. crystallinus Kwon & Fennell
A. elongatus Rai & Agarwal
A. floriformis Samson & Mouchacca
A. granulosus Raper & Thom
A. janus Raper & Thom
A. janus var. *brevis* Raper & Thom
A. lucknowensis Rai *et al.*
A. malodoratus Kwon & Fennell
A. microcysticus Sappa
A. peyronelii Sappa
A. protuberus Muntanola-Cvetkovic
A. pseudodeflectus Samson & Mouchacca
A. pulvinus Kwon & Fennell
A. silvaticus Fennell & Raper

A. *speluneus* Raper & Fennell
A. *sydowii* (Bain. & Sart.) Thom & Church
A. *varians* Wehmer
A. *versicolor* (Vuill.) Tiraboschi

4.5.3. Section *Usti* Gams *et al.* (= *A. ustus* Group Thom & Raper)

This section has remained unchanged since Raper and Fennell (1965) accepted A. *ustus,* A. *puniceus,* A. *panamensis,* A. *conjunctus,* and A. *deflectus.*

4.5.4. Section *Terrei* Gams *et al.* (= *A. terreus* Group Thom & Raper)

Aspergillus terreus has been used in several biotechnological applications, and isolates are for the most part readily recognizable. In some cases, isolates may be confused with brown-color mutants of A. *fumigatus.* This section has remained unchanged as proposed by Raper and Fennell (1965) and includes A. *terreus* Thom and its two varieties, A. *terreus* var. *africanus* Fennell & Raper and A. *terreus* var. *aureus* Thom & Raper.

4.5.5. Section *Flavipedes* Gams *et al.* (= *A. flavipes* Group Thom & Church)

This small section comprises species that are found in soil and dung and are easily recognizable. The teleomorphic state belongs to *Fennellia,* a genus similar to *Emericella* of Section *Nidulantes.*

Accepted Species

ANAMORPHIC SPECIES

A. *carneus* (v. Tiegh.) Blochwitz

TELEOMORPHIC SPECIES

Fennellia flavipes Wiley & Simmons [anamorph: A. *flavipes* (Bain. & Sart.) Thom & Church]
Fennellia monodii Locquin-Linard (no known *Aspergillus* anamorph)
Fennellia nivea (Wiley & Simmons) Samson (anamorph: A. *niveus* Blochwitz)

4.6. Subgenus *Circumdati* Gams *et al.*

4.6.1. Section *Wentii* Gams *et al.* (= *A. wentii* Group Thom & Raper)

Accepted Species

A. *anthodesmis* Bartoli & Maggi
A. *terricola* Marchal

A. terricola var. *americana* Marchal
A. terricola var. *indicus* (Mehrotra & Agnihotri) Raper & Fennell
A. wentii Wehmer

4.6.2. Section *Flavi* Gams *et al.* (*A. flavus* Group Raper & Fennell)

Members of Section *Flavi* are important because of their prominent role in biotechnology, but also because some species can produce mycotoxins. Taxonomy based purely on morphological criteria showed that isolates are often difficult to recognize, and therefore several taxonomic studies have utilized nonconvential approaches. However, several of these studies obtained various results with different taxonomic conclusions. For example, on the basis of a high degree of DNA complementarity, Kurtzman *et al.* (1986) reduced the aflatoxin-producing species *A. parasiticus* to a subspecies of *A. flavus*. *Aspergillus oryzae* and *A. sojae*, used in food fermentation, were also reduced to varieties of the same species. Using electrophoretic comparison of enzymes and ubiquinone systems, Yamatoya *et al.* (1990) found that *A. flavus, A. oryzae, A. parasiticus,* and *A. sojae* could be accommodated in two species, *A. flavus* and *A. parasiticus.*

On the contrary, Liljegren *et al.* (1988) and Cruickshank and Pitt (1990) showed that *A. flavus, A. parasiticus, A. tamarii,* and *A. nomius* can be clearly differentiated on the basis of enzyme electrophoretic patterns. Also, Klich and Pitt (1988b) differentiated isolates of *A. flavus* from those of *A. parasiticus* by conidial ornamentation. In addition, they found that isolates of *A. flavus* produce aflatoxin B_1 and B_2 or cyclopiazonic acid, or both, while *A. parasiticus* produces aflatoxin G_1 and G_2 as well as B1 and B_2, but never cyclopiazonic acid. Other studies have shown that isolates of *A. flavus, A. parasiticus, A. nomius,* and *A. sojae* consistently produce aspergillic acid, while isolates of *A. oryzae* never produce this toxin (Samson and Frisvad 1991a). In a recent study, S. Peterson (unpublished data) found that *A. tamarii,* another species important for biotechnology, was identical with *A. flavofurcatis* when nuclear DNA complementarities of the type cultures were compared. Also, *A. terricola,* a species placed in Section *Wentii,* was closely related to *A. tamarii.* Both species belong in the section *Flavi* and not in a separate "*A. tamarii*-group," as proposed by Kozakiewicz (1989).

Accepted Species

A. avenaceus Smith
A. clavato-flavus Raper & Fennell
A. flavus Link
A. leporis States & M. Christensen

A. nomius Kurtzman *et al.*
A. oryzae (Ahlb.) Cohn
A. parasiticus Speare
A. sojae Sakaguchi & Yamada
A. subolivaceus Raper & Fennell
A. tamarii Kita (= *A. flavo-furcatis* Batista & Maia)
A. zonatus (Kwon & Fennell) Raper & Fennell

4.6.3. Section *Nigri* Gams *et al.* (= *A. niger* Group Thom & Church)

The black aspergilli have always been an important group in biotechnology, and their taxonomy was based primarily on morphological criteria. The systematics of this section has been problematic, however, due to the lack of relevant type material for many described taxa and descriptions of species representing deviant forms of industrial strains. Of the many species described, Raper and Fennell (1965) reduced the number of species to 12. The taxonomic distances between representative isolates of these taxa were assessed by Al-Musallam (1980), using cluster analysis involving all available morphological parameters, after both equal and iterative weighting of characters. She recognized five readily distinguishable species and an *A. niger* aggregate subdivided into seven varieties. Kusters-Van Someren *et al.* (1990) separated the black aspergilli into *A. japonicus, A. carbonarius, A. heteromorphus, A. ellipticus,* and an *A. niger* aggregate using pectin lyase production and a pectin lyase probe to screen RFLPs. In a further study of the *A. niger* aggregate, Kusters-Van Someren *et al.* (1991) studied ribosomal DNA banding patterns supported by Southern blots using several pectin lyase genes isolated from an *A. niger* isolate. The isolates, including many of the (neo)types, could be divided into two groups centered around *A. niger* and *A. tubigensis*. Because heterokaryon formation between the isolates of both groups was not possible, it was suggested that the two groups represent two different species. The taxonomies as proposed by Raper and Fennell (1965), Al-Musallam (1980), and Kusters-Van Someren *et al.* (1990, 1991) are summarized in Table II.

E. J. Mullaney and M. A. Klich (unpublished data) verified the proposed *A. niger*/*A. tubigensis* division using an *A. niger* phytase and an *A. niger* acid phosphatase gene as probes. They also found that the RFLP pattern was not uniform for isolates of *A. ficuum,* another species of the Section *Nigri* and one not included in the investigations of Kusters-Van Someren *et al.* (1991). Some of these *A. ficuum* isolates have a pattern distinct from that reported for either *A. niger* or *A. tubigensis*. Also, their isolates of *A. cinnamoneus* NRRL 348 and *A. schiemanni* NRRL 361 appear

Table II. Classification of Black Aspergilli According to Various Workers

Raper and Fennell (1965)	Al-Musallam (1980)	Kusters–Van Someren *et al.* (1990, 1991)
A. aculeatus Iizuka	*A. japonicus* var. *aculeatus*	*A. japonicus*
A. awamori Nakazawa	*A. niger* var. *awamori*	*A. niger*
A. carbonarius (Bain.) Thom	*A. carbonarius*	*A. carbonarius*
A. ellipticus Raper & Fennell	*A. ellipticus*	*A. ellipticus*
A. ficuum (Reich.) Hennings	*A. niger* var. *niger*	
A. foetidus (Naka.) Thom & Raper	*A. foetidus*	*A. niger*
A. foetidus var. *acidus* Naka *et al.*	*A. foetidus*	*A. niger*
A. foetidus var. *pallidus* Naka *et al.*	*A. foetidus*	*A. niger*
	A. helicothrix Al-Musallam	
A. heteromorphus Batista & Maia	*A. heteromorphus*	*A. heteromorphus*
A. japonicus Saito	*A. japonicus* var. *japonicus*	*A. japonicus*
A. niger van Tieghem	*A. niger* var. *niger*	*A. niger*
	A. niger var. *intermedius*	
	A. niger var. *nanus*	
	A. niger var. *niger* f. *hennebergii*	
A. phoenicis (Cda.) Thom	*A. niger* var. *phoenicis*	*A. niger*
A. pulverulentus (McAlp.) Thom	*A. niger* var. *phoenicis* f. *pulverulentus*	*A. niger*
A. tubingensis (Schöber) Mosseray	*A. niger* var. *niger*	*A. tubingensis*

to yield the *A. niger* RFLP pattern instead of the *A. tubigensis* pattern when digested with *Sma*I. The primary interest of these authors was to determine how widespread and similar the phytase gene and acid phosphatase gene as probes are in Section *Nigri*. If they are common to this group, then the transfer of a modified phytase gene into one of the industrial strains in this group may further enhance the enzyme yield.

S. Peterson (unpublished data) performed nuclear DNA reassociations and segregated two major groups that were 40–50% related to each other. One group was composed mostly of *A. niger*, *A. awamori*, and *A. ficuum* strains; the second group contained isolates that included mostly strains of *A. phoenicis*, *A. pulverulentus*, and *A. tubingensis*. These molecular studies show that morphological classification does not reflect the genetic relatedness of strains. Although separation is clear at the molecular level, the morphological differences cannot be observed microscopically, and identification will remain problematic.

4.6.4. Section *Circumdati* (= *A. ochraceus* Group Thom & Church)

Although Christensen (1982) has revised and provided a synoptic key to the taxa of this section, its taxonomy still requires further study, particularly because the species concepts of the mycotoxinogenic taxa is not clear. Samson and Gams (1985) followed Subramanian (1971) in using the name *A. alutaceus* Berk. & Curt. for *A. ochraceus,* but this nomenclature has been questioned by Klich and Pitt (1988a). Subramanian (1971) proposed the new name *A. fresenii* based on a herbarium specimen and because the name *A. sulphureus* already existed. Until the species concept of this section has been clarified, the name *A. sulphureus* should be retained. In this section, two *Aspergillus* species (*A. alliaceus* and *A. albertensis*) have a *Petromyces* teleomorph.

Accepted Species

Anamorphic Species

A. auricomus (Guéguen) Saito
A. bridgeri M. Christensen
A. campestris M. Christensen
A. dimorphicus Mehrotra & Prasad
A. elegans Gasperini
A. insulicola Montemayor & Santiago
A. lanosus Kamal & Bhargava
A. melleus Yukawa
A. ochraceoroseus Bartoli & Maggi
A. ochraceus Wilhelm
A. ostianus Wehmer
A. petrakii Vörös
A. robustus M. Christensen & Raper
A. sclerotiorum Huber
A. sepultus Tuthill & M. Christensen
A. sulphureus (Fres.) Thom & Church

Teleomorphic Species

Petromyces albertensis Tewari (anamorph: *A. albertensis* Tewari)
Petromyces alliaceus Malloch & Cain (anamorph: *A. alliaceus* Thom & Church)

4.6.5. Section *Candidi* Gams *et al.* (= *A. candidus* Group Thom & Raper)

This section consists of only the common species *A. candidus,* mostly occurring as a contaminant of food.

4.6.6. Section *Cremei* Gams *et al.* (= *A. cremeus* Group Raper & Fennell)

Accepted Species

ANAMORPHIC SPECIES

A. flaschentraegeri Stolk
A. itaconicus Kinoshita
A. ivoriensis Bartoli & Maggi

TELEOMORPHIC SPECIES

Chaetosartorya chrysella (Kwon & Fennell) Subramanian (anamorph: *A. chryseidis* Samson & Gams)
Chaetosartorya cremea (Kwon & Fennell) Subramanian (anamorph: *A. cremeoflavus* Samson & Gams)
Chaetosartorya stromatoides Wiley & Simmons (anamorph: *A. stromatoides* (Raper & Fennell)

4.6.7. Section *Sparsi* Gams *et al.* (= *A. sparsus* Group Raper & Fennell)

This section includes five species: *A. sparsus, A. biplanus, A. diversus, A. funiculosus,* and *A. gorakhpurensis* Kamal & Bhargava.

4.7. Subgenus *Stilbothamnium* (Hennings) Samson & Seifert

This subgenus was established by Samson and Seifert (1985) for *Aspergillus* species with prominent synnematous taxa such as *A. togoensis* (= *Stilbothaminium nudipes*) (Roquebert and Nicot, 1985). Species that are (sub)tropical occur on seeds and are similar to members of Section *Flavi* and could be significant mycotoxin producers. The subgenus now contains five species: *A. amazonenses* (Henn) Samson & Seifert, *A. dybowskii* (Pat.) Samson & Seifert, *A. erythrocephalus* Berk. & Seifert, *A. togoensis* (Henn.) Samson & Seifert, and *A. vitellinus* (Ridley) Samson & Seifert.

ACKNOWLEDGMENTS. The author thanks Drs. J. P. Latgé, Helene Girardin, S. Petersen, Maren Klich, and Ed Mullaney for including some new and partly unpublished data in this chapter.

REFERENCES

Al-Musallam, A., 1980, *Revision of the Black Aspergillus Species,* Dissertation, University of Utrecht, Utrecht, The Netherlands.

Bothast, R. J., and Fennell, D. I., 1974, A medium for rapid identification and enumeration of *Aspergillus flavus* and related organisms, *Mycologia* **66:**365–369.

Christensen, M., 1982, The *Aspergillus ochraceus* group: Two new species from Western soils and synoptic key, *Mycologia* **74:**210–225.

Cousin, M. A., Notermans, S., Hoogerhout, P., and Boom, J. H. Van, 1989, Detection of beta-galactofuranosidase production by *Penicillium* and *Aspergillus* species using 4-nitrophenyl beta-D-galactofuranoside, *J. Appl. Bacteriol.* **66:**311–317.

Cruickshank, R. H., and Pitt, J. I., 1990, Isoenzyme patterns in *Aspergillus flavus* and closely related taxa, in: *Modern Concepts in Penicillium and Aspergillus Classification* (R. A. Samson and J. I. Pitt, eds.), Plenum Press, New York and London, pp. 259–264.

Frisvad, J. C., and Samson, R. A., 1990, Chemotaxonomy and morphology of *Aspergillus fumigatus* and related taxa, in: *Modern Concepts in Penicillium and Aspergillus Classification* (R. A. Samson and J. I. Pitt, eds.), Plenum Press, New York and London, pp. 201–208.

Frisvad, J. C., and Samson, R. A., 1991a, Filamentous fungi in foods and feeds: Ecology, spoilage and mycotoxin production, in: *Handbook of Applied Mycology,* Vol. 3, *Foods and Feeds* (D. K. Arora, K. G. Mukerji, and E. H. Marth, eds.), Marcel Dekker, New York, pp. 31–68.

Frisvad, J. C., and Samson, R. A., 1991b, Mycotoxins produced by species of *Penicillium* and *Aspergillus* occurring in cereals, in: *Cereal Grain, Mycotoxin, Fungi and Quality in Drying and Storage* (J. Chelkowski, ed.), Elsevier, Amsterdam, pp. 441–476.

Frisvad, J. C., Hawksworth, D. L., Kozakiewicz, Z., Pitt, J. I., Samson, R. A., and Stolk, A. C., 1990, Proposal to conserve important species names in *Aspergillus* and *Penicillium,* in: *Modern Concepts in Penicillium and Aspergillus Systematics* (R. A. Samson and J. I. Pitt, eds.), Plenum Press, New York and London, pp. 83–90.

Gams, W., Christensen, M., Onions, A. H. S., Pitt, J. I., and Samson, R. A., 1985, Intrageneric taxa of *Aspergillus,* in: *Advances in Penicillium and Aspergillus Systematics* (R. A. Samson and J. I. Pitt, eds.), New York and London, Plenum Press, pp. 55–62.

Girardin, H., Morrow, B., Shrikantah, S., Latge, J. P., and Soll, D., 1993, The development of probes for DNA fingerprinting of *Aspergillus fumigatus* strains, *J. Clin. Microbiol.* (in press).

Gomi, K., Tanaka, A., Iimvra, Y., and Takahashi, K., 1989, Rapid differentiation of four related species of koji molds by agarose gel electrophoresis of genomic DNA digested units *Sma*I restriction enzyme, *J. Gen. Appl. Microbiol.* **35:**225–232.

Greuter, W., 1988, International Code of Botanical Nomenclature Adopted by the Fourteenth International Botanical Congress, Berlin, July-August, 1987, Koeltz Scientific Books, Konigstein, West Germany.

Kamphuis, H., and Notermans, S., 1992, Development of a technique for immunological detection of fungi, in: *Modern Methods in Food Mycology* (R. A. Samson, A. C. Hocking, J. I. Pitt, and A. D. King, eds.), Elsevier, Amsterdam, pp. 197–203.

Kamphuis, H. J., Notermans, S., Veeneman, G. H., Boom, J. H. Van, and Rombouts, F. M., 1989, A rapid and reliable method for the detection of moulds in foods: Using the latex agglutination assay, *J. Food Prot.* **52:**244–247.

Klich, M. A., and Mullaney, E. J., 1987, DNA restriction enzyme fragment polymorphism as a tool for rapid differentiation of *Aspergillus flavus* from *Aspergillus oryzae, Exp. Mycol.* **11:**170–175.

Klich, M. A., and Mullaney, E. J., 1989, Use of bleomycin-containing medium to distinguish *Aspergillus parasiticus* from *A. sojae, Mycologia* **81:**159–160.

Klich, M. A., and Pitt, J. I., 1988a, *A Laboratory Guide to Common Aspergillus Species and Their Telemorphs,* CSIRO Division of Food Processing, North Ryde, New South Wales.

Klich, M. A., and Pitt, J. I., 1988b, Differentiation of *Aspergillus flavus* from *A. parasiticus* and other closely related species, *Trans. Brit. Mycol. Soc.* **91:**99–108.

Kozakiewicz, Z., 1989, *Aspergillus* species on stored products, *Mycol. Papers* **161:**1–188.

Kozakiewicz, Z., Frisvad, J. C., Hawksworth, D. L., Pitt, J. L., Samson, R. A., and Stolk, A. C., 1992, Proposal for nomina conservanda and rejicienda in *Aspergillus* and *Penicillium, Taxon* **41:**109–113.

Kozlowski, M., and Stepien, P. R., 1982, Restriction enzyme analysis of mitochondrial DNA of members of the genus *Aspergillus* as an aid in taxonomy, *J. Gen. Microbiol.* **128:**471–476.

Kuraishi, H., Itoh, M., Tsuzaki, N., Katayama, Y., Yokoyama, T., and Sugiyama, J., 1990, Ubiquinone system as a taxonomic tool in *Aspergillus* and its teleomorphs, in: *Modern Concepts in Penicillium and Aspergillus Classification* (R. A. Samson and J. I. Pitt, eds.), Plenum Press, New York and London, pp. 407–420.

Kurtzman, C. P., Smiley, M. J., Robnett, C. J., and Wicklow, D. T., 1986, DNA relatedness among wild and domesticated species in the *Aspergillus flavus* group, *Mycologia* **78:**955–959.

Kurtzman, C. P., Horn, B. W., and Hesseltine, C. W., 1987, *Aspergillus nomius,* a new aflatoxin-producing species related to *Aspergillus flavus* and *Aspergillus tamarii, Anton. Leewen.* **53:**147–158.

Kurzeja, K. C., and Garber, E. D., 1973, A genetic study of electrophoretically variant extracellular amylolytic enzymes of wild-type strains of *Aspergillus nidulans, Can. J. Gen. Cytol.* **15:**275–287.

Kusters-Van Someren, M., Kester, H. C. M., Samson, R. A., and Visser, J., 1990, Variation in pectinolytic enzymes of the black aspergilli: A biochemical and genetic approach, in: *Modern Concepts in Penicillium and Aspergillus Classification* (R. A. Samson and J. I. Pitt, eds.), Plenum Press, New York and London, pp. 321–334.

Kusters-Van Someren, M., Samson, R. A., and Visser, J., 1991, The use of RFLP analysis in classification of the black aspergilli: Reinterpretation of the *Aspergillus niger* aggregate, *Curr. Gen.* **19:**21–261.

Liljegren, K., Svendsen, A., and Frisvad, J. C., 1988, Mycotoxin and exoenzyme production by members of *Aspergillus* section *Flavi:* An integrated taxonomic approach to their classification, *Proc. Jpn. Assoc. Mycotoxicol. Suppl.* **1:**35–36.

Matsuda, H., Kohno, S., Maesaki, S., Yamada, H., Koga, H., Tamura, M., Kuraishi, H., and Sugiyama, J., 1992, Application of ubiquinone systems and electrophoretic comparison of enzymes to identification of clinical isolates of *Aspergillus fumigatus* and several other species of *Aspergillus, J. Clin. Microbiol.* **3:**1999–2005.

Moody, S. F., and Tylen, B. M., 1990a, Restriction enzyme and mitochondrial DNA of the *Aspergillus flavus* group, *Aspergillus flavus, Aspergillus parasiticus* and *Aspergillus nomius, Appl. Environ. Microbiol.* **56:**2441–2452.

Moody, S. F., and Tylen, B. M., 1990b, Use of nuclear DNA restriction length polymorphisms to analyse the diversity of the *Aspergillus flavus* group, *Aspergillus flavus, Aspergillus parasiticus* and *Aspergillus nomius, Appl. Environ. Microbiol.* **56:**2453–2461.

Nasuno, S., 1971, Polyacrylamide gel disc electrophoresis of alkaline proteinases from *Aspergillus* species, *Agric. Biol. Chem.* **35:**1147–1150.

Nasuno, S., 1972a, Differentiation of *Aspergillus sojae* from *Aspergillus oryzae* by polyacrylamide gel disc electrophoresis, *J. Gen. Microbiol.* **71:**29–33.

Nasuno, S., 1972b, Electrophoretic studies of alkaline proteinases from strains of *Aspergillus flavus* group, *Agric. Biol. Chem.* **36:**684–689.

Nasuno, S., 1974, Further evidence on differentiation of *Aspergillus sojae* from *Aspergillys oryzae* by electrophoretic patterns of cellulase, pectin-lyase, and acid proteinase, *Can. J. Microbiol.* **20:**413–416.

Nealson, K. H., and Garber, E. D., 1967, An electrophoretic survey of esterases, phosphatases, and leucine amino-peptidases in mycelial extracts of species of *Aspergillus, Mycologia* **59:**330–336.

Notermans, S., and Heuvelman, C. J., 1985, Immunological detection of moulds in food by using the enzyme-linked immunosorbent assay (ELISA): Preparation of antigens, *Int. J. Food Microbiol.* **2:**247–258.

Notermans, S., Heuvelman, C. J., Beumer, R. R., and Maas, R., 1986, Immunological detection of moulds in food: Relation between antigen production and growth, *Int. J. Food Microbiol.* **3:**253–261.

Notermans, S., Veeneman, G. H. J., Zuylen, C. W. E. M. Van, Hoogerhout, P., and Boom, J. H. Van, 1988, (1-5)-Linked beta-D-galactofuranosides are immunodominant in extracellular polysaccharides of *Penicillium* and *Aspergillus* species, *Mol. Immunol.* **25:**975–979.

Pitt, J. I., 1985, Nomenclatural and taxonomic problems in the genus *Eurotium,* in: *Advances in Penicillium and Aspergillus Systematics* (R. A. Samson and J. I. Pitt, eds.), Plenum Press, New York and London, pp. 383–395.

Pitt, J. I., 1989, Recent developments in the study of *Penicillium* and *Aspergillus* systematics, *J. Appl. Bacteriol. Symp. Suppl.* **1989:**37S–45S.

Pitt, J. L., and Hocking, A. D., 1985, *Fungi and Food Spoilage,* Academic Press, Sydney.

Pitt, J. I., and Samson, R. A., 1990a, Approaches to *Penicillium* and *Aspergillus* systematics, *Stud. Mycol. (Baarn)* **32:**77–91.

Pitt, J. I., and Samson, R. A., 1990b, Taxonomy of *Aspergillus* Section *Restricta,* in: *Modern Concepts in Penicillium and Aspergillus Classification* (R. A. Samson and J. I. Pitt, eds.), Plenum Press, New York and London, pp. 249–257.

Pitt, J. I., and Samson, R. A., 1993, Draft list of species names in current use (NCU) in the family Trichocomaceae (Fungi, Eurotiales), *Regnum Veg.* **128:**13–57.

Pitt, J. I., Hocking, A. D., and Glenn, D. R., 1983, An improved medium for the detection of *Aspergillus flavus* and *A. parasiticus, J. Appl. Bacteriol.* **54:**109–114.

Raper, K. B., and Fennell, D. I., 1965, *The Genus Aspergillus,* Williams and Wilkins, Baltimore.

Roquebert, M. F., and Nicot, J., 1985, Similarities between the genera *Stilbothaminium* and *Aspergillus,* in: *Advances in Penicillium and Aspergillus Systematics* (R. A. Samson and J. I. Pitt, eds.), Plenum Press, New York and London, pp. 221–228.

Samson, R. A., 1979, A compilation of the aspergilli described since 1965, *Stud. Mycol. (Baarn)* **18:**1–40.

Samson, R. A., 1992, Current taxonomic schemes of the genus *Aspergillus* and its telemorphs, in: *Aspergillus: The Biology and Industrial Applications* (J. W. Bennett and M. A. Klich, eds.), Butterworth, Heineman, Stoneham, London, pp. 353–388.

Samson, R. A., and Frisvad, J. C., 1991a, Taxonomic species concepts of Hyphomycetes related to mycotoxin production, *Proc. Jpn. Assoc. Mycotoxicol.* **32:**3–10.

Samson, R. A., and Frisvad, J. C., 1991b, Current taxonomic concepts in *Penicillium* and *Aspergillus,* in: *Cereal Grain, Mycotoxins, Fungi and Quality in Drying and Storage* (J. Chelkowski, ed.), Elsevier, Amsterdam, pp. 405–439.

Samson, R. A., and Gams, W., 1985, Typification of the species of *Aspergillus* and associated teleomorphs, in: *Advances in Penicillium and Aspergillus Systematics* (R. A. Samson and J. I. Pitt, eds.), Plenum Press, New York and London, pp. 31–54.

Samson, R. A., and Pitt, J. I., 1985, *Advances in Pencillium and Aspergillus Systematics,* Plenum Press, New York and London.

Samson, R. A., and Pitt, J. I. 1990. *Modern Concepts in Pencillium and Aspergillus Classification.* Plenum Press, New York and London.

Samson, R. A., and Seifert, K. A., 1985, The ascomycete genus *Penicilliopsis* and its anamorphs, in: *Advances in Pencillium and Aspergillus Systematics* (R. A. Samson and J. I. Pitt, eds.), Plenum Press, New York and London, pp. 397–426.

Samson, R. A., and Van Reenen-Hoekstra, E. S., 1988, *Introduction to Food-borne Fungi,* 3rd ed., Centraalbureau voor Schimmelcultures, Baarn, The Netherlands.

Samson, R. A., Nielsen, P. V., and Frisvad, J. C., 1990, The genus *Neosartorya:* Differentiation by scanning electron microscopy and mycotoxin profiles, in: *Modern Concepts in Penicillium and Aspergillus Systematics* (R. A. Samson and J. I. Pitt, eds.), Plenum Press, New York and London, pp. 455–467.

Stynen, D., Meulemans, L., Goris, A., Brandlin, N., and Symons, N., 1992, Characteristics of a latex agglutination test, based on monoclonal antibodies, for the detection of mould antigen in foods, in: *Modern Methods in Food Mycology* (R. A. Samson, A. C. Hocking, J. I. Pitt, and A. D. D. King, eds.), Elsevier, Amsterdam, pp. 213–219.

Subramanian, C. V., 1971, *Hyphomycetes: An Account of Indian Species except Cercosporae,* Indian Council for Agricultural Research, New Delhi.

Sugiyama, J., and Yamatoya, T., 1990, Electrophoretic comparison of enzymes as a chemotaxonomic aid among *Aspergillus* taxa (1). *Aspergillus* sects. *Ornati* and *Cremei,* in: *Modern Concepts of Penicillium and Aspergillus Classification* (R. A. Samson and J. I. Pitt, eds.), Plenum Press, New York and London, pp. 385–394.

Thom, C., and Church, M. B., 1926, *The Aspergilli,* Williams and Wilkins, Baltimore.

Thom, C., and Raper, K. B., 1945, *A Manual of the Aspergilli,* Williams and Wilkins, Baltimore.

Tzean, S. S., Chen, J. L., Liou, G. Y., Chen, C. C., and Hsu, W. H., 1990, *Aspergillus and Related Teleomorphs of Taiwan,* Mycological Monograph No. 1, Culture Collection and Research Center, Taiwan.

Yamatoya, T., Sugiyama, J., and Kuraishi, H., 1990, Electrophoretic comparison of enzymes as a chemotaxonomic aid among *Aspergillus* taxa (2). *Aspergillus* sect. *Flavi,* in: *Modern Concepts of Penicillium and Aspergillus Classification* (R. A. Samson and J. I. Pitt, eds.), Plenum Press, New York and London, pp. 395–406.

Physiology of *Aspergillus* 2

Z. KOZAKIEWICZ and D. SMITH

1. INTRODUCTION

Members of the genus *Aspergillus* occur in a wide variety of habitats. Some are common as saprophytes in soil, while others are on stored food and feed products and in decaying vegetation (Domsch *et al.*, 1980). They are particularly abundant in tropical and subtropical regions (Raper and Fennell, 1965). Together with *Penicillium,* they are the dominant genera on stored products, being able to thrive in situations of low water activity and high temperatures.

Since the aspergilli are filamentous fungi, extrinsic factors such as water, temperature, pH, and gas composition have a profound influence on their growth and mycotoxin biosynthesis. Thus, a knowledge of the effects of these physiobiochemical parameters can provide methodology for preventing their growth and mycotoxin production in foods and animal feeds. However, it should be pointed out that interactions among several of these factors are often of more importance than individual factors acting in isolation.

2. WATER AVAILABILITY

Water availability is usually expressed as equilibrium relative humidity (E.R.H.), water activity (a_w), or water potential (ψ). E.R.H. is the relative humidity of the intergranular atmosphere in equilibrium with the water in a substrate; a_w is the ratio of the vapor pressure of water over a substrate to that of pure water at the same temperature and

Z. KOZAKIEWICZ and D. SMITH • International Mycological Institute, Bakeham Lane, Egham, Surrey TW20 9TY, England.

Aspergillus, edited by J. E. Smith. Plenum Press, New York, 1994.

pressure; E.R.H. and a_w are the same, with E.R.H. being expressed as a percentage and a_w as a decimal of one; and ψ is the sum of osmotic, matric, and turgor potentials. The relation between a_w and water potential is given by the following equation:

$$\psi = (RT/V) \text{ logn } a_w$$

where R is the ideal gas constant, T the absolute temperature, and V the volume of 1 mole of water. W. J. Scott (1957) first introduced the concept of a_w, which is widely used in the food industry as a measure of water availability for microbes, while ψ is more commonly used in soil microbiology.

Table I. Water Activities for Growth of *Aspergillus* Species[a]

	Water activity (a_W)	
Species	Minimum	Optimum
Eurotium amstelodami	0.70–0.78	0.93–0.96
E. chevalieri	0.72–0.78	0.93–0.95
E. echinulatum	0.70–0.74	0.93–0.95
E. herbariorum	0.70–0.74	0.93
E. repens	0.72–0.74	0.93–0.96
E. rubrum	0.73–0.77	0.95
E. xerophilicum	0.74–0.78	
A. penicillioides	0.76	0.91–0.93
A. restrictus	0.70–0.76	0.82–0.93
A. candidus	0.74–0.79	0.98
A. ochraceus	0.76–0.84	
A. melleus	0.86	
A. niger	0.74–0.85	0.98
A. wentii	0.73–0.84	0.94–0.96
A. nidulans	0.79–0.83	0.96–0.98
A. terreus	0.74–0.82	0.95–0.97
A. clavatus	0.88	
A. floriformis	0.88	
A. pseudodeflectus	0.84–0.88	
A. sydowii	0.76–0.79	0.96
A. versicolor	0.71–0.79	
A. flavus	0.71–0.74	0.98
A. oryzae	0.74	
A. parasiticus	0.71	
Neosartorya fischeri	0.84	0.97
A. fumigatus	0.84–0.86	0.97

[a]Taken in part from Lacey and Magan (1991).

Aspergillus species differ in their a_w requirements, and therefore the presence of a particular species in a food or feed is often a good indicator of previous storage conditions. *Eurotium* species, better known as members of the *A. glaucus* species group, are the most xerophilic group of fungi, capable of growing at the lowest limits of a_w, namely, 0.71 a_w (Table I). *Aspergillus restrictus* and *A. penicillioides* are also considered to be xerophilic. *Aspergillus candidus* and *A. ochraceus* require 0.75 a_w, with *A. versicolor* and *A. flavus* requiring 0.78a_w and *A. fumigatus* requiring 0.85 a_w for growth. Table I gives minimum and optimum a_w requirements for several *Aspergillus* species.

2.1. Effect of a_w on Germination and Sporulation

Water activity may affect germination in a number of different ways: (1) the minimum a_w for germination, (2) the lag time before germination, and (3) the rate of germ-tube extension (Table II).

For example, germination can occur at low a_w but without subsequent mycelial growth. Conidia of *A. flavus* germinated at 0.75 a_w and 29°C, but did not grow (Teitell, 1958). Similarly, at less than 0.75 a_w, conidia remained dormant but viable (Teitell, 1958).

With decreasing a_w, lag time before germination increases. At high water activities (>0.98 a_w), lag time can range from a few hours to several days, while at low a_w, this can extend to several months. For example, at 0.62 a_w, *A. brunneus* spores took 730 days to germinate, producing abnormal germ tubes but no mycelial growth (Snow, 1949).

Germ tubes elongate rapidly at high a_w until 150–250 μm in length,

Table II. Minimum Water Activities for Three Growth Phases of Some *Aspergillus* Species[a]

Species	Water activity (a_w)		
	Germination	Growth	Sporulation (anamorph)
Eurotium amstelodami	0.72	0.73	0.78
E. repens	0.72	0.75	0.78
Emericella nidulans	0.83	0.80	0.80
A. candidus	0.78	0.80	0.83
A. versicolor	0.76	0.78	0.80
A. fumigatus	0.94	0.94	0.95

[a]Taken in part from Magan and Lacey (1984).

Table III. Minimum Water Activities for Anamorph and Teleomorph Formation in Some *Aspergillus* Species[a]

Species	Water activity (a_W)	
	Anamorph (conidia)	Teleomorph
Emericella aurantiobrunnea	0.85	0.98
E. nidulans	0.85	0.95
Eurotium species	0.75	0.77[b]–0.86
Fennellia flavipes	0.90	0.86

[a]Taken in part from Lacey and Magan (1991).
[b]Immature after 120 days.

but then growth rate decreases with a_w and elongation ceases to be linear (Magan and Lacey, 1988).

The effect of a_w on anamorphic and teleomorphic sporulation also differs (Table III). Anamorphic sporulation occurs either at the minimum a_w for germination and mycelial growth or at slightly higher a_w (Table II). But the ascomata production and acospore development are often much slower than conidial production and appear to be less tolerant of low a_w (Table III). For example, *Emericella aurantiobrunnea* produced immature ascomata at 0.77 a_w, but mature ascospores were not formed even after 120 days (Lacey, 1986).

3. TEMPERATURE

Pitt (1980) has used the response of different *Penicillium* species to temperature as part of his taxonomic scheme for this genus. The majority of fungi are mesophilic, growing at temperatures within the range 10–40°C. The genus *Aspergillus* is typical, growing readily at temperatures between 15 and 30°C. However, some species are known to grow at temperatures outside this range. Table IV lists the temperature ranges for growth of some *Aspergillus* species.

The more commonly encountered thermotolerant species are: *A. fischerianus*, *A. fumigatus*, and *A. nidulans*. Species of *A. fumigatus* have been isolated from hot desert soil, hay, lungs of birds and mammals, aborted bovine fetuses, and aviation fuel. Although these organisms grow at temperatures higher than normal, they are also able to grow readily in the mesophilic range.

Table IV. Temperature Ranges for Growth of Some *Aspergillus* Species[a]

Species	Temperature (°C)	
	Range	Optimum
Eurotium amstelodami	5–46	33–35
E. chevalieri	5–43	30–35
E. repens	7–40	25–27
E. rubrum	5–40	25–27
E. xerophilicum	10–36	24
A. restrictus	9–40	25–30
A. clavatus	5–42	25
Neosartorya fischeri	12–65	37–43
A. fumigatus	12–65	37–45
Emericella nidulans	6–48	35–37
A. flavipes	6–40	26–28
A. terreus	11–48	35–40
A. ustus	6–42	25–28
A. sydowii	>5–40	25
A. versicolor	4–40	21–30
A. candidus	3–44	25–32
A. ochraceus	12–37	27
A. flavus	6–45	25–37
A. niger	9–60	17–42

[a]Taken in part from Lacey and Magan (1991).

3.1. Low Temperature

Other representatives of the genus grow at low temperatures. *Aspergillus malignus (A. fischerianus)* grows at 4°C, but strains of this species can also grow well at 37°C (Table V). The survival and tolerance of *Aspergillus* species at low temperatures means they can be preserved by frozen storage at subzero temperatures.

3.2. Freezing Injury

Many organisms are damaged by the effects of freezing, some to such an extent that they are unable to recover. *Aspergillus* is affected by freezing, but is able to survive the stresses imposed by this event.

Mazur (1970) proposed a two-factor hypothesis for freezing injury. When organisms are cooled slowly, injury is caused by prolonged exposure to concentrated extracellular solution or cell dehydration. Such injury may be due to many factors, including volumetric and area con-

Table V. Thermotolerant *Aspergillus* Species in the IMI Culture Collection Giving Cardinal Temperatures for Growth

Species	Temperature (°C)
A. fischerianus	45
A. neoglaber	37
A. spinosus	37
A. fumigatus	45
A. fumigatus var. *acolumnaris*	45
A. neoellipticus	45
A. fumigatus var. *helvolus*	40
A. funiculosus	45
A. malignus[a]	37
A. nidulans	45
A. terreus	45

[a]Strains may also grow at 4°C.

traction, concentration of inter- and extracellular solutes, possible pH changes because of different solubilities of buffers, crystallization, and possible removal of water of hydration from macromolecules. Fast rates of cooling cause damage by the formation of intracellular ice. However, it is suggested by some workers that intracellular ice is a consequence of injury to the plasma membrane (Steponkus, 1984). Organisms are not normally affected by only one of these events, but usually by an interaction of several stresses that result in injury. Fungi have been studied by cryogenic light microscopy, and mechanisms of cryoinjury have been investigated (Coulson *et al.*, 1986; Smith *et al.*, 1986; Morris *et al.*, 1988). Shrinkage is seen in several fungi when they are cooled. Both *Phytophthora nicotianae* and *Lentinus edodes* shrink at all rates of cooling up to −120°C/min, whereas the hyphae of most *Aspergillus* species tested shrank only at slow rates of cooling, less than −10°C/min. The formation of ice results in the concentration of the extracellular solution, and there is a subsequent loss in cell volume. The membrane is not elastic; it does not fold to any great extent, and it therefore loses material. Membrane deletions have been seen in the plasma and nuclear membranes of *L. edodes* (Roquebert, 1993).

Faster cooling does not allow the cell to lose water rapidly enough to prevent supercooling, and the cytoplasm consequently freezes. Although this is generally considered to be lethal for vegetative cells, the hyphae of many fungi survive (Morris *et al.*, 1988). Hyphomycetes are

Table VI. Cryogenic Light Microscopy of *Aspergillus* Species

Species	IMI strain	Critical cooling rate (°C/min)[a]	Nucleation temperature (°C)[b]	Age (days)[b]
A. amstelodami	212944	8	−10 to −15	5
A. amstelodami	313749	No ice	NA	9
A. appendiculatus	278374	8.3	−6	NR
A. nidulans	157753	30[c]	NR	13
A. nidulans	314388	No ice	NA	12
A. nidulans	320721	25	−12 to −20	7
A. repens	072050	4.5	−3 to −7	11
A. repens	073207	5	−6 to −9	4
A. repens	298307	2.5	−5 to −7	NR

[a]Critical cooling rate where 50% of hyphae observed have intracellular ice.
[b](NA) Not applicable; (NR) not recorded.
[c]Only 20% of hyphae froze internally.

particularly resistant to intracellular ice nucleation, including species of *Aspergillus* (Table VI). The integrity of the structure of the conidiophore of *A. repens* is lost following freezing and thawing. A breakage of membrane and cell wall has been observed where the vesicle and conidiogenous cells join, with a subsequent loss in cell cytoplasm. However, the majority of hyphae in the mycelium of *Aspergillus* remain intact during freezing and thawing.

3.3. Tolerance to Freezing

Isolates of *Aspergillus* species are extremely tolerant to freezing injury. They survive storage at many different subzero temperatures, from −20 to −196°C. At the International Mycological Institute, 382 strains belonging to 136 species have been frozen and stored successfully in the vapor above liquid nitrogen (Table VII). Only three strains of three species failed to survive, *A. chevalieri, A. nidulans,* and *A. quadricingens,* and in each case other strains of the same species have survived. Of the 136 named species listed in Table VII, 34 species have strains that have survived over 20 years of storage in liquid nitrogen vapor, 44 species have strains that have survived between 15 and 20 years, and a further 37 species have survived between 10 and 15 years. The storage of these strains continues at IMI, the organisms showing an intrinsic ability to survive ultralow temperatures.

The ability to survive stresses imposed by freezing may be due to many factors. The ability to withstand the osmotic stress imposed by slow

Table VII. Frozen Storage in Liquid Nitrogen of Strains of *Aspergillus* Species at the IMI

Species	Number of strains	Longest survival period (yr)	Shortest survival period all remain viable (yr)
A. acanthosporus	1	19	19
A. aculeatus	6	23	1
A. allahabadii	1	21	21
A. alliaceus	3	20	1
A. ambiguus	1	21	21
A. amstelodami	1	10	10
A. anthodesmis	1	13	13
A. arenarius	1	17	17
A. aurantiobrunneus	2	25	20
A. auratus	1	14	14
A. aureolatus	1	7	7
A. aureolus	1	12	12
A. auricomus	4	18	1
A. awamori	2	14	14
A. bicolor	1	12[a]	12
A. biplanus	1	1	1
A. brevipes	1	12	12
A. bridgeri	1	10	10
A. brunneo-uniseriatus	1	13	13
A. caesiellus	3	18	1
A. campestris	1	1	1
A. candidus	4	16	1
A. carbonarius	2	23	17
A. carneus	1	1	1
A. carnoyi	1	18	18
A. chevalieri	1	14	14
A. chrysellus	1	12	12
A. citrisporus	1	1	1
A. clavatonanica	1	11	11
A. clavatus	2	1	1
A. conicus	1	19	19
A. conjunctus	1	20	20
A. coremiiformis	1	14	14
A. corrugatus	1	13	13
A. cremeus	1	12	12
A. cristatus	1	18	18
A. crustosus	1	20	20
A. crystallinus	1	21	21
A. deflectus	4	17	1
A. diversus	2	20	1

(*continued*)

Table VII. (*Continued*)

Species	Number of strains	Longest survival period (yr)	Shortest survival period all remain viable (yr)
A. duricaulis	1	18	18
A. eburneocremeus	2	12	12
A. echinulatus	1	17	17
A. egyptiacus	1	20	20
A. elegans	3	1	1
A. ellipticus	2	18	4
A. fischerianus	11	14	2
A. flaschentraegeri	1	18	18
A. flavipes	1	1	1
A. flavus	15	23	1
A. foetidus	9	18	1
A. fruticulosus	2	20	12
A. fumigatus	19	18	1
A. funiculosus	1	17	17
A. giganteus	4	23	1
A. glaber	1	2	2
A. gorakhpurensis	1	21	21
A. gracilis	1	14	14
A. granulosus	1	14	14
A. halophilicus	2	20	17
A. heteromorphus	1	18	18
A. iizukae	1	21	21
A. ivoriensis	1	14	14
A. japonicus	6	14	1
A. kanagawaensis	7	7	1
A. lanosus	1	19	19
A. longivesica	1	20	20
A. malignus	1	14	14
A. malodoratus	1	17	17
A. manginii	4	17	11
A. medius	1	17	17
A. melleus	1	1	1
A. mellinus	1	18	18
A. microcysticus	1	20	20
A. minimus	1	16	16
A. montevidensis	1	18	18
A. multicolor	1	1	1
A. nidulans	23	22	1
A. niger	24	21	2
A. niveoglaucus	2	19	10
A. niveus	5	18	1
A. nomius	2	16	2

(*continued*)

Table VII. (*Continued*)

Species	Number of strains	Longest survival period (yr)	Shortest survival period all remain viable (yr)
A. nutans	2	12	1
A. ochraceoroseus	1	14	14
A. ochraceus	4	14	1
A. ornatus	2	23	13
A. oryzae	5	14	2
A. ostianus	4	18	1
A. panamensis	1	18	18
A. paradoxus	4	19	1
A. parasiticus	3	17	2
A. parvulus	2	19	1
A. parvathecius	1	20	20
A. penicillioides	3	14	1
A. petrakii	4	18	1
A. peyronelii	2	20	20
A. phoenicis	2	14	14
A. proliferans	1	21	21
A. pulverulentus	1	14	14
A. pulvinus	1	20	20
A. puniceus	4	7	1
A. quadricinctus	2	3	2
A. quadrilineatus	1	3	3
A. quercinus	1	8	8
A. recurvatus	2	21	14
A. repens	6	18	13
A. restrictus	2	20	14
A. robustus	1	14	14
A. ruber	2	14	14
A. rugulosus	1	21	21
A. sclerotiorum	4	20	1
A. silvaticus	1	2	2
A. sojae	9	16	7
A. sparsus	2	1	1
A. spectabilis	1	14	14
A. speluneus	2	16	14
A. stellatus	5	17	3
A. stramenius	1	14	14
A. striatus	1	19	19
A. stromatoides	4	18	17
A. subsessilis	5	20	1
A. sulphureus	3	18	14
A. sydowii	2	14	3
A. tamarii	1	13	13

(*continued*)

Table VII. (*Continued*)

Species	Number of strains	Longest survival period (yr)	Shortest survival period all remain viable (yr)
A. terreus	7	23	1
A. terricola	6	18	1
A. testaceocolorans	1	18	18
A. tubingensis	1	18	18
A. umbrosus	1	14	14
A. unguis	2	3	1
A. ustus	4	22	1
A. varians	1	18	18
A. versicolor	11	24	3
A. violaceus	1	13	13
A. viridinutans	1	13	13
A. wentii	6	24	1
Aspergillus species	4	16	3

[a]Poor recovery.

cooling may depend on the presence of osmoregulatory chemicals. Glycerol, erythritol, and trehalose are some chemicals that can be synthesized by fungi in their natural habitat when responding to osmotic stress (Smith, 1993). The presence of such chemicals in the cell has been shown to give protection (Kelley and Budd, 1991; van Laere, 1989).

The ability of the plasma membrane to transport water quickly will enable the cell to respond by losing water at faster rates of cooling, and therefore less ice may form within the cell, resulting in less damage. The size of the ice crystals formed may also be a reason that fungi survive intracellular ice formation at fast rates of cooling. The fine granular appearance of the cytoplasm following intracellular ice formation indicates a small crystal size (Smith *et al.*, 1986). Despite these factors, *Aspergillus* has been shown to be extremely resistant to freezing injury, remaining viable. However, sublethal damage may result in the loss of properties, perhaps due to loss of nuclear material following injury to the nuclear membrane. It is therefore essential that optimum preservation protocols be employed for storage of strains, avoiding the stresses of freezing in order to retain the stability of their properties.

4. pH

The pH level is difficult to control, since fungal growth itself results in a change in the pH of the substrate. Indeed, the addition of preservatives to foods and feeds drastically alters their pH. These changes can markedly affect metabolic processes, modifying the limits of a_w at which germination and fungal growth occur (Gottlieb, 1978).

Aspergillus niger can grow over a pH range of 1.5–9.8, *A. candidus* over 2.1–7.7, and *E. repens* over 1.8–8.5 at 1.0 a_w (Panasenko, 1967). Other studies have shown that some *Aspergillus* species can grow at lower a_w's at pH 7.0, indeed better than at pH 3.0 or 5.0 (von Schelhorn, 1950). Work by Pitt and Hocking (1977) indicated that germination of *A. flavus, A. ochraceus,* and *A. chevalieri* was equally rapid at pH 4.0 and 6.5. However, Magan and Lacey (1984) found that when the pH was altered from 6.5 to 4.0, the lag time before germination for a number of *Aspergillus* and *Penicillium* species increased by 1–2 days at high a_w and 6–7 days at low a_w. The same pH change also caused the minimum a_w for germination to increase by 0.02 a_w at optimum temperature and 0.05 a_w at marginal temperatures.

5. GAS COMPOSITION

Growth of *Aspergillus* species can also be altered by changes in concentrations of CO_2 and O_2. For germination and mycelial growth to cease, CO_2 levels must be increased and O_2 levels decreased. However, these effects are more marked if a_w and temperature levels interact as well. For example, *A. versicolor* was unaffected by decreasing O_2 to 2%, while *A. flavus* was inhibited only by O_2 concentrations of less than 1% (Landers *et al.,* 1967). However, Magan and Lacey (1984) have shown that the lag time prior to germination was increased by 10–20 days at 0.90–0.85 a_w and 1% O_2, but by only a few days at 0.98 a_w in ambient air.

Storage fungi such as *Aspergillus* appear to be tolerant of high CO_2, being able to grow at concentrations up to 79%, with the *A. glaucus* species group the most tolerant (Pelhâte, 1980). *Aspergillus ochraceus* could grow in 60% CO_2 but not 80% CO_2 (Paster *et al.,* 1983).

6. MYCOTOXIN PRODUCTION

Mycotoxin production is determined by the availability of nutrients, environmental factors such as water and temperature, host resistance, and interactions with other fungi. Indeed, toxigenic fungi seldom occur in isolation, and their ability to compete with other microorganisms will affect toxin production.

6.1. Fungal Factors

The production of a particular mycotoxin is often limited to a few species, and within a species can be strain-specific.

6.1.1. Species and Strain Specificity

Aflatoxin (a decaketide-derived metabolite) has been reported to be produced by a number of *Aspergillus* species, as well as by *Penicillium* and other genera (Steyn *et al.*, 1980). However, there are only three *Aspergillus* species known to be aflatoxin producers: *A. flavus, A. parasiticus,* and *A. nomius. Aspergillus flavus* produces type B aflatoxin, while *A. parasiticus* and *A. nomius* produce both type B and G aflatoxins. However, patulin (a simpler tetraketide-derived metabolite) is produced by *A. clavatus* and *A. terreus,* as well as by *Byssochlamys nivea* and more than 11 *Penicillium* species (P. M. Scott, 1977). Indeed, Moss (1991) has stated that the more complex a biosynthetic pathway, the fewer the fungal species that produce that toxin.

Strains of a particular species also exhibit variability in their ability to produce mycotoxins. Thus, while *Aspergillus* and *Penicillium* are worldwide in distribution, *Aspergillus* is more common in the tropics and *Penicillium* in temperate areas. Extensive studies on aflatoxin production in strains of *A. flavus* and *A. parasiticus* have shown that 74–100% of strains isolated from groundnuts and cottonseed produced aflatoxin, while in those isolated from rice, aflatoxin was produced by only 20–55% of strains (Moss, 1991).

Furthermore, with routine handling, laboratory isolates often exhibit a reduction in aflatoxin production, or even lose this ability entirely (Bennett *et al.*, 1981). Media of low nutrient content have been recommended for routine culture maintenance, and media rich in nutrients for mycotoxin production studies (Moss, 1991).

6.2. Environmental Factors

Most studies on mycotoxin production have been undertaken under laboratory conditions at optimum a_w and temperature, using laboratory media or various autoclaved cereals as substrate. Few studies have considered environmental factors such as water stress.

6.2.1. Temperature and Water Interactions

Although *A. flavus* is considered to be a storage species, under conditions of drought and temperature stress, the species has been isolated

Table VIII. Comparison of Minimum a_W for Growth and Toxin Production in Some *Aspergillus* Species[a]

Species	Mycotoxin	Water activity (a_W)	
		Growth	Toxin production
A. flavus	Aflatoxin	<0.80	0.82
A. parasiticus	Aflatoxin	0.84	0.87
A. ochraceus	Ochratoxin A	0.77	0.85
A. ochraceus	Penicillic acid	0.77	0.88
A. clavatus	Patulin	0.88	0.99

[a]Taken in part from Lacey and Magan (1991).

and aflatoxin produced in groundnuts in the field. A geocarposphere temperature of 30.5°C was optimum for aflatoxin production; none was produced above 32°C or below 25°C, although *A. flavus* still grew at these temperatures (Hill *et al.*, 1985).

The effects of a_w and temperature on mycotoxin production are often different from those on germination and growth. In addition, their effects can be different for two toxins produced by the same species or indeed for the same toxin produced by two separate species. *Aspergillus ochraceus* produces ochratoxin A and penicillic acid. Optimum ochratoxin A production was at 0.98 a_w and 30°C, but at 0.90 a_w and 22°C for penicillic acid (Bacon *et al.*, 1973). No ochratoxin A was produced at 0.74 a_w, the minimum a_w for growth. Similarly, at 0.78 a_w, sporulation was greatest at 30°C and not 22°C (Bacon *et al.*, 1973). Table VIII compares minimum a_w's for growth and toxin production in some toxigenic *Aspergillus* species.

6.2.2. Host–Fungus Interactions

Maize and groundnut cultivars differ in their susceptibility to *A. flavus*. In maize, there can be large differences in aflatoxin content between adjacent kernels in a cob. Some cultivars support more aflatoxin production than others, but these differences are not consistent and cannot be related to kernel hardness or endosperm type (Widstrom *et al.*, 1984). However, sugar content is an important factor (Widstrom *et al.*, 1984).

In groundnuts, surface waxiness, tannic and amino acid content, permeability, cell structure, and arrangement all affect susceptibility to *A. flavus* (Mixon, 1981).

6.2.3. Interactions between Microorganisms

There are many conflicting reports on the effects of competing fungi and bacteria on mycotoxin production. Both enhancement and inhibition of aflatoxin production in *A. parasiticus* by *P. rubrum* have been reported (Fabbri *et al.*, 1984; Moss and Frank, 1985). Inhibition of aflatoxin production, but not of growth, of *A. parasiticus* by *A. candidus, A. niger,* and *Eurotium chevalieri* have also been reported (Lacey, 1986). Similarly, aflatoxin formation was inhibited by the bacteria *Brevibacterium linens* and *Streptococcus lactis,* but stimulated by *Acetobacter aceti* and *Bacillus amyloliquefaciens* (Lacey and Magan, 1991). However, a_w was not controlled in some of these studies. Cuero *et al.* (1987, 1988) showed that a yeast, *Hyphopichia burtonii,* stimulated aflatoxin production by *A. flavus* under all conditions tested except 0.90 a_w at 16°C. But *A. oryzae* and *A. niger* inhibited aflatoxin production at 0.98 a_w, had little effect at 0.95 a_w, and stimulated it at 0.90 a_w.

6.2.4. Effect of Fungicides

The uniform distribution of fungicides in a substrate is very effective in controlling fungal growth. The presence of 1% propionic acid in high-moisture maize completely inhibited the production of aflatoxin and ochratoxin (Vandegraft *et al.*, 1975). However, undertreatment can lead to aflatoxin formation in cereals (Pettersson *et al.*, 1986). Laboratory studies have shown that small concentrations of propionic acid can stimulate aflatoxin production in *A. flavus* (Al-Hilli and Smith, 1979). It is considered that this enhanced toxin production is a response to stress, rather than due to the acid providing precursors for metabolite synthesis (Moss, 1991).

Acetone, benzene, cyclohexane, dioxan, ethanol, ethyl acetate, hexane (Fanelli *et al.*, 1985), phenobarbitone (Bhatnager *et al.*, 1982), carbon tetrachloride (Fanelli *et al.*, 1984), and carbon disulfide and phosphine (Vandegraft *et al.*, 1973) have all been shown to stimulate aflatoxin production.

7. SUMMARY

Growth of *Aspergillus* species is determined by water availability. However, temperature also plays a role, in that growth cannot occur without water, but the rate of growth will be determined by the temperature levels. *Aspergillus* species can also tolerate adverse conditions in one variable, provided other variables are optimal. These variables, such as moisture, temperature, time, O_2/CO_2 levels, substrates, and fungal species, do not function in isolation but are interactive. It may therefore be possible, using a combination of these controlling parameters rather than using extremes of any one, to control the growth of *Aspergillus* species more effectively in the future.

The application of cryogenic light microscopy to examine the response of *Aspergillus* to freezing and thawing has revealed useful information. It not only increases knowledge of the physiology of this fungus, but also has practical benefits. It is essential that organisms used as reference, teaching, industrial, and research strains be maintained without loss in viability or their properties. Such studies provide information to develop appropriate cryopreservation techniques. Fungi preserved by poor technique can suffer freezing damage, which may result in properties being lost or in a delay while repair is carried out. Citric aid production in *A. niger* is strain-specific, and if this property is altered or lost during freezing or subsequent storage, the result could be devastating to the industrial process. However, simply a delay in recovery of a property such as this for days or weeks can be of economic importance. The use of optimized cryopreservation protocols will minimize damage and help ensure long-term stability of strains.

REFERENCES

Al-Hilli, A. L., and Smith, J. E., 1979, Influence of propionic acid on growth and aflatoxin production by *Aspergillus flavus, FEMS Microbiol. Lett.* **6:**367–370.

Bacon, C. W., Sweeney, J. G., Ronnins, J. D., and Burdick, D., 1973, Production of penicillic acid and ochratoxin A on poultry feed by *Aspergillus ochraceus, Appl. Microbiol.* **26:**155–160.

Bennett, J. W., Silverstein, R. B., and Kruger, S. T., 1981, Isolation and characterization of two non aflatoxigenic classes of morphological variants of *Aspergillus parasiticus, J. Am. Oil Chem. Soc.* **58:**952A.

Bhatnager, R. K., Ahmad, S., Kohli, K. K., Mukerji, K. G., and Venkitasubramanian, T. A., 1982, Induction of polysubstrate monooxygenase and aflatoxin production by phenobaritone in *Aspergillus parasiticus, Biochem. Biophys. Res. Commun.* **104:**1287–1292.

Coulson, G. E., Morris, G. J., and Smith, D., 1986, A cryomicroscopic study of *Penicillium expansum* hyphae during freezing and thawing, *J. Gen. Microbiol.* **132:**183–190.

Cuero, R. G., Smith, J. E., and Lacey, J., 1987, Stimulation by *Hyphopichia burtonii* and *Bacillus amyloliquefaciens* of aflatoxin formation by *Aspergillus flavus* in irradiated maize and rice grains, *Appl. Environ. Microbiol.* **53:**1142–1146.

Cuero, R. G., Smith, J. E., and Lacey, J., 1988, Mycotoxin formation by *Aspergillus flavus* and *Fusarium graminearum* in irradiated maize grains in the presence of other fungi, *J. Food Prot.* **51:**452–456.

Domsch, K. H., Gams, W., and Anderson, T. H., 1980, *Compendium of Soil Fungi,* Academic Press, London.

Fabbri, A. A., Panfilli, G., Fanelli, C., and Visconti, A., 1984, Effect of T-2 toxin on aflatoxin production, *Trans. Br. Mycol. Soc.* **83:**150–152.

Fanelli, C., Fabbri, A. A., Finotti, E., Fasella, P., and Passi, S., 1984, Free radical and aflatoxin biosynthesis, *Experientia* **40:**191–194.

Fanelli, C., Fabbri, A. A., Pieretti, S., Panfilli, G., and Passi, S., 1985, Effect of organic solvents on aflatoxin production in cultures of *Aspergillus parasiticus, Trans. Br. Mycol. Soc.* **84:**591–593.

Gottlieb, D., 1978, *The Germination of Fungus Spores,* Meadowfield Press, Durham.

Hill, R. A., Wilson, D. M., McMillan, W. W., Widstrom, N. W., Cole, R. J., Sanders, T. H., and Blankenship, P. D., 1985, Ecology of the *Aspergillus flavus* group and aflatoxin formation in maize and groundnut, in: *Trichothecenes and Other Mycotoxins* (J. Lacey, ed.), John Wiley & Sons, Chichester, pp. 79–95.

Kelley, D. J. A., and Budd, K., 1991, Polyol metabolism and osmotic adjustment in the mycelial ascomycete *Necosmospora vasinfecta* (E. F. Smith), *Exp. Mycol.* **15:**55–64.

Lacey, J., 1986, Water availability and fungal reproduction: Patterns of spore production, liberation and dispersal, in: *Water, Fungi and Plants* (P. G. Ayres and L. Boddy, eds.), Cambridge University Press, Cambridge, pp. 65–86.

Lacey, J., and Magan, N., 1991, Fungi in cereal grains: Their occurrence and water and temperature relationships, in: *Cereal Grain Mycotoxins, Fungi and Quality in Drying and Storage* (J. Chelkowski, ed.), Elsevier, Amsterdam, pp. 77–118.

Landers, K. E., Davis, N. D., and Diener, U. L., 1967, Influence of atmospheric gases on aflatoxin production by *Aspergillus flavus* in peanuts, *Phytopathology* **57:**1086–1090.

Magan, N., and Lacey, J., 1984, The effect of temperature and pH on the water relations of field and storage fungi, *Trans. Br. Mycol. Soc.* **82:**71–81.

Magan, N., and Lacey, J., 1988, Ecological determinants of mould growth in stored grain, *Int. J. Food Microbiol.* **7:**245–256.

Mazur, P., 1970, Cryobiology: The freezing of biological systems, *Science* **168:**939–949.

Mixon, A. C., 1981, Reducing aflatoxin contamination in peanut geotypes by selection and breeding, *J. Am. Oil Chem. Soc.* **58:**961A–966A.

Morris, G. J., Smith, D., and Coulson, G. E., 1988, A comparative study of the morphology of hyphae during freezing with the viability upon thawing of 20 species of fungi, *J. Gen. Microbiol.* **134:**2897–2906.

Moss, M. O., 1991, The environmental factors controlling mycotoxin formation, in: *Mycotoxins and Animal Foods* (J. E. Smith and R. S. Henderson, eds.), CRC Press, Boca Raton, Florida, pp. 37–56.

Moss, M. O., and Frank, J. M., 1985, The influence on mycotoxin production of interactions between fungi and their environment, in: *Trichothecenes and Other Mycotoxins* (J. Lacey, ed.), John Wiley & Sons, Chichester, pp. 257–268.

Panasenko, V. T., 1967, Ecology of microfungi, *Bot. Rev.* **33:**189–215.

Paster, N., Lisker, N., and Chet, I., 1983, Ochratoxin A production by *Aspergillus ochraceus* Wilhelm grown under controlled atmospheres, *Appl. Environ. Microbiol.* **45:**1136–1139.

Pelhâte, J., 1980, Oxygen depletion as a method in grain storage, in: *Controlled Atmosphere Storage of Grains* (J. Shejbal, ed.), Elsevier, Amsterdam, pp. 133–146.

Pettersson, H., Holmberg, T., Kaspersson, A., and Larsson, K., 1986, Occurrence of aflatoxin M1 in milk due to aflatoxin formation by *A. flavus-parasiticus* in acid treated grain, *Proc. Nord. Mykot.,* Uppsala, Sweden (October), p. 8.

Pitt, J. I., 1980, *The Genus Penicillium and Its Teleomorphic States Eupenicillium and Talaromyces,* Academic Press, London.

Pitt, J. I., and Hocking, A. D., 1977, Influence of solutes and hydrogen ion concentration on the water relations of some xerophilic fungi, *J. Gen. Microbiol.* **101**:35–40.

Raper, K. B., and Fennell, D. I., 1965, *The Genus Aspergillus,* Williams & Wilkins, Baltimore.

Roquebert, M. F., 1993, Freezing of *Lentinus edodes, Mycol. Res.* (in press).

Scott, P. M., 1977, *Penicillium* mycotoxins, in: *Mycotoxic Fungi, Mycotoxins, Mycotoxicoses,* Vol. I (T. D. Wyllie and L. G. Morehouse, eds.), Marcel Dekker, New York, pp. 283–356.

Scott, W. J., 1957, Water relations of food spoilage micro-organisms, *Adv. Food Res.* **7**:83–127.

Smith, D., 1993, Tolerance of fungi to freezing and thawing, in: *Stress Tolerance of Fungi* (D. H. Jennings, ed.), Marcel Dekker, New York, pp. 145–171.

Smith, D., Coulson, G. E., and Morris, G. J., 1986, A comparative study of the morphology and viability of hyphae of *Penicillium expansum* and *Phytophthora nicotianae* during freezing and thawing, *J. Gen. Microbiol.* **132**:2014–2021.

Snow, D., 1949, The germination of mould spores at controlled humidities, *Ann. Appl. Biol.* **36**:1–17.

Steponkus, P. L., 1984, Role of the plasma membrane in freezing injury and cold acclimation, *Ann. Rev. Plant Physiol.* **35**:543–584.

Steyn, P. S., Vleggaar, R., and Wessels, P. L., 1980, The biosynthesis of aflatoxin and its congeners, in: *The Biosynthesis of Mycotoxins* (P. S. Steyn, ed.), Academic Press, New York, pp. 105–155.

Teitell, L., 1958, Effects of relative humidity on viability of conidia of aspergilli, *Am. J. Bot.* **45**:748–753.

Vandegraft, E. E., Shotwell, O. L., Smith, M. L., and Hesseltine, C. W., 1973, Mycotoxin production affected by insecticide treatment of wheat, *Cereal Chem.* **50**:264–270.

Vandegraft, E. E., Hesseltine, C. W., and Shotwell, O. L., 1975, Grain preservatives: Effect on aflatoxin and ochratoxin production, *Cereal Chem.* **52**:79–84.

van Laere, A., 1989, Trehalose, reserve and/or stress metabolite, *FEMS Microbiol. Rev.* **63**:201–210.

von Schelhorn, M., 1950, Spoilage of water proof food by osmophilic micro-organisms. II. Substrate concentration for the osmophilic fungus *Aspergillus glaucus* in relation to pH, *Z. Lebensm. Unters. Forsch.* **91**:318–342.

Widstrom, N. W., McMillian, W. W., Wilson, D. M., Garwood, D. L., and Glover, D. V., 1984, Growth characteristics of *Aspergillus flavus* on agar infused with maize kernel homogenates and aflatoxin contamination of whole kernel samples, *Phytopathology* **74**:887–890.

Improvement of Industrial *Aspergillus* Fungi

3

SHIGEOMI USHIJIMA

1. BREEDING BY PROTOPLAST FUSION OF KOJI MOLD, *Aspergillus Sojae*

The koji molds play important roles as producers of various hydrolyzing enzymes in the production of various fermented foods, i.e., soy sauce, sake, miso, and shochu (a Japanese spirit). For example, the protease of koji molds contributes to the solubilization of the nitrogen constituents, and the glutaminase causes the liberation of glutamate, one of the most important flavor components in soy sauce. The breeding of koji molds has long been investigated in attempts to increase their enzyme productivities, using mainly two kinds of methods. One kind involves mutation (Sekine *et al.*, 1969) with several mutagenic agents such as X rays, ultraviolet (UV) light, *N*-methyl-*N'*-nitro-*N*-nitrosoguanadine (MNNG), and so on, and the other involves crossing (Oda and Iguchi, 1963). The latter method was developed after the finding that koji molds, *Aspergillus oryzae* and *Aspergillus sojae,* have a so-called "parasexual" life cycle (Pontecorvo *et al.*, 1953).

Generally, among the koji molds used for the production of shoyu (Japanese fermented soy sauce), sake, and miso, protease hyperproducers show low glutaminase activity and, conversely, glutaminase hyperproducers show insufficient protease activity. Therefore, Ushijima and Nakadai (1987) attempted to obtain a new koji mold showing well-balanced enzyme activities through protoplast fusion of a protease hyperproducer and a glutaminase hyperproducer. They dealt with the

SHIGEOMI USHIJIMA • Research and Development Division, Kikkoman Corporation, 399 Noda, Noda City, Chiba Prefecture 278, Japan.

Aspergillus, edited by J. E. Smith. Plenum Press, New York, 1994.

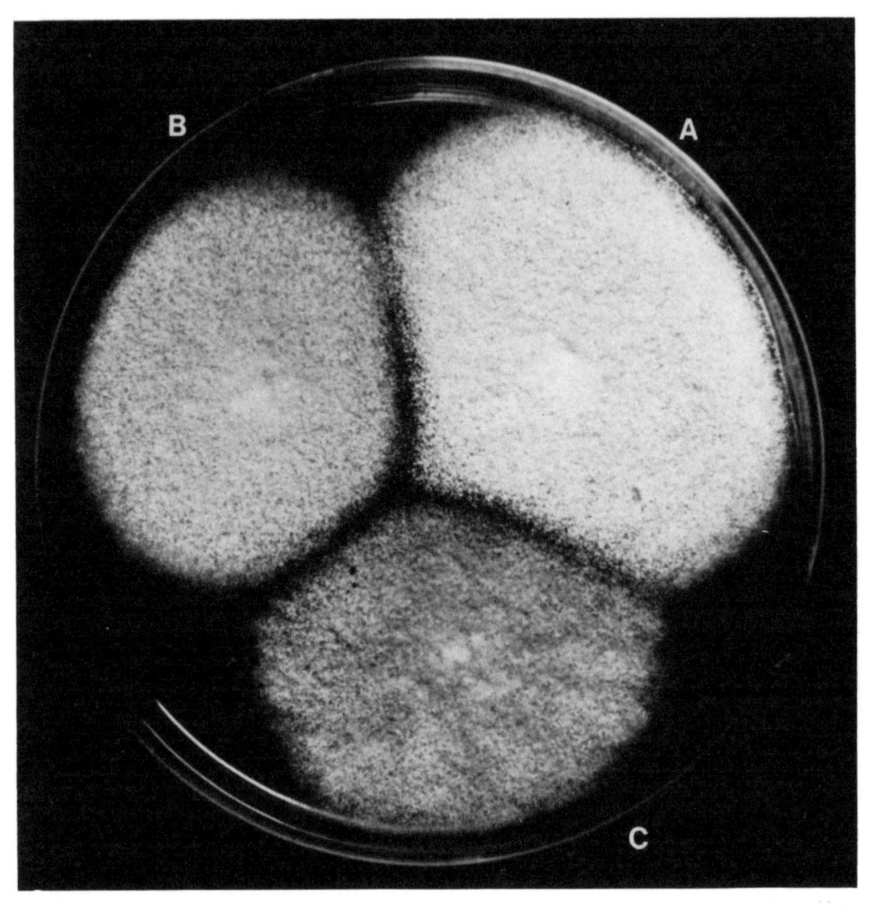

Figure 1. Colonies of *Aspergillus sojae* on the complete medium. (A) Parental strain W19-7 (*whi, bio*); (B) parental strain M12-2-61 (*ylo, nia*); (C) heterokaryon; (D) stable green diploid.

development of an effective method for the protoplast fusion of koji molds, and also with some properties of the resultant fusants. The protease hyperproducer is *A. sojae* 2048; the glutaminase hyperproducer is *A. sojae* 2165, the genealogy of which is unrelated to that of *A. sojae* 2048. Conidia of the parental strains were treated with UV rays or MNNG (Adelberg *et al.*, 1965) in the usual manners. Strains W19-7 (*whi, bio*)* and M-12-2-61 (*ylo, nia*)* are mutants of *A. sojae* 2048 and 2165, respectively (Fig. 1).

1.1. Formation of Protoplasts

Protoplasts were obtained from 12-hr-cultured germlings of each strain. The preincubation medium contained 3 g glucose, 0.1 g

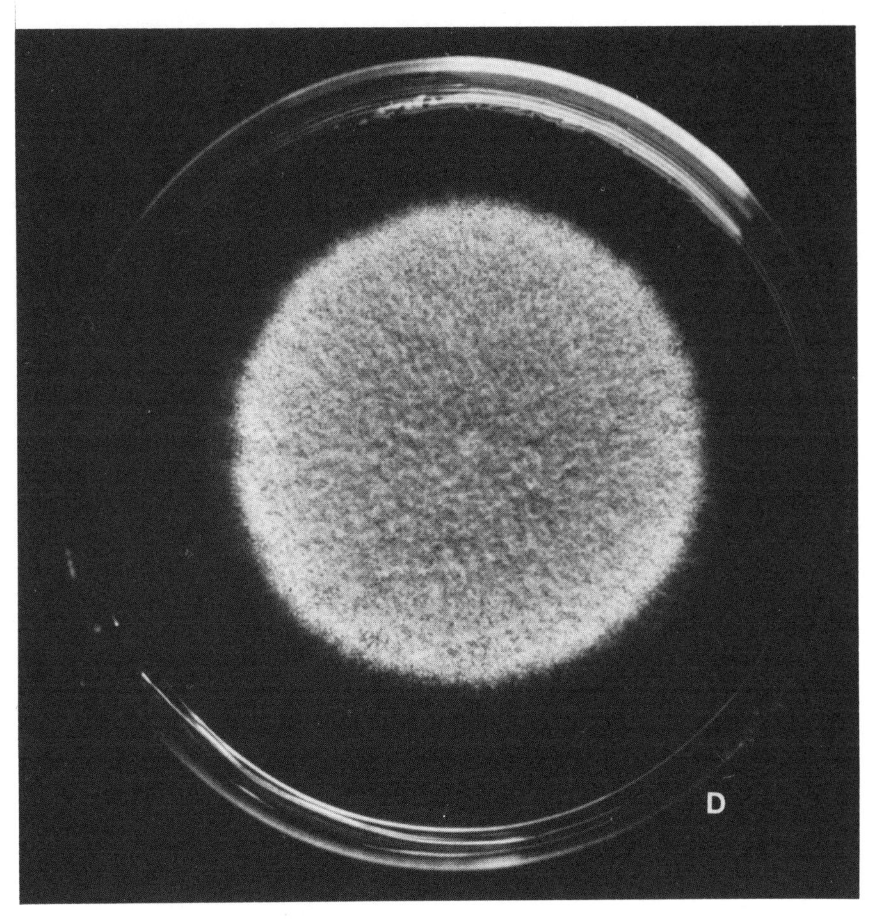

Figure 1. *(Continued)*

KH_2PO_4, 0.05 g $MgSO_4 \cdot 7H_2O$, 0.5 g KCl, 0.2 g Casamino acid (Difco), and 0.1 g Solgen TW-60 (Daiichi Kogyo Seiyaku Co., Ltd., Kyoto, Japan) per liter. The medium was adjusted to pH 6.5 before autoclaving. A 30-ml portion of the medium was inoculated with 1 ml of a conidial suspension containing approximately 10^7 spores/ml. The young germlings were harvested by centrifugation (4500*g*, 20 min), washed twice with 0.01 M phosphate buffer, pH 6.5, and finally resuspended in the same buffer containing an osmotic stabilizer. The suspension was mixed with each of the lytic enzymes, followed by incubation at 25–30°C with slow shaking for 2 hr. The protoplasts were separated from the mycelial debris by filtration through a sintered glass filter (3G3) and then washed twice with the stabilized buffer. When purification of the protoplasts was not satisfactory, they were purified again by centrifugation in 30% sucrose. With the use of the crude enzyme from either *Bacillus circulans* IAM 1165 or *Streptomyces* sp. 0143, sufficient protoplasts were released to give proper results in the following fusion steps ($1–5 \times 10^6$/ml). The addition of 5 mg/ml of chitinase (ICN) to the aforementioned crude enzymes resulted in a marked increase in protoplast formation compared with the single use of one of the crude enzymes. The maximum yield of protoplasts on the combined use of the *B. circulans* IAM 1165 enzyme and chitinase was obtained at pH 6.5. Sorbitol (0.8 M), mannitol (0.8 M), and KCl (0.6 M) showed a good protective effect as osmotic stabilizers for the protoplast formation. The optimum concentration of sorbitol was about 0.7–0.8 M.

1.2. Protoplast Fusion

Protoplasts from the two parental strains to be fused were mixed ($5–10 \times 10^6$ cells/ml of each strain) and then centrifuged (2000 rpm, 15 min). The pelleted protoplast mixtures were resuspended in 1 ml 20% (wt./vol.) polyethylene glycol 6000 in 0.01 M $CaCl_2$ and 0.05 M glycine, which was adjusted to pH 7.5. After incubation for 10–30 min at room temperature, the suspension was diluted with 6 ml of the minimal medium (MM) containing 0.8 M sorbitol, followed by centrifugation. The protoplasts were washed twice with 8 ml 0.8 M sorbitol and then resuspended in 5 ml 0.8 M sorbitol. To select nutritionally complemented fused protoplasts, an aliquot of a suspension was put on solid MM (2% agar) in a Petri dish, and then 5 ml soft MM agar (0.5% agar, 45°C) was poured onto the plate and immediately mixed with the protoplasts. The overlayered plates were incubated at 30°C. The regeneration ratio for the protoplasts was more than 30% for both strains. The frequency of fusion after the fusing treatment was about 0.1%.

The two double-marked strains used for fusion exhibit discriminating conidial color and auxotrophic requirements as genetic markers. Many colonies showing complementation as to both auxotrophic requirements arose on the minimal regeneration plates. They appeared to be fusants, there being green, white, and yellow conidial heads in a single colony (Fig. 1), and the parental marker strains were frequently segregated. Thus, these fusants were thought to be heterokaryons (Ishitani *et al.*, 1956; Ishitani and Sakaguchi, 1956).

1.3. Isolation of Heterozygous Diploids

Ishitani *et al.* (1956) and Oda and Iguchi (1963) reported that UV irradiation was effective for the induction of stable green diploids from heterokaryons. According to their methods, conidial suspensions of heterokaryons were irradiated with UV light from a 15 W lamp at a distance of 39 cm for 3 min. The irradiated conidial suspensions were plated on the complete medium (CM) at appropriate dilutions. The stable green colonies that appeared were picked up (Fig. 1). Stable green colonies were obtained under irradiation conditions that led to a 2–4% survival ratio.

1.4. Properties of the Fused Green Strain

In several fusion trials, 130 fused diploids were isolated, and they were examined as to their protease and glutaminase activities. A mixture of 30 g cooked soybean grits, 30 g roasted wheat, and 40 ml water was then inoculated with a pure culture of a koji mold strain. This mixture was incubated for 3 days at 30°C. A 10-g sample of the koji thus obtained was extracted with water. After filtration, the extract was used as the enzyme solution for the protease assay. The koji homogenate was used for the glutaminase assay. In most cases, the activities of both protease and glutaminase were between those of the parents or lower. A few diploids (D-047, D-063, and D-105), however, were found to show excellent activities of both enzymes with good balances (Fig. 2).

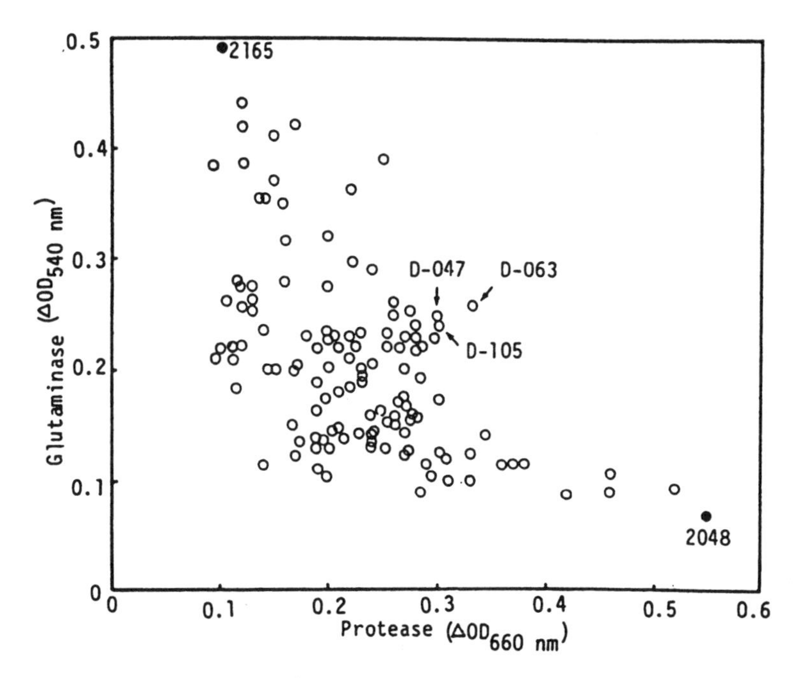

Figure 2. Distribution of the protease and glutaminase activities of 130 fused green strains. Fused green strains (D-047, D-063, and D-105) showing excellent activities with a good balance.

With these experiments, Ushijima and Nakadai (1987) demonstrated the usefulness of this technique for the improvement of koji molds used in the production of soy sauce, through the hybridization of genealogically unrelated *A. sojae* strains.

Breeding of koji molds with the aim of increasing their enzyme productivities has long been attempted, these attempts involving mainly mutation (Sekine *et al.*, 1969; Nasuno *et al.*, 1971) or crossing (Oda and Iguchi, 1963; Uchida *et al.*, 1958). The strains obtained on mutation usually showed poor growth and poor conidiation.

Furuya *et al.* (1983) successfully used this technique for the breeding of *A. oryzae,* through the fusion of a protease-improved mutant and its well-growing parent.

Protoplast fusion has also been investigated in *A. nidulans* (Kevei and Peberdy, 1977, 1979); this protoplast fusion technique was used not only for breeding a certain desired strain but also for analyzing genetic defects of the organism. In *A. parasiticus* (Bennett, 1979) and *A. flavus* (Papa, 1976), heterokaryons and diploids were obtained by crossing and used to investigate their aflatoxin genetics. Protoplast fusion will be a useful tool for studying the genetics of enzyme productivity of *Aspergillus.*

2. IMPROVEMENT OF ENZYME PRODUCTIVITIES THROUGH MUTATION OR HAPLOIDIZATION OF HETEROZYGOUS DIPLOIDS OBTAINED BY PROTOPLAST FUSION OF *Aspergillus sojae*

The protoplast fusion technique, first described by Anné and Peberdy (1976) and Ferenczy *et al.* (1975) in fungi, has proved to be a valuable method for the breeding of new organisms or for investigating genetic problems in *Aspergillus* (Furuya *et al.*, 1983; Kevei and Peberdy, 1977), *Penicillium* (Ferenczy *et al.*, 1975), *Cephalosporium* (Hamlyn *et al.*, 1985), *Mucor* (Goto-Hamamoto *et al.*, 1986), and *Trichoderma* (Toyama *et al.*, 1984).

Ushijima and Nakadai (1987) reported that the stable heterozygous diploids of *A. sojae* were obtained through protoplast fusion of a protease hyperproducer and a glutaminase hyperproducer. Enzyme assaying of these diploids revealed that the productivities of either enzyme in most diploids were distributed around the mean value for the parents or below. They attempted to improve the enzyme productivities of fused diploids through mutation or haploidization. Through usual mutation of a heterozygous diploid, both the protease and glutaminase productivities were simultaneously improved to a certain extent.

Ushijima *et al.* (1987) investigated haploidization of fused diploids of *A. sojae* with benomyl or *p*-fluorophenylalanine (FPA), attempting to obtain haploidized recombinants with maximized enzyme productivities, i.e., ones that had inherited the favorable properties of both parents. The ploidy of the resulting haploidized recombinants was also determined by DNA content assaying.

2.1. Mutation of Heterozygous Diploids

A conidial suspension of a stable heterozygous diploid with green conidia was irradiated with UV light from a 15 W lamp at a distance of 39 cm for 15 min. The irradiated suspension was then plated on casein medium (Sekine *et al.*, 1969) after appropriate dilution. About 200 mutants showing increased protease productivities were obtained from 10 heterozygous diploids through a screening procedure. These mutants were examined as to their productivities of protease and glutaminase in koji cultures. Most of the screened mutants exhibiting protease productivity higher than that of the parent heterozygous diploids showed lower glutaminase productivity than the parent. Seven mutants shown in Fig. 3, obtained from diploid D-013, were found to exhibit improved productivities of the two enzymes. Another 6 mutants shown in Fig. 3 were also obtained from diploid D-063.

In the case of the mutation of usual haploid strains, no mutant showing simultaneous increases in the productivities of the two enzymes has ever been obtained. It was interesting that in the diploid mutants, both enzyme activities increased simultaneously (Fig. 3).

2.2. Haploidization of Heterozygous Diploids

In an attempt to investigate haploidization of the heterozygous diploids and, if possible, to improve their enzyme productivities, the two diploids selected were treated with a haploidizing agent, FPA or benomyl. Haploidization of heterozygous diploids was performed as described by Hastie (1970), Upshall *et al.* (1977), and Bennett (1979) with the following modifications: The diluted conidial suspension of heterozygous diploids was plated on CM agar plates containing 100–300 μg/ml of FPA or 1.0–5.0 μg/ml of benomyl ("Benlate"; Du Pont Co., Delaware), followed by incubation for 7–21 days at 30°C. Color segregants (Fig. 4) that appeared were picked up and examined as to their properties. Spontaneous segregation of the heterozygous diploids was not observed in this experiment, and color segregants occurred only in the cultures with haploidizing agents. The phenotypes of the segregants obtained on

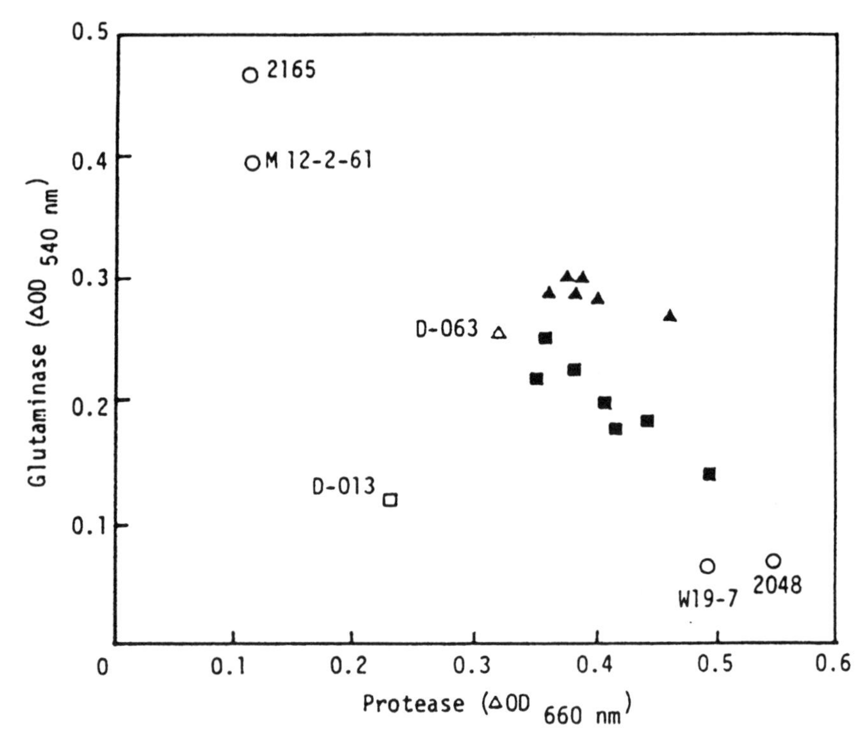

Figure 3. Distribution of the protease and glutaminase activities of mutants obtained from heterozygous diploids. (■) Mutant obtained from diploid D-013; (▲) mutant obtained from diploid D-063.

benomyl treatment (2.5 ppm) are shown in Table I. These phenotypes showed great variety. Most segregants showed the same phenotype as one of the fusion parents, but some showed recombinant phenotypes, e.g., *whi, nia* and *ylo, bio*. This result suggested that genetic recombination had occurred through haploidization of heterozygous diploids bred by protoplast fusion of two *A. sojae,* as observed in *A. nidulans* (Kevei and Peberdy, 1977) and *A. parasiticus* (Bennett, 1979; Bradshaw *et al.,* 1983). The fact that segregants with phenotypes of the fusion parents were separated at a relatively high frequency suggests that whole genomes coding for the phenotypes of the parents were integrated into the heterozygous diploids.

The effect of benomyl on the growth of a heterozygous diploid is shown in Table II. The survival ratio decreased with increasing benomyl concentration.

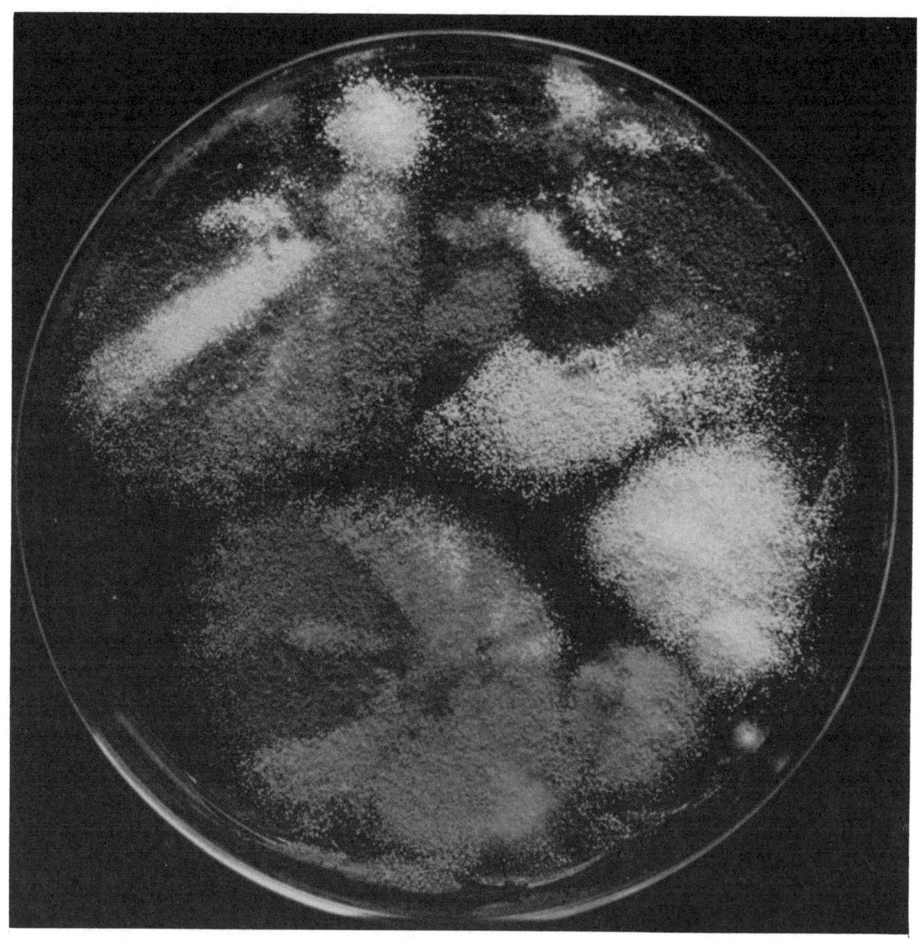

Figure 4. Appearance of color segregants with benomyl treatment.

2.3. Enzyme Productivities of Haploidized Strains

In several haploidization trials, many haploidized strains were isolated from heterozygous diploids. On treatment with 2.5 ppm benomyl, 160 strains were randomly isolated and then examined as to their protease and glutaminase activities (Fig. 5). On the basis of their productivities of the two enzymes, these strains were divided into five types, i.e., two parental types and a lower, an intermediate, and a higher type. Further studying the higher type, Ushijima *et al.* (1987) found a strain that fulfilled their requirements for improvement of the enzyme pro-

Table I. Number of Segregants Obtained on Benomyl Treatment of Heterozygous Diploid D-063[a]

Conidial color	Prototrophs	Auxotrophs			Total
		bio	*nia*	bio nia	
whi	27	4	28	22	81
ylo	35	1	23	2	61
g	16	0	1	1	18

[a]Heterozygous diploid D-063 was treated with 2.5 ppm benomyl for 14 days.

ductivities. Through several haploidization trials with benomyl or FPA, some excellent strains exhibiting high productivities of both enzymes were obtained (Table III).

Segregation induced by benomyl or FPA has permitted the isolation of numerous recombinants in *A. nidulans* (Kevei and Peberdy, 1977; Upshall *et al.*, 1977), *A. flavus* (Papa, 1979, 1982, 1984), *A. parasiticus* (Bennett, 1979; Bradshaw *et al.*, 1983; Bennett *et al.*, 1980), and yeasts (Flores, 1970; Sipiczki and Ferenczy, 1977; Maraz *et al.*, 1978). In these works, auxotrophic markers or spore-color markers were recombined, and new strains with properties different from those of the parents were obtained. Haploidizing experiments on fused diploids of *A. sojae* with FPA or benomyl produced many recombinants with a variety of color and nutritional markers, in accordance with the reports cited above. On screening these recombinants, Ushijima *et al.* (1987) succeeded in obtaining haploid recombinants with improved enzyme productivities, i.e., strains that exhibited high productivities of both protease and glutaminase. Specific nutrition requirements or conidial colors for these recombinants were not observed. A haploidized recombinant, Ben-638Y, was found to produce as much protease as the hyperprotease producer

Table II. Effect of Benomyl on the Survival of the Diploid *Aspergillus sojae* D-063

Benomyl concentration	Survival ratio (%)
Control (no benomyl)	100
1.0 ppm	73.9
1.5 ppm	64.4
2.5 ppm	19.9
3.5 ppm	5.8
5.0 ppm	10^{-5}

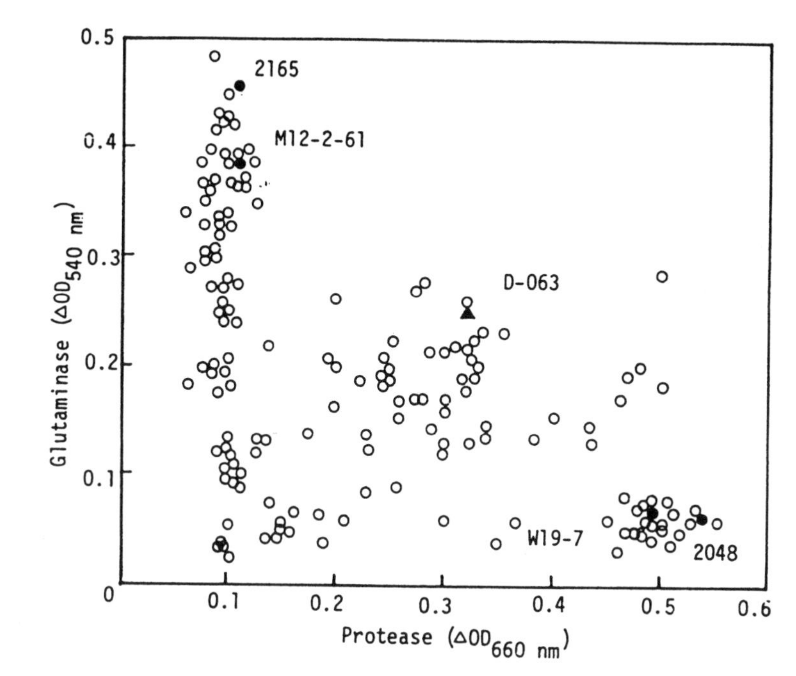

Figure 5. Distribution of the protease and glutaminase activities of segregants obtained on haploidizing treatment. Heterozygous diploid D-063 (▲) was treated with 2.5 ppm benomyl.

(2048) and 4-fold as much glutaminase as this parent (2048). This finding would suggest the presence of a certain regulation system for the production of these two enzymes in the koji mold, *A. sojae.*

Through this study, it has been shown that mutation or haploidization of heterozygous diploids with intermediate enzyme-producing properties enables us to obtain greatly improved strains for industrial purposes.

The fusion of haploid cells, followed by haploidization of the resulting fused diploids, has been used to determine the linkage groups in some aspergilli (Bennett, 1979; Bradshaw *et al.,* 1983; Bennett *et al.,* 1980; Kafer, 1958; Lhoas, 1967) that have no sexual life cycle. It would be feasible, through the use of this artificial sexual cycle, and of more genetic markers to be allocated, to begin the mapping of genes on linkage groups in koji molds.

2.4. Estimation of DNA Content

The DNA content and number of nuclei per conidium were determined. The DNA contents of conidia were estimated by the methods of

Table III. Properties of Segregants Obtained on Haploidizing Treatment of Heterozygous Diploid D-063

Strain No.	Phenotype	Protease ($\Delta OD_{660\,nm}$)	Glutaminase ($\Delta OD_{540\,nm}$)
Parents			
W19-7	*whi, bio*	0.488	0.069
M12-2-61	*ylo, nia*	0.108	0.379
Heterozygous diploid			
D-063	*g,* —	0.325	0.257
Protease hyperparental type			
Ben-56	*whi,* —	0.493	0.072
Ben-136	*whi, nia*	0.512	0.049
FPA-88	*whi, nia*	0.488	0.050
Glutaminase hyperparental type			
Ben-23	*whi, nia*	0.105	0.385
Ben-96	*whi,* —	0.105	0.354
FPA-118	*ylo,* —	0.116	0.356
Lower type			
Ben-154	*whi,* —	0.082	0.051
Ben-158	*whi, nia*	0.088	0.054
FPA-30	*whi,* —	0.119	0.045
Intermediate type			
Ben-41	*ylo, bio*	0.317	0.259
Ben-140	*whi, nia*	0.334	0.232
Ben-468	*ylo,* —	0.351	0.231
Higher type			
Ben-60	*whi,* —	0.478	0.201
Ben-114	*whi,* —	0.468	0.193
Ben-638Y	*ylo,* —	0.497	0.265
Ben-467	*ylo,* —	0.506	0.190

Herbert *et al.* (1971) and Burton (1956). For microscopic examination, nuclei in fresh conidia were stained by means of the basic fuchsin technique of DeLamater (1950). The DNA contents per nucleus of the parents, heterozygous diploids, and haploidized recombinants are shown in Table IV. The recombinants obtained through the haploidization procedure showed nearly the same nuclear DNA contents as those of the fusion parents, but the value for the heterozygous diploid was found to be about twice that of the parents.

The numbers of nuclei and DNA contents determined indicate that the stable diploid fusants with green conidia are diploids and that the recombinants segregated from these heterozygous diploids are haploids.

Table IV. DNA Content per Conidium and per Nucleus of Parental, Diploid, and Haploidized Strains of *Aspergillus sojae*

Strain	DNA content per conidium (10^{-7} μg)	Number of nuclei per conidium[a]	DNA content per nucleus (10^{-7} μg)
Parents			
W19-7	4.36 ± 0.61	4.7	0.93
M12-2-61	3.87 ± 0.54	4.5	0.86
Diploid			
D-063	3.44 ± 0.48	2.1	1.64
Haploidized strains			
Ben-638Y	5.07 ± 0.57	4.8	1.06
Ben-467	4.74 ± 0.52	4.7	1.01
Ben-114	4.18 ± 0.30	4.5	0.93

[a]The contents of 100 conidia were counted.

3. INTERSPECIFIC PROTOPLAST FUSION BETWEEN *Aspergillus oryzae* AND *Aspergillus sojae*

Aspergillus oryzae and *A. sojae* are industrially important for the production of various fermented foods. Generally, *A. oryzae* is characterized by a high productivity of α-amylase and *A. sojae* by a high productivity of proteinase. To breed a variety of new and more talented koji molds, it would be useful to introduce some foreign genes from organisms of diverse taxa. In regard to this objective, interspecific fusion between *A. oryzae* and *A. sojae* is the most urgent object in this field. The interspecific fusion technique was first used in genetic mapping in *A. nidulans* (Kevei and Peberdy, 1977). Heterokaryons or diploids obtained by crossing among the strains of *A. parasiticus* (Bennett *et al.*, 1980) or *A. flavus* (Papa, 1984) were used to locate some auxotrophic markers, conidial color determinants, or aflatoxin genes.

Ushijima *et al.* (1990a) reported on interspecific protoplast fusion between *A. oryzae* and *A. sojae* and described the behaviors of some phenotypic characters, not only in the expression of morphological characters but also in genetic behaviors of the productivity of hydrolyzing enzymes, through the process of protoplast fusion and subsequent haploidization.

The parental strains are *A. oryzae* 1299 and *A. sojae* 2048. Their color and auxotrophic double mutants were derived by UV irradiation or MNNG treatment. Strain W19-7 (*whi, bio*) and 6Y-1 (*ylo, met*) are mutants of *A. sojae* 2048 and *A. oryzae* 1299, respectively. Regenerating frequencies of these protoplast preparations were 10–30%.

3.1. Protoplast Fusion

Altogether, 30 prototrophic colonies complementing the auxotrophic characteristics were obtained from several trials of protoplast fusion between *A. oryzae* 6Y-1 and *A. sojae* W19-7. The frequency of prototrophic colonies was 5×10^{-5} per protoplast pairs used for a trial. During the polyethylene glycol (PEG) treatment of a mixture of the two lines of protoplasts, some protoplasts aggregated with each other and then fused into a cell.

Prototrophs appearing on an MM plate formed green, white, and yellow conidial heads in each single colony, and they frequently segregated lines the same as the parental-marker strains. Thus, these prototrophic colonies were thought to be fused heterokaryons. The frequency of interspecific fungal fusion was lower than that of intraspecific fusion.

3.2. Isolation of Heterozygous Diploids

It is known that UV irradiation or *d*-camphor treatment is effective for induction of stable diploids from fused heterokaryons (Ishitani *et al.*, 1956; Ogawa *et al.*, 1988a). In this fusion experiment, however, these methods were not effective. Stable green fusants were obtained by successive subcultures of the heterokaryons on MM as reported with *A. parasiticus* (Bennett *et al.*, 1980) or *A. flavus* (Papa, 1984).

3.3. Phenotypes of Heterozygous Diploids

All the fusants were prototrophs. Seven of the nine heterozygous diploids formed rough conidia (*A. oryzae* type) and two formed echinulate conidia (*A. sojae* type). None showed intermediate roughness.

3.4. Haploidization of Fusants

Nine fusants with green conidia were treated with a haploidizing agent, benomyl. Color segregants that appeared on agar plates containing 1.5, 2.5, and 3.5 μg/ml benomyl, respectively, were selected and examined for their phenotypic properties. No color segregant occurred on the plates without the haploidizing agent.

Table V. DNA Content per Conidium and per Nucleus of Parental, Diploid, and Haploidized Strains

Strain	DNA content per conidium (10^{-7} μg)[a]	Number of nuclei per conidium[b]	DNA content per nucleus (10^{-7} μg)
Parents			
A. oryzae 6Y-1	3.12	2.8	1.11
A. sojae W19-7	3.81	4.7	0.81
Fusant			
OS-176	3.92	2.5	1.57
OS-96	4.31	2.4	1.79
Haploidized strains			
OS-176-1W	2.63	4.2	0.86
OS-176-2W	5.60	4.8	1.17
OS-96-1Y	4.41	4.6	0.96

[a]Mean of 4 independent assays. The standard deviations are about 0.36.
[b]The contents of 100 conidia were counted.

3.5. Conidial DNA Content

DNA contents and number of nuclei per conidium were measured. DNA contents per nucleus of the parents, fusant, and haploidized strains are shown in Table V. The fusant with green conidia showed a DNA content per nucleus nearly double those of the parents, though the value was somewhat low. Two segregants obtained through the haploidization showed almost the same nuclear DNA contents as those of the fusion parents. These data indicated that the fusant with green conidia was a diploid and that the segregants from this diploid were haploids.

3.6. Electrophoretic Discrimination of Alkaline Proteinase*

The cultural methods and preparation of the electrophoretic samples were reported in detail by Nasuno (1972). Figure 6 shows electrophoretic mobilities of alkaline proteinases from the fusion parents, fused heterozygous diploids, and haploidized segregants. The heterozygous diploid OS-176 showed mobility of the *A. oryzae* type. A haploidized derivative OS-176-1W, which was derived from OS-176, however, showed the *A. sojae* type. This confirmed that both the alkaline pro-

*This section is based on the results of experiments by Ushijima *et al.* (1990b).

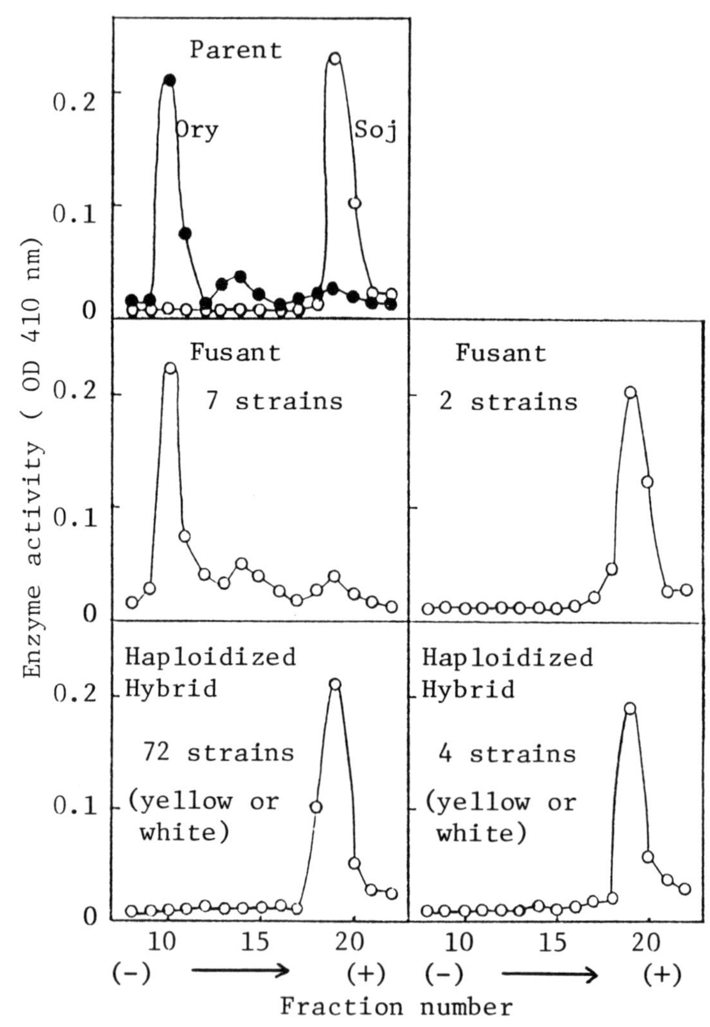

Figure 6. Electrophoretic mobility of alkaline proteinase from parents, fusants, and haploidized strains. Electrophoresis on pH 9.4 gels for 120 min.

teinase genomes of *A. oryzae* and *A. sojae* simultaneously existed in the diploid OS-176.

Throughout the haploidization trials, 76 white- or yellow-spored segregants were isolated from 9 stable green-spored fusants. Electrophoretic examination of enzyme preparations from 72 color segregants, which were derived from the 7 fusants expressing *A. oryzae*-type proteinase, showed that all the segregants produced *A. sojae*-type enzyme with exception. All of the 4 segregants induced from the 2 fusants having *A. sojae*-type enzyme were also found to produce the A. sojae-type proteinase. It is very noteworthy that even though about half the color segregants obtained in this study had phenotypic markers of the *A. oryzae* type, all of them formed exclusively the alkaline proteinase with the mobility of one selected side of the parents, and that none of them formed any intermediate type.

3.7. Phenotypes of Haploidized Strains

Benomyl treatment of green-conidial heterozygous diploids permitted the isolation of various color, auxotrophic, and prototrophic segregants.

Some properties of 72 segregants derived from 7 phenotypically *A. oryzae*-type fusants are presented in Table VI. The ratio of white conidia segregants to yellow ones in 72 isolates was 34:38. Of the 34 white segregants, 10 were auxotrophs and 24 were prototrophs. Among these 10 auxotrophs, 7 strains were phenotypically recombinant in relation to the markers *whi* and *met*. In addition to some back-segregated strains, 13 recombinant-like strains in relation to *ylo* and *bio* were also obtained in the 38 yellow segregants. The facts that segregants with mutant spore color and nutritional requirements were obtained at a relatively high frequency, and that segregants with phenotypes of the fusion parents

Table VI. Conidial-Color and Nutrition Requirements of Benomyl-Induced Segregants from "*Oryzae*-Type" Fused Diploids with Green Conidia

Conidial color[a]	Requirements[b]		Number of segregants
	met	*bio*	
whi	+	+	24
[p]*whi*	+	−	3
whi	−	+	7
whi	−	−	0
ylo	+	+	13
ylo	+	−	9
[p]*ylo*	−	+	13
ylo	−	−	3
			Total 72

[a](*p*) Parental type; (*whi*) white; (*ylo*) yellow.
[b](+) Not requiring; (−) requiring.

were obtained, suggest that whole genomes originated from both the parent were integrated once into the heterozygous fusants.

Sixteen (7 + 9) segregants showed recombinant types, e.g., *whi, met* or *ylo, bio*. This result suggests that certain genetic recombinations had occurred in the course of the interspecific protoplast fusion of *A. oryzae* and *A. sojae* and subsequent haploidization of the heterozygous fusants. The haploidized recombinants were obtained with a frequency of 0.97–1.25×10^{-1}.

Some haploidized strains were considered to be phenotypic recombinants with respect to their morphological properties and artificial markers.

Aspergillus sojae has usually been distinguished from *A. oryzae* by the roughness of their conidial walls. The fact that this difference has been used as a key character in the taxonomy would mean that variations of this character were discrete rather than continuous. Among the interspecific fused diploids obtained in this study, there could be found no fusant showing intermediate roughness of the conidial wall, although their conidial color and auxotrophy were both complemented. That is, they were either *A. oryzae* type or *A. sojae* type. It is very interesting that only a set of characteristics from one side of the parents was expressed in the fused heterozygous diploids, though both the genomes of parents including those of conidial walls and alkaline proteinase were evidently integrated into the diploid.

Since purified alkaline proteinases from the two species do not contain any carbohydrate, it is quite possible that this difference is based on some differences in the amino acid sequences of the enzymes and, accordingly, on the sequences of their structural genes. Therefore, the behavior of the alkaline proteinase is very interesting in connection with a possible factor that controls the selective expression of the gene.

3.8. Hydrolyzing-Enzyme Productivities of the Fusants and the Haploidized Segregants

Productivities of some extracellular hydrolyzing enzymes in two typical fusants and two haploidized segregants are compared with those of the parents in Table VII (Ushijima *et al.*, 1990b). The parents were significantly different in their enzyme productivities, especially of protease, glutaminase, or pectinase. Fusants or haploidized segregants showed enzyme productivities fairly different from the parents. It is likely, however, that the higher activities in proteinase and peptidase, and the lower ones in glutaminase, pectinase, and CMC-ase, both of which seemed to have originated from the *A. sojae* parents, behaved

Table VII. Enzymatic Activities of Parents, Fusants, and Haploidized Hybrids[a]

Strain	Protease	Glutaminase	Peptidase	Pectinase[b]	CMC-ase[b]	Amylase × 10^2 DB/g
Parent						
A. oryzae 6Y-1	431	3.23	1.45	432	218	611
A. sojae W19-7	1122	0.82	3.86	84	180	586
Fusant						
OS-176	536	3.17	4.31	263	320	745
OS-96	1268	1.24	3.11	74	110	751
Haploidized segregants						
OS-176-1W	580	1.75	2.66	137	218	394
OS-96-1Y	1259	0.83	6.69	87	122	734

[a]Enzymatic activities are expressed as units/g koji.
[b](Pectinase) Pectin liquefying activity; (CMCase) carboxymethyl cellulose liquefying activity.

together through the fusion and haploidization. These enzyme productivities were not always consistent with the other phenotypic characteristics such as conidial color or electrophoretic mobility of proteinase. A haploidized segregant OS-96-1Y showed a high activity of peptidase compared with that of the parents or fusants.

By this combined process of artificial fusion and haploidization including genetic recombination, some characteristics of different species could be successfully introduced into a strain as described above. Protoplast fusion and subsequent haploidization, like intraspecific fusion, has thus been shown to be a useful technique for the breeding of koji molds.

4. INTERSPECIFIC ELECTROFUSION BETWEEN PROTOPLASTS OF *Aspergillus oryzae* AND *Aspergillus sojae*

Protoplast fusion has become a valuable technique for strain improvement, e.g., in *Aspergillus, Penicillium, Mucor,* and *Trichoderma.* PEG has been used almost exclusively as the fusogenic agent for the fusion of these fungus protoplasts. This method, however, has a number of disadvantages in comparison with the newly developed electrical fusion methods. Ushijima et al. (1991) described the electrofusion of protoplasts between *A. oryzae* and *A. sojae* to increase the frequency of occurrence of fusants.

Strains 1Y-90 (*ylo, pant*) and 2Y (*ylo, met*) are marker mutants of *A. oryzae* 1830. Strains W19-7 (*whi, bio*) and W5-11 (*whi, ade*) are mutants of *A. sojae* 2048. Preparation of protoplasts was done as described in Sec-

Table VIII. Optimum Electrical Conditions for Koji-Mold Fusion

AC field strength	400 V/cm
AC frequency	1.1 MHz
DC pulse field strength	4.0 kV/cm
DC pulse width	60 μsec
DC pulse number	3 times

tion 1.1, with a slight modification. The 12-hr-cultured germlings were suspended in the buffer containing 0.6 M KC1 (pH 5.5). The suspension was mixed with the lytic enzyme Novozyme 234 (final 10 mg/ml). The protoplasts were washed three times with 1.1 M sorbitol. An equal number of protoplasts of each of two lines (1×10^6/ml of each) was then mixed for the fusion experiments. After electrofusion, the protoplasts were regenerated on the stabilized MM.

From results of these examinations, the optimum electrical conditions for koji-mold fusion, such as *A. oryzae, A. sojae,* are presented in Table VIII. The frequency of fusant formation via electrofusion was 2 or 4 times higher than with PEG fusion.

Many stable fusants with green conidia were obtained in this study, and three fusants were selected for subsequent haploidization. Three fusants with green conidia were treated with benomyl, 3.5 μg/ml; color segregants appearing on agar plates were selected and examined for their phenotypic properties.

4.1. Phenotypic Characters of Fusants and Haploidized Segregants

All of the 15 fusants with green conidia were prototrophs. They formed rough conidia (*A. oryzae* type) and showed the gel electrophoretic mobility of alkaline proteinase of *A. oryzae* type. None of them showed intermediate phenotypic characteristics as to roughness of conidial walls or electrophoretic mobility of alkaline proteinase.

Protease and α-amylase activities of koji cultures of each strain are shown in Table IX. One fusant, F4161, showed intermediate protease productivity. It is noteworthy that fusants F1762 and F1918 showed high protease productivities like the *A. sojae* parent, but showed the mobility of the alkaline proteinase of *A. oryzae* type. The productivities and mobility of proteinase behaved independently of each other in the fusants. Three fusants, however, had low productivity of amylase.

After treatment with 3.5 μg/ml of benomyl, five selected segregants from three fusants were assayed for their two enzyme productivities

Table IX. Properties of the Parents, Fused Green Strains, and Haploidized Strains

Strain	Combination of fusion[a]	Phenotype	Protease (ΔOD)	α-Amylase 10^3 DB/g	Type of conidial wall[b]	Type of ALP[b]
Parents						
A. oryzae 1830		*g*, —	0.096	87.8	Ory	Ory
1Y-90		*ylo, pant*	0.067	104.4	Ory	Ory
A. sojae 2048		*g*, —	0.328	31.8	Soj	Soj
W19-7		*whi, bio*	0.320	31.1	Soj	Soj
Fused green strains						
F4161	D	*g*, —	0.226	28.6	Ory	Ory
F1762	C	*g*, —	0.303	24.2	Ory	Ory
F1918	C	*g*, —	0.286	26.4	Ory	Ory
Haploidized strains						
F4161-1W	D	*whi, pant*	0.043	88.8	Ory	Ory
F4161-4Y	D	*ylo, pant*	0.075	59.7	Ory	Ory
F1762-2Y	C	*ylo*, —	0.321	24.3	Ory	Ory
F1762-1W	C	*whi*, —	0.343	23.6	Ory	Ory
F1918-27Y	C	*ylo, pant*	0.047	48.6	Ory	Ory

[a](D) 1Y90 (*ylo, pant*) × W5-11 (*whi, ade*); (C) 1Y90 (*ylo, pant*) × W19-7 (*whi, bio*).
[b]Type of conidial wall and electrophoretic type of alkaline proteinase (ALP): (Ory) *A. oryzae*; (Soj) *A. sojae*, combination of fusion.

(Table IX). None showed the excellent activities of both protease and amylase manifested by the parents. This result would suggest the presence of a certain regulation system for the production of these two enzymes in *A. oryzae* and *A. sojae*.

Through this study, it was shown that interspecific electrofusion of *Aspergillus koji* molds enables us to obtain improved or specific strains for industrial purposes.

5. INTERSPECIFIC HYBRIDIZATION OF *Aspergillus awamori* AND *Aspergillus oryzae* BY PROTOPLAST FUSION

A new breeding technique for *A. awamori* and *A. oryzae* involving protoplast fusion was studied to obtain more useful interspecific hybrids for making spirits for drinking. A strain obtained from the heterokaryon using *d*-camphor showed high stability and prolific sporulation. The amylase productivity of the heterozygous diploid was shown to be comparable to that of *A. oryzae,* but the productivity of citric acid in the diploid strain was about 7-fold that of *A. oryzae* (Ogawa *et al.,* 1988b).

6. INTRASPECIFIC PROTOPLAST FUSION OF *Aspergillus niger*

Intraspecific protoplast fusion was done between different citric-acid-producing strains of *A. niger.* Heterodiploids were induced from a heterokaryon by *d*-camphor treatment. Citric acid productivity of one heterodiploid was intermediate between those of parent strains in both shaking and solid cultures (Kirimura *et al.,* 1986).

REFERENCES

Adelberg, E. A., Mandel, C., and Chen, C. C., 1965, Optimal conditions for mutagenesis by *N*-methyl-*N'*-nitro-*N*-nitrosoguanidine in *Escherichia coli* K12, *Biochem. Biophys. Res. Commun.* **18:**788–795.

Anné, J. J., and Peberdy, J. F., 1976, Induced fusion of fungal protoplasts following treatment with polyethene glycol, *J. Gen. Microbiol.* **92:**413–417.

Bennett, J. W., 1979, Aflatoxin and anthraquinones from diploids of *Aspergillus parasiticus, J. Gen. Microbiol.* **113:**127–136.

Bennett, J. W., Vinnett, C. H., and Goynes, W. R., Jr., 1980, Aspects of parasexual analysis in *Aspergillus parasiticus, Can. J. Microbiol.* **26:**706–713.

Bradshaw, R. E., Bennett, J. W., and Peberdy, J. F., 1983, Parasexual analysis of *Aspergillus parasiticus, J. Gen. Microbiol.* **129:**2117–2123.

Burton, K., 1956, A study of the conditions and mechanism of the diphenylamine reaction for the colorimetric estimation of deoxyribonucleic acid, *Biochem. J.* **62**:315–323.

DeLamater, E. D., 1950, The nuclear cytology of the vegetative diplophase of *Saccharomyces cerevisiae*, *J. Bacteriol.* **60**:321–332.

Ferenczy, L., Kevei, F., and Szegedi, M., 1975, Increased fusion frequency of *Aspergillus nidulans* protoplasts, *Experientia* **31**:50–52.

Flores da Cunha, M., 1970, Mitotic mapping of *Schizosaccharomyces pombe*, *Genet. Res. Cambr.* **16**:127–144.

Furuya, T., Ishige, M., Uchida, K., and Yoshino, H., 1983, Koji-mold breeding by protoplast fusion for soy sauce production, *Nippon Nogeikagaku Kaishi* **57**:1–8.

Goto-Hamamoto, M., Ohnuki, T., Uozumi, T., and Beppu, T., 1986, Intraspecific hybridization by protoplast fusion in *Mucorales* producing milk-clotting proteases, *Agric. Biol. Chem.* **50**:1467–1473.

Hamlyn, P. F., Birkett, J. A., Perez, G., and Peberdy, J. F., 1985, Protoplast fusion as a tool for genetic analysis in *Cephalosporium acremonium*, *J. Gen. Microbiol.* **131**:2813–2823.

Hastie, A. C., 1970, Benlate-induced instability of *Aspergillus* diploids, *Nature (London)* **226**:771.

Herbert, D., Phipps, P. J., and Srange, R. E., 1971, Chemical analysis of microbial cells. VII. Determination of nucleic acids, in: *Methods in Microbiology*, Vol. 5B (J. R. Norris and D. W. Ribbons, eds.), Academic Press, New York, pp. 324–328.

Ishitani, C., and Sakaguchi, K., 1956, Hereditary variation and recombination in *koji*-molds: Heterokaryosis, *J. Gen. Appl. Microbiol.* **2**:345–400.

Ishitani, C., Ikeda, Y., and Sakaguchi, K., 1956, Hereditary variation and genetic recombination in *koji*-molds. VI. Genetic recombination in heterozygous diploids, *J. Gen. Appl. Microbiol.* **2**:401–430.

Kafer, E., 1958, An eight-chromosome map of *Aspergillus nidulans*, *Adv. Genet.* **9**:105–145.

Kevei, F., and Peberdy, J. F., 1977, Interspecific hybridization between *Aspergillus nidulans* and *Aspergillus rugulosus* by fusion of somatic protoplasts, *J. Gen. Microbiol.* **102**:255–262.

Kevei, F., and Peberdy, J. F., 1979, Induced segregation in interspecific hybrids of *Aspergillus rugulosus* obtained by protoplast fusion, *Mol. Gen. Genet.* **170**:213–218.

Kirimura, K., Yaguchi, T., and Usami, S., 1986, Intraspecific protoplast fusion of citric acid-producing strains of *Aspergillus niger*, *J. Ferment. Technol.* **64**:473–479.

Lhoas, P., 1967, Genetic analysis by means of the parasexual cycle in *Aspergillus niger*, *Genet. Res.* **10**:45–61.

Maraz, A., Kiss, M., and Ferenczy, L., 1978, Protoplast fusion in *Saccharomyces cerevisiae* strains of identical and opposite mating types, *FEMS Lett.* **3**:319–322.

Nasuno, S., 1972, Differentiation of *Aspergillus sojae* from *Aspergillus oryzae* by polyacrylamide gel disc electrophoresis, *J. Gen. Microbiol.* **71**:29–33.

Nasuno, S., Ohara, T., and Iguchi, N., 1971, Successive isolation of proteinase hyperproductive mutants of *Aspergillus sojae*, *Agric. Biol. Chem.* **35**:291–293.

Oda, K., and Iguchi, N., 1963, Genetic and biochemical studies on the formation of protease in *Aspergillus sojae*. Part I. Genetic studies, *Agric. Biol. Chem.* **27**:758–766.

Ogawa, K., Ohara, H., and Toyama, N., 1988a, Intraspecific hybridization of *Aspergillus awamori* var. *kawachi* by protoplast fusion, *Agric. Biol. Chem.* **52**:337–342.

Ogawa, K., Ohara, H., and Toyama, N., 1988b, Interspecific hybridization of *Aspergillus awamori* var. *kawachi* and *Aspergillus oryzae* by protoplast fusion, *Agric. Biol. Chem.* **52**:1985–1991.

Papa, K. E., 1976, Linkage groups in *Aspergillus flavus*, *Mycologia* **68**:159–165.

Papa, K. E., 1979, Genetics of *Aspergillus flavus:* Complementation and mapping of aflatoxin mutants, *Genet. Res.* **34:**1–9.
Papa, K. E., 1982, Norsolorinic acid mutant of *Aspergillus flavus, J. Gen. Microbiol.* **128:**1345–1348.
Papa, K. E., 1984, Genetics of *Aspergillus flavus:* Linkage of aflatoxin mutants, *Can. J. Microbiol.* **30:**68–73.
Pontecorvo, G., Roper, J. A., Hemmons, D. W., MacDonald, K. D., and Bufton, A. W. J., 1953, The genetics of *Aspergillus nidulans, Adv. Genet.* **5:**141–238.
Sekine, H., Nasuno, S., and Iguchi, N., 1969, Isolation of highly proteolytic mutants from *Aspergillus sojae, Agric. Biol. Chem.* **33:**1477–1482.
Sipiczki, M., and Ferenczy, L., 1977, Protoplast fusion of *Schizosaccharomyces pombe* auxotrophic mutants of identical mating-type, *Mol. Gen. Genet.* **151:**77–81.
Toyama, H., Yokoyama, T., Shinmyo, A., and Okada, H., 1984, Interspecific protoplast fusion of *Trichoderma, J. Biotechnol.* **1:**25–35.
Uchida, K., Ishitani, C., Ikeda, Y., and Sakaguchi, K., 1958, An attempt to produce interspecific hybrids between *Aspergillus oryzae* and *Asp. sojae, J. Gen. Appl. Microbiol.* **4:**31–38.
Upshall, A., Giddings, B., and Mortimore, I. D., 1977, The use of benlate for distinguishing between haploid and diploid strains of *Aspergillus nidulans* and *Aspergillus terreus, J. Gen. Microbiol.* **100:**413–418.
Ushijima, S., and Nakadai, T., 1987, Breeding by protoplast fusion of *koji* mold, *Aspergillus sojae, Agric. Biol. Chem.* **51:**1051–1057.
Ushijima, S., Nakadai, T., and Uchida, K., 1987, Improvement of enzyme productivities through mutation or haploidization of heterozygous diploids obtained by protoplast fusion of *Aspergillus sojae, Agric. Biol. Chem.* **51:**2781–2786.
Ushijima, S., Nakadai, T., and Uchida, K., 1990a, Breeding of new *koji*-molds through interspecific hybridization between *Aspergillus oryzae* and *Aspergillus sojae* by protoplast fusion, *Agric. Biol. Chem.* **54:**1667–1676.
Ushijima, S., Nakadai, T., and Uchida, K., 1990b, Further evidence on the interspecific protoplast fusion between *Aspergillus oryzae* and *Aspergillus sojae* and subsequent haploidization, with special reference to their production of some hydrolyzing enzymes, *Agric. Biol. Chem.* **54:**2393–2399.
Ushijima, S., Nakadai, T., and Uchida, K., 1991, Interspecific electrofusion of protoplasts between *Aspergillus oryzae* and *Aspergillus sojae, Agric. Biol. Chem.* **55:**129–136.

Molecular Genetics and Expression of Foreign Proteins in the Genus *Aspergillus*

J. R. KINGHORN and S. E. UNKLES

1. INTRODUCTION

The advent of recombinant DNA procedures in the early 1970s provided methods to isolate genes from almost any organism. Such genes could be transferred by plasmid-mediated transformation procedures or by bacteriophage infection to the bacterium *Escherichia coli*, in which such heterologous gene(s) could be synthesized in large quantities and studied, taking advantage of the replicative abilities of plasmids within the *E. coli* cell.

The 1980s brought the development of genetic engineering procedures for the filamentous fungi, in particular aspergilli. There are several reasons for such attention. First, to study the regulation of gene expression was perhaps the major driving force, since a number of genes and their regulatory circuits had been identified by previous classic genetic studies, especially in the "model" ascomycetous fungus *Aspergillus nidulans*. Second, aspergilli have a long history in producing materials and substances beneficial to man, such as food, alcoholic beverages, and enzymes. Third, certain aspergilli are of medical interest, such as species that produce aflatoxins. Fourth, although *E. coli* and the budding yeast *Saccharomyces cerevisiae* were proving, in many cases, to be attractive hosts for the recombinant protein industry, filamentous fungi offered alternative systems (see Section 3).

We review in this chapter the current state of fungal molecular

J. R. KINGHORN and S. E. UNKLES • Plant Science Laboratory, School of Biological and Medical Science, University of St. Andrews, Fife KY 16 9TH, Scotland.

Aspergillus, edited by J. E. Smith. Plenum Press, New York, 1994.

genetic technology as it relates to *Aspergillus,* as well as an application, namely, the production of foreign proteins.

2. METHODOLOGY

2.1. Isolation of *Aspergillus* Genes

Several routes are available to permit the cloning of genes in aspergilli. The principal and most routinely used methods are cloning via (1) function, (2) antibody recognition, (3) DNA homology, and (4) chromosomal position. The first reported *Aspergillus* gene, *aromA* from *A. nidulans* (Kinghorn and Hawkins, 1982), was isolated on the basis of expression and complementation of a corresponding *E. coli* mutant. This approach has since been successful for a number of other fungal genes such as *trpC* (Yelton *et al.,* 1984), but suffers the disadvantage that most fungal genomic clones cannot be faithfully expressed in a prokaryote due to the presence of introns in the fungal genes. Similar approaches were made using mutants of *S. cerevisiae,* and this has led to the isolation of the *A. nidulans argB* gene (Berse *et al.,* 1983). However, although several *S. cerevisiae* genes contain introns, yeast splicing mechanisms fail, or have difficulty at least, in faithfully splicing introns of the filamentous fungi. Following the successful development of efficient transformation systems, self-cloning by complementation of *Aspergillus* mutants has been a better alternative. This is particularly an advantage for *A. nidulans* in which there are many genes represented by mutations. An *A. nidulans* genomic library prepared in a plasmid or bacteriophage lambda vector can be used to transform a mutant to prototrophy. The wild-type complementing gene can then be isolated by so-called "marker rescue," in which total DNA from the fungal transformant is used to transform *E. coli.* Selection for ampicillin-resistant *E. coli* colonies allows recovery of plasmids that also contain the gene complementing the fungal mutation. *Aspergillus* genomic libraries in cosmid vectors have also been successful in self-cloning. The gene of interest within a complementing cosmid can be isolated by cotransforming restriction endonuclease fragments and testing for complementation ability (Yelton *et al.,* 1985). A major disadvantage of marker rescue cloning approaches, however, is the possibility of rearrangement of genes during the procedure. This can be overcome using sib selection, a refinement described for the isolation of *Neurospora crassa* genes (Vollmer and Yanofsky, 1986), in which DNA clones are first divided into pools and each pool is tested for complementation of the fungal mutation. The complementing pool is then subdivided and retested, and so on until a single complementing clone is identified. Recombinant clones obtained in this fashion therefore have been amplified only in *E. coli* without passage through and possible rearrangement in the fungus. Although self-cloning has been applicable for the isolation of several *A. nidulans* genes, including several regulatory genes such as *creA* (Dowzer and Kelly, 1989), the technique has been less useful with other aspergilli, particularly those used in industry, in which well-defined mutations are difficult to isolate and characterize. It has been possible, however, to use mutants of *A. nidulans* to allow isolation of the corresponding gene from other *Aspergillus* species. For example, the *argB* gene of *A. niger* was isolated by complementation of the corresponding *A. nidulans* mutant.

Another successful method of cloning has been antibody recognition of clones in cDNA expression libraries. This route has been particularly useful for the isolation of genes encoding extracellular proteins that are often produced in large quantity and hence are easier to purify for antibody production than low-abundance cytosolic or membrane proteins. Examples in this category include the *alp* and *nep* genes of *A. oryzae* encoding extracellular proteases (Tatsumi *et al.*, 1989, 1991) and the *glaA* gene of *A. niger* var. *awamori* encoding glucoamylase (Nunberg *et al.*, 1984).

Many genes have been isolated using one of a number of DNA:DNA hybridization techniques. These include differential hybridization, in which DNA clones are identified by virtue of their expression in cells grown under one condition and not another. Refinements of this technique are possible using subtractive hybridization, which allows enrichment of DNA clones specific to one condition such as those involved in cellular differentiation (Timberlake, 1980). Isolation of clones is also commonly achieved by hybridization to a gene library using a DNA probe. Probes can consist of a DNA fragment from a gene previously isolated from another organism or a degenerate oligonucleotide mixture, the sequence of which is derived either from a portion of amino acid sequence of the protein product of the gene or from the amino acid sequence of regions of that protein conserved between species. For example, the *pelD* gene encoding pectin lyase was isolated from an industrial strain of *A. niger* by hybridization to a gene library with synthetic oligonucleotide probes derived from the N-terminal amino acid sequence (Gysler *et al.*, 1990). A fragment of this gene was used as a probe to identify clones containing the corresponding gene from a better-characterized, laboratory *A. niger* strain, and fragments of the latter gene were used in relaxed hybridization conditions to obtain clones representing another five genes of a pectin lyase family (Harmsen *et al.*, 1990). Recently, gene cloning has also been possible using the polymerase chain reaction (PCR) to amplify fragments of a gene using degenerate oligonucleotide primers derived either from the amino acid sequence of the protein or from sequence comparison of the protein from other species. These PCR fragments are then used as homologous probes to identify the full-length genomic clone. This procedure has been successful for the isolation of the *A. fumigatus* alkaline protease gene (Tang *et al.*, 1992).

A less common and more laborious procedure that is possible in *A. nidulans,* due to the number of existing cloned genes and the availability of an excellent genetic map, is chromosome walking from a known to an unknown gene using a bacteriophage lambda or cosmid genomic library. Overlapping clones are tested for complementation of a mutation in the desired gene. The disadvantage of this method is primarily the number of clones that need to be screened, as what may seem a short distance in classic genetic map units may in fact be several tens or hundreds of kilobase pairs in actual physical distance. For instance, the *amdA* gene of *A. nidulans* required screening of approximately 20 phage clones representing some 120 kb in a walk from the gene *gatA* (R. Lints and M. J. Hynes, unpublished data). The development of chromosome-specific subcollections of cosmid clones by Timberlake and colleagues (Brody *et al.*, 1991) and the organization of overlapping clones into large contiguous chromosomal regions or maps known as "contigs" should greatly facilitate position-based cloning.

To date, the DNA sequences of at least 70 *Aspergillus* genes (reviewed by Ballance, 1986; Gurr *et al.*, 1987; Rambosek and Leach, 1987; Unkles, 1992) have been determined, around 50 from *A. nidulans* alone, and many more are reported to have been isolated. The isolation and characterization of genes provides invaluable information on gene regulation and structure, which is necessary for the biotechnological manipulation of these organisms.

2.2. Transformation of the Model Fungus *Aspergillus nidulans*

It has been almost 10 years since the first reports of genetic transformation of *A. nidulans* (Ballance *et al.*, 1983; Tilburn *et al.*, 1983; John and Peberdy, 1984; Yelton *et al.*, 1984) following earlier success with *Neurospora crassa* (Case *et al.*, 1979). There are a number of excellent fungal transformation reviews, which include aspergilli, available for further information (e.g., Arst and Scazzocchio, 1985; Turner and Ballance, 1985; Turner *et al.*, 1985; Fincham, 1989; Timberlake, 1991; May, 1992).

2.2.1. Selectable Markers

Among the earliest reports of *A. nidulans* transformation was that using the *N. crassa pyr-4* gene to reverse the auxotroph requirement for

Table I. Nutritional Markers Used to Transform *Aspergillus nidulans*

Marker	Selection	Reference
amdS	Acetamide utilization	Tilburn *et al.* (1983)
argB	Arginine prototrophy	John and Peberdy (1984)
qutE	Quinic acid utilization	Streatfield *et al.* (1992)
trpC	Tryptophan prototrophy	Yelton *et al.* (1984)

uracil of the *A. nidulans pyrG89* mutation, allowing uridine biosynthesis (Ballance *et al.*, 1983). Other contemporary selection systems utilized *A. nidulans* genes for complementation—*argB* [arginine biosynthesis (John and Peberdy, 1984], *trpC* [tryptophan biosynthesis (Yelton *et al.*, 1984], and *amdS* [acetamide utilization (Tilburn *et al.*, 1983)]. Now, many selection systems have been developed for *A. nidulans*. Those in routine use can be conveniently divided into two classes: (1) nutritional selective markers (Table I), which were the basis of the earlier successful transformation attempts, and (2) dominant selection markers (Table II). Further possible selection systems, which at present are not widely used but nevertheless are of potential interest depending on particular studies, include those based on almost any *A. nidulans* gene that has been isolated, for example, *prn* [proline utilization (Hull *et al.*, 1992)], which comes into the first category of nutritional transformation markers.

One of the drawbacks with most nutritional markers is that corresponding mutants need to be isolated for complementation to be successful. This process is usually laborious, time-consuming, and not always successful. However, in this regard, the *pyr-4* and *niaD* systems, for example, are attractive, since loss-of-function mutants can be isolated directly on the basis of resistance to 5-fluoroorotic acid (Ballance *et al.*, 1983) or chlorate (Johnstone *et al.*, 1990; Malardier *et al.*, 1989), respec-

Table II. Dominant Selectable Markers Used to Transform *Aspergillus nidulans*

Marker	Selection	Reference
Phleomycin[a]	Resistance to phleomycin	Austin *et al.* (1990)
benA[b]	Resistance to benomyl	May *et al.* (1985)
oliC[b]	Resistance to oligomycin	M. Ward *et al.* (1986, 1988)

[a]Bacterial genes flanked by fungal transcription signals.
[b]*Aspergillus nidulans* genes.

Table III. Bidirectional Selectable Markers for *Aspergillus nidulans*

Marker	Isolation of mutants on the basis of resistance	Transformation selection	Reference
acuD[a]	Fluoroacetate	Acetate utilization	Gainey *et al.* (1992)
pyr-4[b]	5-Fluoroorotic acid	Uridine prototrophy	Ballance *et al.* (1983)
niaD[a]	Chlorate	Nitrate utilization	Johnstone *et al.* (1990)
sG[a]	Selenate	Sulfate utilization	Buxton *et al.* (1989)

[a]Homologous gene.
[b]Heterologous gene.

tively. Such resistant mutants have a concomitant nutritional growth defect (uridine auxotrophy or nitrate nonutilization, respectively), the phenotype of which can be used for complementation purposes. There are several other selection systems in this category (Table III).

Dominant selection markers (see Table II) are encoded by antibiotic-resistance genes and confer to transformant cells the ability to grow in the presence of an antibiotic. These antibiotic-resistance genes, for example, the hygromycin phosphotransferase gene, can be derived from bacteria, in which case expression of the gene is governed by cloned fungal promoter and terminator sequences, or they can be of fungal origin, the gene containing a mutation that confers resistance to an antimicrobial agent. The latter systems, for benomyl and oligomycin C resistance, are therefore properly described as semidominant, since the population of molecules within a resistant transformant will consist of wild-type as well as mutant varieties, and so the level of resistance will depend on the relative contributions of the two types of molecules (Turner *et al.*, 1985; M. Ward *et al.*, 1988; May, 1992). One obvious advantage of systems employing antibiotic-resistance genes is that they circumvent the requirement for a fungal recipient mutant. Such markers have been widely used in fungal species for which mutants are not available for various reasons (see Section 2.3).

2.2.2. Transformation Procedures

Although successful transformation of filamentous fungi has been achieved using lithium acetate treatment of conidia, electroporation (M. Ward *et al.*, 1989, 1990) and tungsten microprojectiles or "biolistics" (Armaleo *et al.*, 1990), by far the most common procedure for transformation of this group of organisms, including *Aspergillus,* involves the uptake of DNA by protoplasts and their subsequent regeneration to give transformed colonies. The procedure necessitates first the removal of the cell wall in order to generate protoplasts. This can be achieved using a range of lytic enzymes, of which the commercial preparation Novozyme 234 is currently the most common in use. This cocktail of enzymes prepared from *Trichoderma viride* contains principally glucanases and chitinase to which other lytic enzymes such as driselase or β-glucuronidase may be added depending on the laboratory practice. Usually, very young mycelial cells or germlings are used as starting material, and an osmoticum such as sorbitol or salts is used to prevent lysis of protoplasts. Protoplasts in osmoticum are transformed in the presence of calcium ions by the addition of DNA followed by polyethylene glycol, which causes clumping of protoplasts and facilitates DNA uptake. The protoplasts are then placed on regeneration and selection medium allowing growth of transformed colonies. This very basic outline of the transformation procedure is usually refined and optimized by individual laboratories. Comprehensive descriptions of procedures are included in the earlier publications (Ballance *et al.*, 1983; Tilburn *et al.*, 1983; John and Peberdy, 1984; Yelton *et al.*, 1984) and discussed in reviews by Turner and Ballance (1985), Fincham (1989), and May (1992).

2.2.3. Integration Events

Hybridization analysis has shown that transforming DNA is most commonly integrated into the host chromosome in *A. nidulans*. Integrations can take place at either homologous or nonhomologous sites, and several recent reviews detail the proposed mechanisms of these events (Fincham, 1989; Timberlake, 1991; May, 1992).

Multiple integrations appear to be common at both homologous (or resident) and nonhomologous (or ectopic) sites, depending to some extent on the selection system used—*trpC* (Yelton *et al.*, 1984), *argB* (Upshall, 1986), or *niaD* (Malardier *et al.*, 1989) selection tends to produce low copy number transformants, while *amdS* selection can generate high numbers of integrated copies (Wernars *et al.*, 1985). These may be the result of independent integration events at scattered sites, or they may represent tandem integrations into the homologous locus or integration into a vector previously integrated at a nonhomologous site. Ectopic integration usually occurs when there is little or no DNA sequence homology between the vector and the host genome, although a substantial proportion of integration events may take place at ectopic sites even if the transforming DNA has significant sequence homology (Yelton *et al.*, 1984). The frequency with which nonhomologous integrations occur in the latter case seems to be to some extent recipient strain-dependent (Wernars *et al.*, 1985). Additionally, it is not known whether these ectopic integrations reflect recombination between short regions of homology in the vector and host DNA.

Despite the relatively high frequency of nonhomologous integration following transformation in *A. nidulans*, vectors that carry an *A. nidulans* gene most frequently integrate at the resident locus. This property therefore allows the directed manipulation of the genome, including gene replacement and disruption, which in turn permits the creation of desirable mutations or the functional assessment of the effect of mutations created *in vitro*.

Gene replacement occurs when a linear transforming molecule integrates at a homologous locus. Integration of linear fragments takes place by double crossover, and introduction of a desired mutation generated *in vitro* can therefore be made if the mutation is flanked on a linear fragment by regions homologous to the host chromosome (Miller *et al.*, 1985). This procedure of direct gene replacement can also be used to create null mutations if, as in the case of the *pepA* gene of *A. niger* var. *awamori* (see Section 3.1.), the coding region of the gene is replaced with a selective marker, homology with the resident locus being maintained in the 5′ and 3′ noncoding regions. Indirect gene replacement is also possible using circular plasmid containing a gene of interest. Integration of circular DNA at the homologous locus involves a single crossover leading to duplication of the target sequence separated by vector sequences. During meiosis of self-fertilized transformants, intrachromosomal recombination or unequal crossing over can result in plasmid loss and, depending on the position of the recombination event, the retention of the original transforming sequences (Miller *et al.*, 1985). The requirement for a sexual cycle can be circumvented if selection for rare mitotic plasmid loss can be made, for example, using 5-fluoroorotic acid resistance or chlorate resistance to select for eviction of *pyrG*- or *niaD*-containing vector sequences, respectively (May, 1992).

Gene disruptions can be made using circular transforming DNA if that vector contains a fragment of the gene of interest internal to the coding region. Homologous insertion of such a plasmid causes a duplication in which neither copy has the entire coding region. This technique has been used to study, for example, the role of fungal genes in plant pathogenesis and in fungal development processes.

2.2.4. Cotransformation of Nonselectable Genes

For applied molecular purposes, it is frequently necessary to integrate genes for which there is no direct selection. Such is the case with several mammalian gene expression vectors. The commonly used tactic to facilitate this event is to transform a mixture of a vector containing such a gene of commercial interest together with a vector on which is a selectable marker (Section 2.1), usually with the unselectable vector in 5- to 10-fold excess. Cotransformation rates of up to 90% have been observed with *amdS* as selectable marker (Wernars *et al.*, 1987). However, cotransformation frequencies appear to be marker-dependent and can be as low as 15% with the *niaD* system (Campbell *et al.*, 1989).

2.2.5. Replicating Vectors

While sequences that enable replication of plasmids within yeast and *Mucor* cells were fairly straightforward to obtain, it is only recently that such a facility has been available for aspergilli. The main advantages of autonomously replicating vectors is, of course, the substantial increase in transformation frequency from a few to tens of thousands of transformants per microgram of transforming plasmid DNA. Such high frequencies of transformation permit the isolation of genes by "shotgun" complementation of mutants and recovery of plasmids without recourse to excision of the cloned gene from fungal genomic sequences.

Part of the difficulty in obtaining autonomously replicating vectors may have been that there appear to be no native *Aspergillus* plasmids on which to base a cloning plasmid with replicon properties. In addition, attempts to use fungal mitochondrial replicons or to isolate replicons in *S. cerevisiae* failed in terms of replicating activity when returned to *Aspergillus*. However, one such sequence isolated in *S. cerevisiae* (designated *ans-1*) provided a 50- to 100-fold increase in transformation frequency, although little evidence was obtained for its autonomous replication (Ballance and Turner, 1985). Instead, its effect is probably due to increased stability of integration.

A sequence that permits plasmid replication, at least in *A. nidulans*, *A. niger*, and *A. oryzae*, has been isolated recently by Clutterbuck and colleagues (Gems *et al.*, 1991). The fragment was isolated from an *A. nidulans* gene library in an *argB* vector on the basis that it resulted in reduced stability of the transforming *argB* gene with concomitant reduction in growth rate. Plasmid rescue from a slow-growing *argB* transformant allowed recovery of a plasmid designated ARp1, which contained the *argB* gene and a 6.1 kb sequence responsible for autonomous replication, termed *AMA1*. This sequence includes a short unique region flanked by long inverted repeats, and its inclusion in an *argB* vector results in a 250-fold increase in transformation frequency in *A. nidulans*. The ARp1 vector also replicates autonomously in *A. oryzae* and *A. niger* with 30-fold and 25-fold increases in frequencies, respectively. Using as little as 250 ng transforming DNA, 120,000 transformants can be obtained in *A. nidulans*. A further advantage of this vector is that so-called "instant gene" banks can be synthesized (Gems and Clutterbuck, 1993). In this procedure, an *A. nidulans* mutant is cotransformed with ARp1, or its derivative pDHG25, and total genomic DNA from a wild-type strain. Arginine prototrophs are tested for reversal of the mutation of interest and for instability of that trait. Plasmid rescue from such strains can result in recovery of a molecule containing the gene corresponding to the mutation of interest. The basis of the recombination events that result in the instant gene bank are unclear, but presumably the *AMA1* sequence allows the vector to "pick up" cotransforming sequences at random. The drawback of the technique appears to be that rearrangements of sequences are frequent, but it can nevertheless provide fragments of a gene that could be used as a hybridization probe to a conventional gene bank.

2.3. Transformation of Other Aspergilli

Following successful transformation of *A. nidulans,* transformation systems have been developed for a number of other *Aspergillus* species. These include *A. niger, A. oryzae,* and *A. flavus,* which are industrially important due to their ability to produce copious amounts of primary metabolites that are industrially important, such as citric acid, as well as enzymes used in many industrial and pharmaceutical processes. Additionally, there is interest in *A. niger* and *A. oryzae* as vehicles for production of foreign proteins (see Section 3). Development of transformation thus allows the molecular manipulation of commercially important genes or biosynthetic pathways. Some of the selection systems available for transformation are shown in Table IV.

The choice of selection system is to some extent more limited in aspergilli other than *A. nidulans* due to the general lack of basic classic genetic information. Particularly in industrial strains, well-characterized mutants for use as recipients for nutritional markers may not be available. Nevertheless, transformation has been developed in *A. niger* and *A. oryzae* using the *argB* gene of *A. nidulans* to repair arginine auxotrophic mutants (Buxton *et al.*, 1985; Gomi *et al.*, 1987; Hahm and Batt, 1988), and methionine prototrophic transformants of *A. oryzae* have been obtained by introduction of the *A. oryzae metA* gene to a methionine auxotrophic strain (Iimura *et al.*, 1987). Industrial strains, however, are often diploid or aneuploid, making the isolation of nutritional mutants extremely difficult. Additionally, mutagenesis may introduce undesirable secondary mutations in industrial strains, hence the attraction of

Table IV. Examples of Transformation Selection Systems for Other Aspergilli

Species	Marker	Selection	References
A. niger	*amdS*[b]	Acetamide utilization	Kelly and Hynes (1985)
	argB[b]	Arginine prototrophy	Buxton *et al.* (1985)
	Hygromycin	Resistance to hygromycin	Punt *et al.* (1987)
	niaD[a]	Nitrate utilization	Unkles *et al.* (1989a)
	oliC[a]	Resistance to oligomycin C	M. Ward *et al.* (1988)
	pyrG[a]	Uridine prototrophy	van Hartingsveldt *et al.* (1987)
A. oryzae	*amdS*[a]	Acetamide utilization	Gomi *et al.* (1991)
	argB[a,b]	Arginine prototrophy	Hahm and Batt (1988), Gomi *et al.* (1987)
	metA[a]	Methionine prototrophy	Iimura *et al.* (1987)
	niaD[a]	Nitrate utilization	Unkles *et al.* (1989b)
	pyrG[a,b]	Uridine prototrophy	Mattern *et al.* (1987), De Ruiter-Jacobs *et al.* (1989)
A. flavus	*benA*[b]	Benomyl resistance	Seip *et al.* (1990)
	pyr4	Uridine prototrophy	Woloshuk *et al.* (1987)
	niaD[b]	Nitrate utilization	Chevalet *et al.* (1992)
A. parasiticus	*niaD*[a]	Nitrate utilization	Chang *et al.* (1992)
A. fumigatus	Hygromycin	Resistance to hygromycin	Tang *et al.* (1992)

[a]Homologous gene.
[b]Heterologous gene.

systems involving dominant selectable markers such as resistance to hygromycin, phleomycin, benomyl, or oligomycin, which require only that the organism be suitably susceptible to the antibiotic used (see Section 2.2.1). Vectors generated using *A. nidulans* transcription signals have been successfully transformed into other aspergilli; for example, the vector pAN7-1 with the *A. nidulans* glyceraldehyde-3-phosphate dehydrogenase (*gpdA*) promoter and *trpC* terminator allowing transcription of the bacterial hygromycin-resistance gene is used for transformation of *A. niger* (Punt *et al.*, 1987). A useful alternative to dominant selectable markers, however, is those systems in which the corresponding recipient mutant can be easily selected without the need for mutagenesis. Thus, systems based on selection for nitrate utilization or uridine prototrophy have been developed using either a heterologous gene, as in *A. flavus* transformation with the *A. oryzae niaD* gene (Chevalet *et al.*, 1992), or the homologous gene, such as *A. niger niaD* (Unkles *et al.*, 1989a), *pyrG* (van Hartingsveldt *et al.*, 1987), or *A. parasiticus niaD* (Chang *et al.*, 1992).

Transformation procedures are generally based on those developed for *A. nidulans* using calcium and polyethylene glycol-induced uptake of DNA by protoplasts. Some attempts at transformation have also been made using electroporation, for example, in *A. niger* (M. Ward *et al.*, 1990), but this technique has not overtaken the more conventional methods.

2.4. Development of Systems to Analyze Gene Expression

More than 100 *Aspergillus* genes have been isolated since the *aromA* gene (see Section 2.1). Recently, research has focused on studying the 5′ non-protein-coding end of fungal genes—sequences that are responsible for the gene expression. A comprehensive review by Punt and van den Hondel (1992) on this topic has appeared recently. Initial approaches usually involve nucleotide comparisons of sequences between corresponding genes isolated from different fungi on similarly regulated genes, the rationale being that functionally essential sequences would be retained. Similar motifs have been observed in a number of systems; e.g., a 50-nucleotide stretch upstream of the *gpdA* gene, encoding glyceraldehyde-3-phosphate dehydrogenase, has been observed in *A. nidulans* and *A. niger* with a 95% similarity (Punt *et al.*, 1990). The function of such motifs is unclear without functional studies being performed, but such comparisons can provide clues, at least. Furthermore, such sequences can be of applied interest. For example, the insertion of the *gpdA* motif into the 5′ sequences of other fungal genes, including ones of direct commercial interest such as glucoamylase, can (1) deregulate the expression of such genes and (2) increase their expression by up to 50-fold.

In many cases, the fungal upstream region is fused *in vivo* to bacterial reporter systems, such as *lacZ* [β-galactosidase (van Gorcom *et al.*, 1985, 1986)] or *uidA* [β-glucuronidase (Roberts *et al.*, 1989)], which are very easy to assay enzymatically (reviewed by van den Hondel *et al.*, 1985). These systems are particularly useful for measuring expression levels of *Aspergillus* genes the products of which are difficult to assay, e.g., genes involved in mitosis, development, and other functions. A further embellishment of this approach is to subject the 5′ prime ends to *in vitro* mutational analysis, transform mutant constructs in *Aspergillus*, and determine the effect of the mutation in its effect on the reporter gene. Such work has been useful for the determination of upstream nucleotide sequences involved in the regulation of the important industrial enzyme amylase in *A. oryzae*. When transformed into *A. oryzae*, certain deletions of the amylase promoter region fused to the *uidA* gene result in loss of inducibility of β-glucuronidase activity (Tada *et al.*, 1991; Sakaguchi *et al.*, 1992).

A further technique takes advantage of the fact that regulatory gene products (proteins) bind to sequences in the 5′ flanking regions to effect control of the structural gene that the control gene regulates. Several *in vitro* methods have been reported to recognize important binding motifs in the 5′ region. Preliminary experiments identify by gel shift assays a small fragment or oligonucleotide, from the upstream region of the regulated gene, which binds specifically to the regulatory protein. Binding reactions may be carried out with crude nuclear extract or with the putative DNA binding region of the regulatory protein purified from extracts of *E. coli* in which it is expressed in an *E. coli* expression vector. Precise nucleotides that interact with the DNA binding domain of the regulatory protein can then be determined by footprinting. Kulmburg *et al.* (1992) have shown by DNAase1 footprinting and methylation interference that the short nucleotide sequence CCGCA in the upstream region of the *A. nidulans alcA* gene interacts specifically with the zinc finger domain of the *alcR* regulatory product.

2.5. Electrophoretic Karyotypes of Aspergilli

The development of pulse-field gel electrophoresis (PFGE) has permitted the separation of linear chromosomes from a number of organisms, including several aspergilli. In this regard, Brody and Carbon (1989) used the CHEF system, a refinement of the original method, to resolve the genome of *A. nidulans* (Fig. 1). Six chromosomal bands were observed, of which two showed higher intensities representing doublets. The eight chromosomes were therefore in agreement with classic genetic and cytological studies that strongly suggested eight linkage groups. The size of the chromosomes was between 2.9 and 5.0 megabases (Mb), from which an estimate of 31 Mb for the total genome was calculated. Combining electrophoresis with Southern blotting and DNA hybridization using probes of four genes that were previously mapped by formal genetics to individual linkage groups provided equivalence between linkage group (as determined by classic genetics) and chromosome (as determined by electrophoresis) (Fig. 2). *Aspergillus nidulans* mutant strains carrying known chromosomal translocations were used to identify the location of the other four linkage groups by observing changes in gel migration (Brody and Carbon, 1989). In this way, correlation between linkage group and all eight chromosomes was made (Table V).

The technique of chromosome separation has been particularly useful for assignment of genes to chromosomes by Southern hybridization in species in which the lack of a classic genetic system makes this impossible. Thus, linkage may be established using either cloned homo-

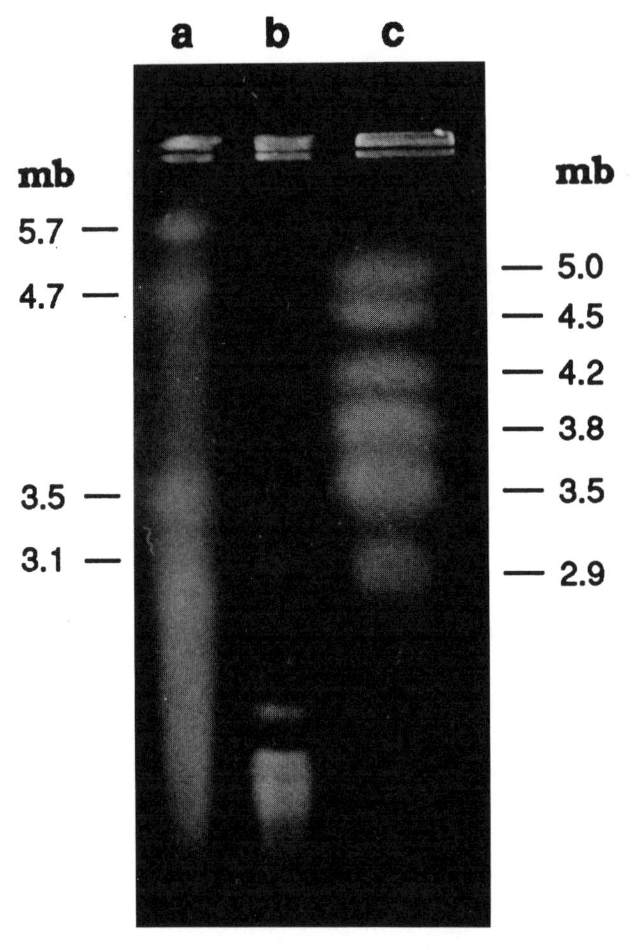

Figure 1. Separation of *A. nidulans* intact chromosomal DNAs using a CHEF gel electrophoresis. Chromosomal DNAs were prepared from *Sc. pombe* strain 334 (lane a), *S. cerevisiae* (lane b), and *A. nidulans* wild-type strain (lane c). The approximate lengths of the resolved chromosome bands are indicated for *Sc. pombe* (lane a) and *A. nidulans* (lane c). Data taken from Brody and Carbon (1989).

logous gene probes or heterologous gene probes. The ability to separate the chromosomes is currently permitting the generation of chromosome-specific gene libraries for several aspergilli. Furthermore, this technique may allow the construction of artificial chromosomes for the aspergilli containing multiple copies of industrially important genes and their essential regulators. Seven chromosomes have been resolved in *A. niger* var. *awarmori* ranging in size from 2.9 Mb to 6.2 Mb (Dunn-Coleman *et*

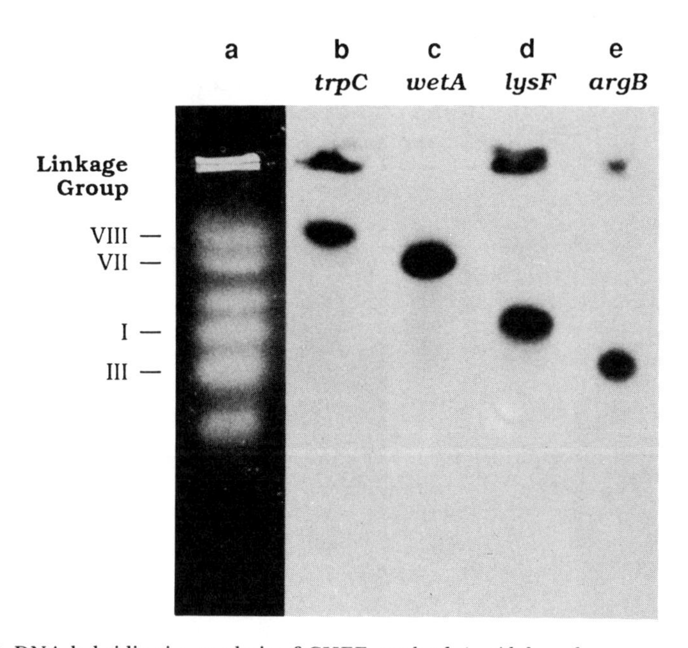

Figure 2. DNA hybridization analysis of CHEF-resolved *A. nidulans* chromosome DNAs. Agarose plugs containing intact chromosomal DNAs from *A. nidulans* wild-type strain FGSC4 (lanes a–e) were electrophoresed and stained with ethidium bromide (lane a), and the DNAs were transferred to a nylon membrane (lanes b–e). Hybridization of separated chromosomes with labeled DNA gene probes specific to four of the eight *A. nidulans* linkage groups (LGs). After transfer of the DNAs to the nylon membrane, the membrane was cut and probed. The genes used as probes are indicated above the autoradiogram. The LGs of the gene probes are as follows: *trpC,* LG VIII; *wetA,* LG VII; *lysF,* LG I; and *argB,* LG III. Data taken from Brody and Carbon (1989).

al., 1992) and seven in *A. niger* from 3.5 to 6.6 Mb (Debets *et al.,* 1990), while eight have been observed for *A. oryzae,* from 2.5 to 6.5 Mb (K. Kitamoto, unpublished data). An industrial application of the technique may be the description of strains, developed by classic or recombinant means, in terms of their overall chromosomal pattern in ethidium-bromide-stained gels.

2.6. Classification and Differentiation of Species and Strains

The ability to recognize unequivocally a fungal species or strain is of importance for two main reasons. The first arises from the similarity between certain species, such as the black aspergilli, some of which—for example, *A. niger*—are commonly used in the food industry and others of which are pathogenic. Distinguishing these different species on the

Table V. Equivalence of *Aspergillus nidulans* Classic Genetic Linkage Group and Chromosomal Electrophoretic Karyotype[a]

Linkage group	Chromosome size (Mb)
I	3.8
II	4.2
III	3.5
IV	2.9
V	3.8
VI	3.5
VII	4.5
VIII	5.0

[a]For further information on the correlation between linkage groups and chromosomal bands by PFGE, see Brody and Carbon (1989). For the characterization of linkage groups by classic genetics, see Clutterbuck (1984).

basis of morphological or biochemical tests is extremely difficult, but taxonomic identification is desirable in view of their biotechnological importance. The second reason for precise differentiation is the potential for proprietary labeling of strains used in industrial processes.

RFLP analysis has been used to reclassify the black aspergilli. Detection of differences in ribosomal DNA banding patterns and in hybridization patterns using cloned *A. niger* genes has allowed the separation of these strains into several reproducibly defined species (Kusters-van Someren *et al.,* 1991). Another technique that may be of use is that of DNA fingerprinting using DNA probes or oligonucleotides that hybridize to hypervariable repetitive sequences or "microsatellite DNA." Such probes have been used successfully in *A. niger* to demonstrate the presence of these repetitive loci (Meyer *et al.,* 1991).

3. EXPRESSION OF FOREIGN PROTEINS

With the development of plasmid-mediated genetic transformation systems and related gene-cloning technology for fungi, there has been considerable interest in applying the technology to the industrial situation. This has led to the isolation of a number of genes of commercial importance, such as the genes required for biosynthesis of penicillin and of extracellular enzymes, e.g., amylases, glucoamylases, pectinases, phytases, proteases, cellulases, and ligninases. By increasing the gene copy, elevated levels of antibiotics and enzymes may be achieved [see Kinghorn and Turner (1992) and reviews therein]. The aspergilli, in particular, have offered alternative systems to *E. coli* and yeast for the expression of recombinant proteins. These include fungal (from other fungi) and foreign (nonfungal, eukaryotic in this review) proteins. In the former regard, fungal enzyme-encoding genes of commercial interest such as the cellulase genes (isolated from *Trichoderma,* for example) may be transferred to the aspergilli, since aspergilli are often easier to handle. In the latter regard, GRAS (Generally Regarded As Safe) status has been bestowed by the United States Food and Drug Administration and the World Health Organization on certain aspergilli such as *A. niger* and *A. oryzae,* making them acceptable hosts for the production of recombinant proteins of pharmaceutical interest. There have been a number of attempts to program filamentous fungi to produce foreign proteins, and this topic is discussed below.

There has been considerable commercial interest in exploiting certain filamentous fungi, especially the aspergilli, as hosts particularly for the production of novel mammalian proteins (van Brunt, 1986). There are several reasons for this interest. First, aspergilli have the ability to secrete copious amounts of their own enzymes. This can be as much as 25 g/liter of glucoamylase by *A. niger* (although there is a general feeling this may well be much higher, since industrial companies tend to be "economical" with dissemination of data). Second, filamentous fungi have a traditional history of industrial use for antibiotics and enzymes, for primary metabolites such as citric acid, and for Oriental alcoholic beverage production. There are therefore already fermentation procedures, conditions, and equipment in place (Solomons, 1980). Third, filamentous fungi such as *A. niger* and *A. oryzae* have so-called GRAS status. Fourth, certain proteins have been found to be difficult to express or are inactive in *E. coli* or *S. cerevisiae,* and consequently the performance of filamentous fungi has been evaluated (van Brunt, 1986). Finally, the fact that many bacterial systems and yeast were encumbered by patents provided additional incentive to assess filamentous fungi.

Aspergillus has been the filamentous fungal genus most exploited for the purpose of foreign protein expression. This choice is perhaps due to the combination of the factors stated above and to the molecular genetics of the "model" *A. nidulans,* which is more advanced than any other filamentous fungus. It must be pointed out that as *A. nidulans* itself is used as a host for heterologous expression of industrial proteins, the distinction between "model" and "industrial" fungus becomes blurred. *A. nidulans* together with *A. niger* have been used most commonly to express mammalian proteins. Recent reviews of heterologous expression by filamentous fungi are available for further detail (Saunders *et al.*, 1989; van den Hondel *et al.*, 1991, 1992; Davies, 1991a,b, 1992; Gwynne *et al.*, 1987a; Gwynne, 1992; Dunn-Coleman *et al.*, 1992).

There are two principal reasons for expressing a protein in a heterologous host, and these to some extent dictate the level at which expression is deemed to be successful. The first is the wish to study the structure or function, or both, of a protein in terms of either biological activity or protein engineering. In this case, the levels of protein expressed need not be extremely high as long as the protein can be obtained in a form that is indistinguishable (or almost so) from the authentic molecule. The second reason for heterologous expression is commercial. The availability of certain proteins such as chymosin (see Section 3.1.) may be erratic, and so expression in a microbial host can fulfill demand. As long as the levels expressed are sufficiently high, heterologous production is therefore commercially lucrative. Alternatively, pharmaceutically useful proteins that previously were obtainable only from biological samples such as serum—among which are interferon α-2 (Section 3.2.) and tissue plasminogen activator (Section 3.3.)—can be profitable even if they are produced at much lower levels.

Before successful expression of a foreign gene within a fungal cell can take place, that gene must be inserted between an efficient fungal transcriptional promoter system and terminator region. A number of filamentous fungal genes have been used commonly for this purpose. Figure 3 outlines a typical plasmid expression construct in which glucoamylase (*glaA*) transcriptional sequences are used to drive the expression of the human interleukin-6 (hIL-6) gene. In addition to this expression cassette, the plasmid molecule may contain a selectable marker, such as *niaD* (Section 2.2.1), for successful transformation into the host (Section 2.2.1). However, it is often unnecessary to incorporate a marker into the expression plasmid, as entry can be effected using cotransformation approaches (Section 2.2.4). Case situations of foreign genes (nonfungal eukaryotic) expressed so far in aspergilli are given in Tables VI and VII and discussed below. Heterologous expression of fungal and bacterial genes is extensively reviewed by van den Hondel *et al.* (1991).

3.1. Calf Chymosin

Researchers at Genencor (San Francisco, California) developed a successful system for the synthesis of recombinant calf chymosin in *A. niger* (var. *awamori*) (reviewed by M. Ward, 1991), making it the first commercially successful mammalian product synthesized by filamentous

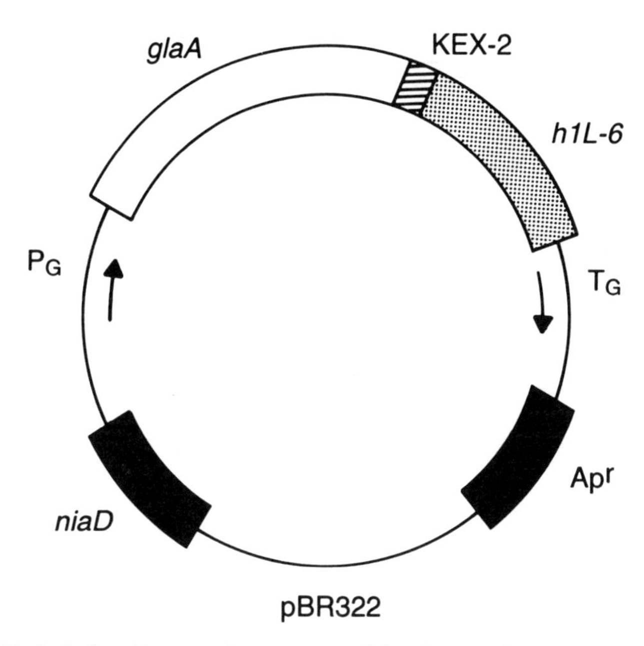

Figure 3. Typical plasmid expression vector used for the transformation and expression of foreign genes in aspergilli. Human interleukin-6 (hIL-6) is fused to the *Aspergillus* glucoamylase gene (*glaA*). A proteolytic processing recognition site such as KEX-2 may be included to facilitate cleavage of product from the fusion protein by host proteases. (P_G, T_G) Promoter and terminator, respectively, of the *glaA* gene. The *niaD* gene permits selection for transformants in an *niaD* mutant *Aspergillus* host. The *Ap*^{r} (for ampicillin resistance) gene is used to maintain the vector in *E. coli.* Plasmid BR322 sequences contain a bacterial origin of replication that allows autonomous replication on *E. coli.*

fungi. Chymosin is an important enzyme used in cheese manufacture, the natural protein often being in short supply. Chymosin is synthesized as preprochymosin from which the 16-amino-acid pre sequence is cleaved during secretion, followed by the 42-amino-acid pro sequence, which is autocatalytically cleaved at low pH.

Early attempts at expression and secretion of chymosin utilized the *A. niger* glucoamylase (*glaA*) gene promoter and secretion signal to direct expression of the bovine prochymosin cDNA clone in *A. nidulans* (Cullen *et al.*, 1987) and *A. niger* var. *awamori* (M. Ward *et al.*, 1990), but yields were well below those necessary for commercial development despite the demonstration of abundant chymosin mRNA. Significant increase in yields was obtained using *A. niger* var. *awamori* as host and the prochymosin cDNA fused to the entire *A. niger* var. *awamori glaA* gene, i.e., promoter, signal sequence, and protein-coding region (M. Ward *et al.*,

Table VI. Foreign Proteins Secreted by *Aspergillus nidulans*[a]

Protein	Yield[b]	Expression signals[c]		Reference
		Promoter	Terminator	
Human interferon α-2	1 mg/liter	*alcA*	*glaA*	Gwynne *et al.* (1987b)
Human lactoferrin	5 mg/liter	*alcA*	*benA*	P. P. Ward *et al.* (1992a)
Human tissue plasminogen activator	100 μg/liter	*thiA*	*tpiA*	Upshall *et al.* (1987)
Human tissue plasminogen activator	1 mg/liter	*alcA*	*tpiA*	Upshall *et al.* (1987)
Human α-1 antitrypsin	1 mg/liter	*alcA*	*thiA*	Upshall *et al.* (1987)
Human granulocyte macrophage colony stimulating factor	1 mg/liter	*alcA*	*tpiA*	Upshall *et al.* (1987)
Human interleukin-6	2 mg/liter	Complete *glaA* gene[d]		Contreras *et al.* (1991)
Cattle tick surface	1.8 mg/liter	*amdS*	*amdS*	Turnbull *et al.* (1990)

[a]Foreign refers to nonfungal eukaryotic proteins.
[b]Highest published yields quoted.
[c]For further details of the fungal genes used in expression cassettes, see the text or primary references.
[d]For construct details, see Section 3.7.

Table VII. Foreign Proteins Secreted by *Aspergillus niger*[a]

Protein	Yield[b]	Expression signals[c]		Reference
		Promoter	Terminator	
Calf chymosin[d]	<1000 mg/liter	Complete *glaA* gene[e]		Dunn-Coleman *et al.* (1991)
Hen egg lysozyme	1 mg/liter	*gpdA*	*trpC*	Archer *et al.* (1990)
Hen egg lysozyme	12 mg/liter	*glaA*	*glaA*	Archer *et al.* (1990)
Porcine pancreatic glycoprotein A^2	10 mg/liter	Complete *glaA* gene[e]		Roberts *et al.* (1992)

[a]Foreign refers to nonfungal eukaryotic proteins.
[b]Highest published yields quoted.
[c]For further details of the fungal genes used in expression cassettes, see the text or primary references.
[d]*Aspergillus niger* var. *awamori* was used as the host.
[e]For plasmid construct details, see Sections 3.4 and 3.5.

1990). The protein was thus secreted as a fusion to glucoamylase and subsequently underwent autocatalytic cleavage by reduction of the medium pH to 2. One problem that was identified as a factor contributing to low yields was the production by *A. niger* of an extracellular aspartyl protease, aspergillopepsin, which caused degradation of the secreted chymosin. The gene encoding this protease, *pepA*, was cloned and a vector generated in which the *pepA* coding region was replaced by an *argB* gene (Berka *et al.*, 1990). This vector allowed deletion of *pepA* by direct gene replacement following transformation and selection for arginine prototrophy. Deletion of the *pepA* gene resulted in a further chymosin yield increase, but still below the level of commercial feasibility. The group therefore embarked on a program of classic mutagenesis and selection for yield improvement. This was facilitated by the availability of a simple milk turbidity assay for chymosin activity and a robotic screening procedure that allowed the testing of 50,000–60,000 mutated viable spores per screen. From expression levels of 250 mg/liter of extracellular chymosin, six rounds of classic mutational strain improvement resulted in 650 mg/liter of extracellular chymosin. Further increase in yield was obtained by selection for 2-deoxyglucose-resistant mutants. It has been suggested that this toxic analogue of glucose may affect carbon regulation and result in mutations that relieve glucose repression. The resulting production strains achieved greater than 1 g/liter of extracellular chymosin (Dunn-Coleman *et al.*, 1991).

Chymosin is now the first heterologous fungal product that has the approval of the United States Food and Drug Administration. These high-chymosin-yielding mutant strains may be useful for production of other heterologous proteins, as it has been shown that they have elevated levels of many extracellular proteins, such as α-amylase, presumably due to the introduction of mutations that result in supersecretion. Such mutations were shown to be associated with the host genome and not the result of mutations within the transforming chymosin vector. This was determined by curing chymosin production strains of the integrated vector, which carried the *pyrG* gene, by selection for 5-fluoroorotic acid resistance (Sections 2.2.1 and 2.2.3). A proportion of those colonies selected were shown by Southern blotting to have completely deleted the original vector. Retransformation of the cured strain with another chymosin-encoding vector resulted in some transformants that produced chymosin levels equivalent to that of the strain before curing. Transformation of a chymosin vector into a high-yielding glucoamylase production strain, however, did not result in levels of chymosin above those of the progenitor of the chymosin production strain, i.e., the original strain before mutagenesis and selection. This evidence suggests that the mutations in the chymosin production strain direct increased secretion particularly of heterologous proteins and that such a strain may be of general use as a host for heterologous protein production (Dunn-Coleman *et al.*, 1991).

3.2. Human Interferon α-2

Interferon α-2 is a secreted protein with antiviral properties. The main source of the protein is human serum, and its purification is thus extremely expensive and difficult. Hence, expression of this protein in a microbial host even at low levels could be commercially attractive. Research workers at Allelix Biopharmaceuticals (Toronto, Canada) have attempted this task using *A. nidulans* as host organism (Gwynne *et al.*, 1987b). The vectors that they employed consisted of the *A. niger glaA* promoter and signal sequence or the *A. nidulans alcA* (encoding alcohol dehydrogenase) promoter and a synthetic signal sequence fused to the mature interferon α-2 cDNA. These vectors were cotransformed using an *argB* selectable marker into *A. nidulans* (Section 2.2.1). Cotransformation was shown to be efficient with up to 85% of transformants showing hybridization to interferon-encoding sequences in Southern blots. Most transformants had multiple tandem copies of the vector, and interferon-specific mRNA was detectable.

Both promoters used in the constructs are inducible by either starch (*glaA*) or threonine (*alcA*) and repressed by glucose. Regulation of the *A. niger glaA* promoter appeared to be normal in *A. nidulans*. Low levels of secreted interferon α-2 were detected in culture supernatants of cells grown in inducing conditions by a viral plaque reduction assay, and following purification on a monoclonal antibody affinity column, a product of around the expected size could be observed on sodium dodecyl sulfate (SDS)–polyacrylamide gels. The *alcA*-based vector system was optimized by creation of a characteristic Kozak sequence around the initiation codon (Kozak, 1984) and by the use of an *A. nidulans* strain that contained multiple copies of *alcR*. This gene encodes a *trans*-acting positive regulatory protein required for induction of *alcA*. In the original transformants, there were insufficient levels of *alcR* product to permit transcription from many *alcA* promoters, but use of an *alcR* overproducing host strain overcame this titration effect and allowed secretion of up to 1 mg interferon α-2/liter.

3.3. Human Tissue Plasminogen Activator

Successful attempts at expressing human tissue plasminogen activator (t-PA) in *A. nidulans* have been carried out at Zymogenetics (Seattle,

Washington). This enzyme can be used in blood clot removal following strokes and heart attacks. The protein is a glycosylated serine protease that in humans is secreted as a propeptide that undergoes proteolytic processing to generate a mature protein with glycine as the N-terminal residue. Amino peptidase activity can remove a further four residues, producing a second single-chain form with a serine at the N-terminus.

Initial expression vectors utilized the promoter and terminator sequences of the constitutive triosephosphate isomerase (*tpiA*) gene of *A. nidulans* in a vector that included *argB* as a selectable marker (Upshall *et al.*, 1987). A precise translational fusion of the *tpiA* promoter to a cDNA encoding pre-pro t-PA (i.e., including signal sequence) was generated and the construct transformed into *A. nidulans.* Arginine prototrophs were screened for the ability to produce t-PA in a fibrin lysis assay using agar plugs of transformant mycelium. The presence of the expression construct was confirmed by Southern hybridization, and some correlation of copy number to the level of secreted t-PA was demonstrated. Secreted t-PA, partially purified from liquid cultures by monoclonal antibody affinity chromatography, showed two species in Western blots of a size similar to authentic t-PA single chains. Thus, it appeared that *A. nidulans* was capable of mimicking the differential glycosylation of single-chain forms leading to a doublet seen when the protein is secreted from mammalian cells. N-terminal amino acid sequence analysis demonstrated that both the mature glycine and serine forms were present, indicating correct processing of the propeptide and subsequent aminopeptidase activity in *A. nidulans,* although the ratios of the two forms were different in the mammalian and fungal products. Levels of up to 100 μg active t-PA/liter were secreted, but this was increased to 1 mg/liter by the use of expression signals from the ethanol-inducible *A. nidulans* gene encoding alcohol dehydrogenase 3 (*alcC*) or a constitutive *A. niger* gene encoding alcohol dehydrogenase (*adhA*). Additionally, using these expression signals, similar levels of secreted human α-1-antitrypsin and human granulocyte-macrophage colony-stimulating factor were obtained (Upshall *et al.,* 1987).

3.4. Hen Egg White Lysozyme

Hen egg white lysozyme (HEWL) was the first enzyme to have its three-dimensional structure determined by X-ray chrystallography and is an excellent model for the study of structure–function relationships of individual amino acid residues within the protein using *in vitro* mutagenesis. However, although the native hen protein may be abundant, another production system is necessary for the analysis of mutant proteins produced by recombinant DNA techniques. The feasibility of *A. niger* as a host for expression of these recombinant proteins was assessed (Archer *et al.,* 1990) after inefficient attempts at expression in *E. coli* and *S. cerevisiae.*

With fusion of the entire HEWL cDNA clone encoding the enzyme and its signal sequence between the *A. niger* var. *awamori glaA* promoter and terminator, levels of up to 12 mg lysozyme/liter were secreted in medium containing starch for induction of the *glaA* promoter. The recombinant protein was purified from culture filtrates by FPLC and was indistinguishable on SDS–polyacrylamide gels from natural HEWL. Likewise, the catalytic activity of the recombinant and natural proteins was identical. Some proteolysis (around 10%) was shown to have occurred by the presence of faint bands of lower molecular weight that reacted with anti-HEWL antibodies. N-terminal amino acid sequence determination of these minor products suggested that the protein was specifically cleaved by an endopeptidase. This was subsequently shown to be a mycelium-associated protease distinct from the product of *pepA* or other secreted proteases (Archer *et al.,* 1992). Amino acid sequence determination of intact recombinant lysozyme showed that the N-terminus was identical to natural lysozyme. *A. niger* is therefore capable of recognizing and correctly processing the hen signal sequence. Further analysis of the recombinant protein using two-dimensional nuclear magnetic resonance spectroscopy demonstrated that the protein was correctly folded in *A. niger* with authentic disulfide bond formation, the spectrum being identical to that of natural HEWL.

3.5. Porcine Pancreatic Prophospholipase A_2

Extensive biochemical and biophysical information is available for prophospholipase A_2 (PLA_2), making it an ideal model for protein engineering (Roberts *et al.,* 1992). As with HEWL, it would be useful to have an efficient heterologous host for production of the enzyme in order to perform structure–function studies. This has been accomplished in an *A. niger* host using a vector containing the *A. niger glaA* promoter and terminator sequences for expression and *pyrG* for transformant selection (Section 2.2.1). The successful construct contained in-frame fusions of the *glaA* promoter to a fragment, obtained using fusion PCR, containing the entire *glaA* signal and coding sequence in frame with the pro-PLA_2 cDNA. Deletion of the *pepA* gene was essential for detection of the recombinant protein. Several forms of the protein were seen on polyacrylamide gels, one of which corresponded in size to natural PLA_2, and PLA_2 enzyme activity could be assayed in culture filtrates of transformed strains. Purification of recombinant PLA_2 by ion-exchange chromatography and FPLC resulted in three major protein peaks indistinguishable in size from authentic PLA_2 on SDS–polyacrylamide gels. N-terminal amino acid sequence determination showed that one of these forms had been processed from the fusion correctly by *A. niger,* cleavage occurring after an arginine residue, while in the other two forms, processing had occurred between two serine residues, which resulted in inclusion of two residues of the propeptide. Yields of active enzyme were up to 10 mg PLA_2/liter.

3.6. Cattle Tick Cell-Surface Glycoprotein (Bm 86)

The cell surface glycoprotein Bm86 from the cattle tick *Boophilus microplus* has been used to elicit a protective immune response against the ectoparasite, but it is difficult to isolate sufficient quantities. Using the *A. nidulans amdS* promoter and terminator sequences, vectors were created that could express the Bm86 propeptide when cotransformed into *A. nidulans* (Turnbull *et al.,* 1990). Cotransformants with high copy numbers of the expression vector (Section 2.2.4) were recognized by their reduced ability to grow on 2-pyrrolidinone due to the extra copies of the *amdS* 5′ region, causing titration of the *amdR* product, which is required for growth on that substrate (Kelly and Hynes, 1987). The presence of the vector was confirmed by Southern hybridization, and Northern blotting showed the presence of abundant Bm86-hybridizing transcript of the expected size. All detectable recombinant protein was secreted, indicating recognition of the Bm86 signal sequence. Western blots showed a broad band with an apparent molecular mass greater than that expected and similar to the native protein. This was probably due to recognition by *A. nidulans* of the several glycosylation sites in the authentic protein. Using complex medium, 1.8 mg Bm86/liter could be detected in the culture medium. *A. niger* was also tested as a host, but the expression levels were much reduced.

3.7. Human Interleukin-6

The cytokine hIL-6 is a secreted glycoprotein with antiviral and B-cell proliferative properties. Preliminary attempts to express hIL-6 in *A. nidulans* were carried out using vectors in which the sequence encoding mature hIL-6 was fused to the starch-inducible *A. niger glaA* promoter and signal sequence or in which transcription of the hIL-6 cDNA including its own signal sequence was driven by the strong constitutive *A. nidulans gpdA* promoter (Carrez *et al.,* 1990). Expression vectors were introduced by cotransformation with an *argB* selectable marker, and several arginine prototrophic transformants were shown by Southern hybridization to contain multiple copies of the hIL-6 gene. The majority of hIL-6 produced was secreted, but maximum levels reached only 20 μg hIL-6/liter.

In order to increase the level of secreted hIL-6, constructs were generated, again using the *glaA* or *gpdA* gene promoter to drive transcription, but including the entire glucoamylase precursor protein-encoding region with a translational fusion at the C-terminus to an oligonucleotide encoding a spacer peptide (see Fig. 3) (Contreras *et al.*, 1991). This peptide included the recognition site for a proteolytic processing signal similar to that encoded by the *KEX-2* gene in *S. cerevisiae*, which recognizes basic dipeptides Lys-Arg, Arg-Arg, or Arg-Lys. KEX-2-like processing activity has been demonstrated in *A. nidulans* (Devchand *et al.*, 1989) and other filamentous fungi (Calmels *et al.*, 1991). Fused in-frame to this spacer was the cDNA clone encoding mature hIL-6 with the *trpC* gene providing transcription termination signals. The cotransformants producing most recombinant protein were identified by polyacrylamide gel electrophoresis of aliquots of culture medium, and further analysis was made of transformants expressing vectors driven by the *gpdA* promoter.

Expression vectors carrying the glucoamylase gene alone under control of the *gpdA* promoter produced the same amount of secreted glucoamylase as those containing the hIL-6 fusion protein, demonstrating that the secretion rate of glucoamylase was unaffected by the carboxy-terminal fusion of hIL-6. Additionally, expression of the fusion protein led to equimolar quantities of secreted glucoamylase and hIL-6.

Following semipurification, the N-terminal amino acid sequence of the recombinant hIL-6 was determined. However, instead of the expected cleavage after the KEX-2 signal Lys-Arg, an additional two residues had been cleaved, indicating the presence in *A. nidulans* of a prolyl dipeptidase. Partial removal of the dipeptide is also observed in humans. The size of the recombinant hIL-6 was similar on Western blots to that of natural hIL-6, and the protein appeared to be nonglycosylated. Maximum levels of biologically active hIL-6 obtained were around 5 mg/liter in a nonoptimized growth system.

3.8. Human Lactoferrin

Human lactoferrin (hLF) is present in milk and other exocrine secretions and is thought to play an important role in iron regulation. In order to study its putative functions, expression at useful levels in a heterologous host was sought (P. P. Ward *et al.*, 1992a). However, as the protein may have antifungal properties, expression via a regulated promoter was thought to be essential to enable growth of the organism to a sufficient cell density before expression of the potentially inhibitory hLF gene. The promoter used was that of the *A. nidulans alcA* gene, from which was transcribed the hLF cDNA, encoding the signal sequence and mature protein, with a *benA* transcriptional terminator. Transformation into *A. nidulans* resulted in integration of 1–20 copies of the expression vector. There appeared to be a correlation between copy number and hLF levels, secreted hLF being detected only in multicopy transformants. Strains were grown initially under glucose-repressing conditions, followed by a switch of medium to ethanol-inducing conditions. Western blot analysis showed that 30% of total recombinant hLF was secreted, and this product had a size similar to that of mature hLF, indicating cleavage of the hLF signal sequence. The secreted protein retained biological activity as assayed by its iron-binding ability and was produced at levels of up to 5 mg/liter.

Lactoferrin cDNA has also been expressed in *A. oryzae* using the *A. oryzae* starch-inducible α-amylase promoter and signal sequence and *A. niger glaA* terminator (P. P. Ward *et al.*, 1992b). Transformations contained between 1 and 35 copies, and on induction, comparable levels of α-amylase and hLF mRNA were observed. However, the concentration of secreted hLF protein (5–25 mg/liter) was 20-fold less than that of α-amylase, suggesting either inefficient translation or secretion of the recombinant protein. Human lactoferrin secreted by *A. oryzae* was indistinguishable from the native protein by size, immunoreactivity, and biological activity and appeared to be glycosylated to a similar extent via *N*-glycosidic linkages.

3.9. Factors That Affect Heterologous Expression

Many factors are likely to affect the successful expression of heterologous proteins. As shown in the previous examples, the choice of promoter may be important. The production of certain heterologous proteins may be detrimental to the host, and so an inducible promoter such as *glaA* or *alcA* is desirable. For most proteins expressed thus far, however, a strong constitutive promoter like that of *gpdA* is sufficient. Fortunately, it appears that most filamentous fungi are capable of efficiently recognizing promoter and terminator sequences of filamentous fungal origin. Therefore, expression cassettes utilizing promoter and terminator signals derived from a gene of one *Aspergillus* species can usually be expressed in another species.

In most instances, there appears to be a correlation, albeit loose, between copy number of the expression cassette and the level of product. This is not universally true, however, as demonstrated in certain chymosin-producing transformants (Cullen *et al.*, 1987). The reasons for the differences in expression levels between transformants with similar vector copy numbers are unclear, but may include differences in transcriptional efficiency dependent on integration site, differences in mRNA stability and processing, or host mutations caused by gene disruptions on integration of expression vectors. Additionally, the availability of *trans*-acting regulatory proteins may place a limitation on transcription.

The influence of the secretion signal sequence in filamentous fungi has not been rigorously assessed. It appears, however, that aspergilli are rather permissive with respect to signal-sequence recognition. For example, levels of secreted chymosin were similar with the signal sequence of either chymosin or *A. niger* glucoamylase (Cullen *et al.*, 1987); indeed, *A. nidulans* was capable of recognizing an entirely synthetic signal for secretion of interferon α-2 (Gwynne *et al.*, 1987b). Studies in *S. cerevisiae* have nevertheless shown that the signal sequence can affect the quantity and quality of the product (Sleep *et al.*, 1990), and so the signal peptide may be important in fine-tuning of secretion in *Aspergillus*.

Significant advances in secretion levels have been obtained by fusion of the desired protein to a fungal protein that is usually secreted at high level, such as glucoamylase. Little is known about the secretion pathway in filamentous fungi, however, and so the precise way in which fusion to glucoamylase aids secretion of a foreign protein is speculative. A factor that undoubtedly contributed to the successful high-level production of active chymosin was the autocatalytic ability of the protein even when fused to glucoamylase (M. Ward *et al.*, 1990). Other chimeric fusion proteins have to rely on the host processing apparatus, about which, again, little is known. Aspergilli do appear to have processing systems similar to those of yeast and humans (Calmels *et al.*, 1991). The glucoamylase–pro-PLA_2 fusion could be correctly cleaved by an *A. niger* protease, although aberrant processing was also observed (Roberts *et al.*, 1992). The engineering of a particular proteolytic cleavage recognition site such as KEX-2 may be useful to stimulate faithful cleavage from a fusion protein.

Many of the proteins of commercial and research interest are glycosylated. Posttranslational modification by glycosylation is probably of importance in immunogenicity, function, and stability of proteins, and so accurate glycosylation by a heterologous host may be relevant particularly for pharmaceutical proteins. Filamentous fungi are capable of *O*-

and *N*-glycosylation, possibly by a mechanism similar to that of animal cells (Salovuori *et al.*, 1987), and extensive hyperglycosylation, which is often observed in yeast, does not appear to be as great a problem in aspergilli. Several foreign proteins are assumed from their size relative to authentic protein to be similarly glycosylated, e.g., t-PA (Upshall *et al.*, 1987), Bm86 (Turnbull *et al.*, 1990), and hLF (P. P. Ward *et al.*, 1992), but no studies have yet empirically demonstrated this.

A major problem for the expression of foreign proteins is the production by aspergilli of numerous extracellular proteases (Cohen, 1976). With *A. nidulans,* this may be less of a problem, as proteases are repressed in cells grown under conditions of nitrogen, carbon, and sulfur repression, and so their production can be manipulated by growth conditions. *A. niger* and other aspergilli, however, produce copious amounts of these extracellular enzymes in complex media. The production of the acid protease aspergillopepsin by *A. niger* probably accounted for the discrepancy between the levels of secreted chymosin expected and those observed (M. Ward *et al.*, 1990). The presence of this protease also prevented detection of secreted PLA_2 (Roberts *et al.*, 1992). Other secreted or mycelium-associated proteases may also be involved in degradation of secreted foreign proteins (Archer *et al.*, 1990, 1992). Currently, protease-deficient strains obtained by classic mutagenesis and screening (Mattern *et al.*, 1992) are being assessed as hosts for foreign protein production.

Undoubtedly, the aspergilli have the potential to secrete correctly folded and glycosylated foreign proteins. However, many if not most of the processes involved in translation, processing, and secretion of proteins in filamentous fungi are unknown. The presence of most of the posttranslational modifications seen in higher eukaryotes has yet to be demonstrated and the similarity of glycosylation patterns to be investigated. Likewise, the role of chaperones and the secretion route, which is likely to be different from that in unicellular organisms like *S. cerevisiae,* have to be determined.

ACKNOWLEDGMENTS. Research studies carried out in the author's laboratory (J.R.K.) have been funded by the Science and Engineering Research Council and the Commission of the European Communities.

REFERENCES

Archer, D. B., Jeenes, D. J., MacKenzie, D. A., Brightwell, G., Lambert, N., Lorne, G., Radford, S., and Dobson, C. M., 1990, Hen egg lysozyme expressed and secreted from *Aspergillus niger* is correctly processed and folded, *Bio/Technology* **8**:741–745.

Archer, D. B., Mackenzie, D. A., Jeenes, D. J., and Roberts, I. N., 1992, Proteolytic degradation of heterologous proteins expressed in *Aspergillus niger, Biotech. Lett.* **14:**357–362.

Armaleo, D., Ye, G. N., Klein, T. M., Shark, K. B., Sanford, J. C., and Johnston, S. A., 1990, Biolistic nuclear transformation of *Saccharomyces cerevisiae* and other fungi, *Curr. Genet.* **17:**97–103.

Arst, H. N., and Scazzocchio, C., 1985, Formal genetics and molecular biology of the control of gene expression in *Aspergillus nidulans,* in: *Gene Manipulations in Fungi* (J. W. Bennett and L. L. Lasure, eds.), Academic Press, New York, pp. 309–320.

Austin, B., Hall, R. M., and Tyler, B. M., 1990, Optimised vectors and selection for transformation of *Neurospora crassa* and *Aspergillus nidulans* to bleomycin and phleomycin resistance, *Gene* **93:**157–162.

Ballance, D. J., 1986, Sequences important for gene expression in filamentous fungi, *Yeast* **2:**229.

Ballance, D., and Turner, G., 1985, Development of a high-frequency transforming vector for *Aspergillus nidulans, Gene* **36:**321–331.

Ballance, D. J., Buxton, E. P., and Turner, G., 1983, Transformation of *Aspergillus nidulans* by the orotidine-5-phosphate decarboxylase gene of *Neurospora crassa, Biochem. Biophys. Res. Commun.* **112:**284–289.

Berka, R. M., Ward, M., Wilson, L. J., Hayenga, K. H., Kodama, K. H., Carlomagna, L. P., and Thompson, S. A., 1990, Molecular cloning and deletion of the aspergillopepsin A gene from *Aspergillus awamori, Gene* **86:**153–162.

Berse, B., Dinochowska, A., Skrzypek, M., Weglenski, P., Bates, M. A., and Weiss, R. L., 1983, Cloning and characterization of the ornithine carbamoyl transferase gene from *Aspergillus nidulans, Gene* **25:**109–117.

Brody, H., and Carbon, J., 1989, Electrophoretic karyotype of *Aspergillus nidulans, Proc. Natl. Acad Sci. U.S.A.* **86:**6260–6263.

Brody, H., Griffith, J., Cuticchia, A. J., Arnold, J., and Timberlake, W. E., 1991, Chromosome-specific recombinant DNA libraries from the fungus *Aspergillus nidulans, Nucleic Acids Res.* **19:**3105–3109.

Buxton, F. P., Gwynne, D. I., and Davies, R. W., 1985, Transformation of *Aspergillus niger* using the *argB* genes of *Aspergillus nidulans, Gene* **37:**207–214.

Buxton, F. P., Gwynne, D. I., and Davies, R. W., 1989, Cloning of a new bidirectionally selectable marker for *Aspergillus* strains, *Gene* **84:**329–334.

Calmels, T. P. G., Martin, F., Durand, H., and Tiraby, G., 1991, Proteolytic events in the processing of secreted proteins in fungi, *J. Biotechnol.* **17:**51–66.

Campbell, E. I., Unkles, S. E., Macro, J., van den Hondel, C. A. M. J. J., and Kinghorn, J. R., 1989, An improved transformation system for *Aspergillus niger, Curr. Genet.* **16:**53–56.

Carrez, D., Janssens, W., Degrave, P., van den Hondel, C. A. M. J. J., Kinghorn, J. R., Fiers, W., and Contreras, R., 1990, Heterologous gene expression by filamentous fungi: Secretion of human interleukin-6 by *Aspergillus nidulans, Gene* **94:**147–154.

Case, M. E., Schweizer, M., Kushner, S. R., and Giles, N. H., 1979, Efficient transformation of *Neurospora crassa* by utilizing hybrid plasmid DNA, *Proc. Natl. Acad. Sci. U.S.A.* **76:**5259–5263.

Chang, P. K., Skory, C. D., and Linz, J. E., 1992, Cloning of a gene associated with aflatoxin B_1 biosynthesis in *Aspergillus parasiticus, Curr. Genet.* **21:**231–233.

Chevalet, L., Tiraby, G., Cabane, B., and Loison, G., 1992, Transformation of *Aspergillus flavus:* Construction of urate oxidase-deficient mutants by gene disruption, *Curr. Genet.* **21:**447–453.

Clutterbuck, A. J., 1984, *Aspergillus nidulans,* in: *Handbook of Genetics 1* (R. C. King, ed.), Plenum Press, New York, pp. 447–510.

Cohen, B. L., 1976, Extracellular proteases, in: *The Genetics and Physiology of Aspergillus* (J. E. Smith and J. A. Pateman, eds.), Academic Press, London, pp. 281–292.

Contreras, R., Carrez, D., Kinghorn, J. R., van den Hondel, C. A. M. J. J., and Fiers, W., 1991, Efficient KEX2-like processing of a glucoamylase–interleukin-6 fusion by *Aspergillus nidulans* and secreting of mature interleukin-6, *Bio/Technology* **9:**378–381.

Cullen, D., Gary, G. L., Wilson, L. J., Hayenga, K. J., Lamsa, M. H., Rey, M. W., Norton, S., and Berka, R. M., 1987, Controlled expression and secretion of bovine chymosin in *Aspergillus nidulans, Bio/Technology* **5:**369–376.

Davies, R. W., 1991a, Molecular biology of a high-level recombinant protein production system in *Aspergillus,* in: *Molecular Industrial Mycology: Systems and Applications for Filamentous Fungi* (S. A. Leong and R. M. Berka, eds.), Marcel Dekker, New York, pp. 45–82.

Davies, R. W., 1991b, Expression of heterologous genes in filamentous fungi, in: *Applied Molecular Genetics of Fungi* (J. F. Perberdy, C. E. Caten, J. E. Ogden, and J. W. Bennett, eds.), Cambridge University Press, Cambridge, England, pp. 103–117.

Davies, R. W., 1992, Genetic engineering of filamentous fungi for heterologous gene expression and protein secretion, in: *Transgenesis—Applications of Gene Transfer* (J. A. H. Murray, ed.), John Wiley, New York.

Debets, A. J. M., Holub, E. F., Swart, K., van den Brock, H. W. J., and Bos, C. J., 1990, An electrophoretic karyotype of *Aspergillus niger, Mol. Gen. Genet.* **224:**264–268.

De Ruiter-Jacobs, Y. M. J. T., Broekhuijsen, M. P., Campbell, E. I., Unkles, S. E., Kinghorn, J. R., Contreras, R., Pouwels, P. H., and van den Hondel, C. A. M. J. J., 1989, A gene transfer system based on the homologous *pyrG* gene and efficient expression of bacterial genes in *Aspergillus oryzae, Curr. Genet.* **16:**159–163.

Devchand, M., Gwynne, D., Baxton, F., and Davies, R., 1989, An efficient cell-free translation system from *Aspergillus nidulans* and *in vitro* translocation of prepro-x-mating factor across *Aspergillus* microsomes, *Curr. Genet.* **14:**561–566.

Dowzer, C. E. A., and Kelly, J. M., 1989, Cloning of the *creA* gene from *Aspergillus nidulans:* A gene involved in carbon catabolite repression, *Curr. Genet.* **15:**457–459.

Dunn-Coleman, N. S., Bloebaum, P., Berka, R. M., Bodie, E., Robinson, N., Armstrong, G., Ward, M., Przetak, M., Carter, G. L., LaCost, R., Wilson, L. J., Kodama, K. H., Baliu, E. F., Bower, B., Lamsa, M., and Heinsohn, H., 1991, Commercial levels of chymosin production by *Aspergillus, Bio/Technology* **9:**976–982.

Dunn-Coleman, N. S., Bodie, E., Carter, G. L., and Armstrong, G., 1992, Stability of recombinant strains under fermentation conditions, in: *Applied Molecular Genetics of Filamentous Fungi* (J. R. Kinghorn and G. Turner, eds.), Blackie Press, Glasgow, pp. 152–174.

Fincham, J. R. S., 1989, Transformation in fungi, *Microbiol. Rev.* **53:**148–170.

Gainey, L. D. S., Connerton, I. F., Lewis, E. H., Turner, G., and Ballance, D. J., 1992, Characterisation of the glyoxysomal isocitrate ligase genes of *Aspergillus nidulans (acuD)* and *Neurospora crassa (acu-3), Curr. Genet.* **21:**43–47.

Gems, D., and Clutterbuck, A. J., 1993, Conversion of a gene bank in an integrating vector to autonomous replication by cotransformation with an autonomously replicating plasmid and its use in cloning the *Aspergillus nidulans* genes *adC* and *adD, Curr. Genet.* (in press).

Gems, D., Johnstone, I. L., and Clutterbuck, A. J., 1991, An autonomously replicating plasmid transforms *Aspergillus nidulans* at high frequency, *Gene* **98:**61–67.

Gomi, K., Limura, Y., and Hara, S., 1987, Integrative transformation of *Aspergillus oryzae* with a plasmid containing the *Aspergillus nidulans argB* gene, *Agric. Biol. Chem.* **51**:2549–2555.

Gomi, K., Kitamoto, K., and Kumagai, C., 1991, Cloning and molecular characterisation of the acetamidase-encoding gene (*amdS*) from *Aspergillus oryzae, Gene* **108**:91–98.

Gurr, S. J., Unkles, S. E., and Kinghorn, J. R., 1987, The structure and organisation of nuclear genes of filamentous fungi, in: *Gene Structure in Eukaryotic Microbes* (J. R. Kinghorn, ed.), IRL Press, Oxford, pp. 93–138.

Gwynne, D. J., 1992, Foreign proteins, in: *Applied Molecular Genetics of Filamentous Fungi* (J. R. Kinghorn and G. Turner, eds.), Blackie Press, Glasgow, pp. 132–151.

Gwynne, D. I., Buxton, F. P., Gleeson, M. A., and Davies, R. W., 1987a, Genetically engineered secretion of foreign proteins from *Aspergillus* species, in: *Protein Purification: Micro to Macro* (R. Burgess, ed.), Alan R. Liss, New York.

Gwynne, D. I., Buxton, F. P., Williams, S. A., Garven, S., and Davies, R. W., 1987b, Genetically engineered secretion of active interferon and a bacterial endoglucanase from *Aspergillus nidulans, Bio/Technology* **5**:713–719.

Gysler, C., Harmsen, J. A. M., Kester, H. C. M., Visser, J., and Heim, J., 1990, Isolation and structure of the pectin lyase D-encoding gene from *Aspergillus niger, Gene* **89**:101–108.

Hahm, Y. T., and Batt, C. A., 1988, Genetic transformation of an *argB* mutant of *Aspergillus oryzae, Appl. Environ. Microbiol.* **54**:1610–1611.

Harmsen, A. M., Kusters-van Someren, M. A., and Visser, J., 1990, Cloning and expression of a second *Aspergillus niger* pectin lyase gene (*pelA*): Indications of a pectin lyase gene family in *A. niger, Mol. Gen. Genet.* **18**:161–166.

Hull, E. P., Green, P. M., Arst, H. N., and Scazzocchio, C., 1992, Cloning and physical characterization of the L-proline catabolism gene cluster of *Aspergillus nidulans, Curr. Genet.* **21**:447–453.

Iimura, Y., Gomi, K., Uzu, H., and Hara, S., 1987, Transformation of *Aspergillus oryzae* through plasmid-mediated complementation of the methionine-auxotrophic mutation, *Agric. Biol. Chem.* **51**:323–328.

John, M. A., and Peberdy, J. F., 1984, Transformation of *Aspergillus nidulans* using the *argB* gene, *Enzyme Microbial Technol.* **6**:386–389.

Johnstone, I. L., McCabe, P. C., Greaves, P., Gurr, S. J., Cole, G. E., Brow, M. A. D., Unkles, S. E., Clutterbuck, A. J., Kinghorn, J. R. Kinghorn and Innis, M. A., 1990, Isolation and characterisation of the *crnA-niiA-niaD* gene cluster for nitrate assimilation in *Aspergillus nidulans. Gene* **90**:181–192.

Kelly, J. M., and Hynes, M. J., 1985, Transformation of *Aspergillus niger* by the *amdS* gene of *Aspergillus nidulans, EMBO J.* **4**:475–479.

Kelly, J. M., and Hynes, M. J., 1987, Multiple copies of the *amdS* gene of *Aspergillus nidulans* cause titration of *trans*-acting regulatory proteins, *Curr. Genet.* **11**:21–31.

Kinghorn, J. R., and Hawkins, A. R., 1982, Cloning and expression in *Escherichia coli* K-12 of the biosynthetic dehydroquinase function of the *arom* cluster gene from the eukaryote, *Aspergillus nidulans, Mol. Gen. Genet.* **186**:145–152.

Kinghorn, J. R., and Turner, G. (eds.), 1992, *Applied Molecular Genetics of Filamentous Fungi*, Blackie Press, Glasgow.

Kozak, M., 1984, Compilation and analysis of sequences upstream from the translational start site in eukaryotic mRNAs, *Nucleic Acids Res.* **12**:857–872.

Kulmburg, P., Sequeval, D., Lenouvel, F., Mathieu, M., and Felenbok, B., 1992, Identification of the promoter region involved in the autoregulation of the transcriptional activator ALCR in *Aspergillus nidulans, Mol. Cell. Biol.* **12**:1932–1939.

Kusters-van Someren, M. A., Samson, R. A., and Visser, J., 1991, The use of RFLP analysis in classification of the black aspergilli: Reinterpretation of the *Aspergillus niger* aggregate, *Curr. Genet.* **19:**21–26.

Malardier, L., Daboussi, M. J., Julian, J., Roussel, F., Scazzocchio, C., and Brygoo, Y., 1989, Cloning of the nitrate reductase gene (*niaD*) of *Aspergillus nidulans* and its use for transformation of *Fusarium oxysporum, Gene* **78:**1247–156.

Mattern, I. E., Unkles, S., Kinghorn, J. R., Pouwels, P. H., and van den Hondel, C. A. M. J. J., 1987, Transformation of *Aspergillus oryzae* using the *A. niger pyrG* gene, *Mol. Gen. Genet.* **210:**460–461.

Mattern, I. E., van Noort, J. M., van den Berg, P., Archer, D. B., Roberts, I. N., and van den Hondel, C. A. M. J. J., 1992, Isolation and characterization of mutants of *Aspergillus niger* deficient in extracellular proteases, *Mol. Gen. Genet.* **234:**332–336.

May, G., 1992, Fungal technology, in: *Applied Molecular Genetics of Filamentous Fungi* (J. R. Kinghorn and G. Turner, eds.), Blackie Press, Glasgow, pp. 1–27.

May, G. S., Gambino, J., Weatherbee, J. A., and Morris, N. R., 1985, Identification and functional analysis of beta-tubulin genes by site specific integrative transformation in *Aspergillus nidulans, J. Cell Biol.* **101:**712–719.

Meyer, W., Koch, A., Niemann, C., Beyermann, B., Epplen, J. T., and Borner, T., 1991, Differentiation of species and strains among filamentous fungi by DNA fingerprinting, *Curr. Genet.* **19:**239–242.

Miller, B. L., Miller, K. Y., and Timberlake, W. E., 1985, Direct and indirect gene replacements in *Aspergillus nidulans, Mol. Gen. Genet.* **5:**1714–1721.

Nunberg, J. H., Meade, J. H., Cole, G., Lawyer, F. C., McCabe, P., Schweickart, V., Tal, R., Wittman, V. P., Flatguard, J. E., and Innis, M. A., 1984, Molecular cloning and characterisation of the glucoamylase gene of *Aspergillus awamori, Mol. Cell. Biol.* **4:**2306–2315.

Punt, P. J., and van den Hondel, C. A. M. J. J., 1992, Analysis of transcription control sequences of fungal genes, in: *Molecular Signals in Plant-Microbe Communications* (D. P. S. Verma, ed.), CRC Press, Boca Raton, Florida.

Punt, P. J., Oliver, R. P., Dingemanse, M. A., Pouwels, P. H., and van den Hondel, C. A. M. J. J., 1987, Transformation of *Aspergillus* based on the hygromycin B resistance marker from *Escherichia coli, Gene* **56:**117–124.

Punt, P. J., Dingemanse, M. A., Kuyvenhoven, A., Soede, R. D. M., Pouwels, P. H., and van den Hondel, C. A. M. J. J., 1990, Functional elements in the promotor region of the *Aspergillus nidulans gpdA* gene, encoding glyceraldehyde-3-phosphate dehydrogenase, *Gene* **93:**101.

Rambosek, J., and Leach, J., 1987, Recombinant DNA in filamentous fungi: Progress and prospects, *CRC Crit. Rev. Biotechnol.* **6:**357–393.

Roberts, I. N., Oliver, R. P., Punt, P. J., and van den Hondel, C. A. M. J. J., 1989, Expression of the *Esherichia coli* β-glucuronidase gene in industrial and phytopathogenic filamentous fungi, *Curr. Genet.* **15:**177–180.

Roberts, I. N., Jeenes, D. J., MacKenzie, D. A., Wilkinson, A. P., Sumner, I. G., and Archer, D. B., 1992, Heterologous gene expression in *Aspergillus niger:* A glucoamylase-porcine pancreatic prophospholipase A_2 fusion protein is secreted and processed to yield mature phospholipase A_2, *Gene* (in press).

Sakaguchi, K., Gomi, K., Takagi, M., and Horiuchi, H., 1992, Fungal enzymes used in oriental food and beverage industries, in: *Applied Molecular Genetics of Filamentous Fungi* (J. R. Kinghorn and G. Turner, eds.), Blackie Press, Glasgow, pp. 54–99.

Salovuori, I., Makarow, M., Rauvala, H., Knoules, J., and Kaariainen, L., 1987, Low molecular weight high-mannose type glycans in a secreted protein of the filamentous fungus *Trichoderma reesei, Bio/Technology* **5:**152–156.

Saunders, G., Picknett, T. M., Tuite, M. F., and Ward, M., 1989, Heterologous gene expression in filamentous fungi, *Trends Biotechnol.* **7**:283–287.

Seip, E. R., Woloshuk, C. P., Payne, G. A., and Curtis, S. E., 1990, Isolation and sequence analysis of a β-tubulin gene from *Aspergillus flavus* and its use as a selectable marker, *Appl. Environ. Microbiol.* **56**:3686–3692.

Sleep, D., Belfield, G. P., and Goodey, A. R., 1990, The secretion of human serum albumin from the yeast *Saccharomyces cerevisiae* using five different leader sequences, *Bio/Technology* **8**:42–46.

Solomons, G. L., 1980, Fermentor design and fungal growth, in: *Fungal Biotechnology* (J. E. Smith, D. R. Berry, and B. Kristiansen, eds.), Academic Press, London, pp. 55–90.

Streatfield, S. J., Toews, S., and Roberts, C. F., 1992, Functional analysis of the expression of the 3′-phosphoglycerate kinase *pgk* gene in *Aspergillus nidulans, Mol. Gen. Genet.* **233**:231–240.

Tada, S., Gomi, K., Kitamoto, K., Takahashi, K., Tamura, G., and Hara, S., 1991, Construction of a fusion comprising the Taka-amylase A promoter and the *Escherichia coli* β-glucuronidase gene and analysis of its expression in *Aspergillus oryzae, Mol. Gen. Genet.* **229**:301–306.

Tang, C. M., Cohen, J., and Holden, D. W., 1992, An *Aspergillus fumigatus* alkaline protease mutant constructed by gene disruption is deficient in extracellular elastase activity, *Mol. Microbiol.* **6**:1663–1671.

Tatsumi, H., Ogawa, Y., Murakami, S., Ishida, Y., Murakami, K., Masaki, A., Kawabe, H., Arimura, H., Nakano, E., and Motai, H., 1989, A full length cDNA clone for the alkaline protease from *Aspergillus oryzae:* Structural analysis and expression in *Saccharomyces cerevisiae, Mol. Gen. Genet.* **219**:33–38.

Tatsumi, H., Murakami, S., Tsuji, R. F., Ishida, Y., Murakami, K., Masaki, A., Kawabe, H., Arimura, H., Nakano, E., and Motai, H., 1991, Cloning and expression in yeast of a cDNA clone encoding *Aspergillus oryzae* neutral protease II, a unique metalloprotease, *Mol. Gen. Genet.* **228**:97–103.

Tilburn, J., Scazzocchio, C., Taylor, G. G., Zabicky-Zissman, J. H., Lockington, R. A., and Davies, R. W., 1983, Transformation by integration in *Aspergillus nidulans, Gene* **26**:205–221.

Timberlake, W. E., 1980, Developmental gene regulation in *Aspergillus nidulans, Dev. Biol.* **78**:497–510.

Timberlake, W. E., 1991, Cloning and analysis of fungal genes, in: *More Gene Manipulations in Fungi* (J. W. Bennett and L. L. Lasure, eds.), Academic Press, London.

Turnbull, I. F., Smith, D. R. J., Sharp, P. J., Cobon, G. S., and Hynes, M. J., 1990, Expression and secretion in *Aspergillus nidulans* and *Aspergillus niger* of a cell Surface Glycoprotein from the Cattle Tick, *Boophilus microplus,* by using the fungal *amdS* promoter system, *Appl. Environ. Microbiol.* **56**:2847–2852.

Turner, G., and Ballance, D. J., 1985, Cloning and transformation in *Aspergillus,* in: *Gene Manipulation in Fungi* (J. W. Bennett and L. L. Lasure, eds.), Academic Press, New York, pp. 259–278.

Turner, G., Ballance, D. J., Ward, M., and Beri, R. K., 1985, Development of cloning vectors and a marker for gene replacement techniques in *Aspergillus nidulans,* in: *Molecular Genetics of Filamentous Fungi* (W. E. Timberlake, ed.), Alan R. Liss, New York, pp. 15–28.

Unkles, S. E., 1992, Gene organisation in industrial filamentous fungi, in: *Applied Molecular Genetics of Filamentous Fungi* (J. R. Kinghorn and G. Turner, eds.), Blackie Press, Glasgow.

Unkles, S. E., Campbell, E. I., Carrez, D., Grieve, C., Contreras, R., Fiers, W., van den

Hondel, C. A. M. J. J., and Kinghorn, J. R., 1989a, Transformation of *Aspergillus niger* with the homologous nitrate reductase gene, *Gene* **78:**157–166.

Unkles, S. E., Campbell, E. I., de Ruiter-Jacobs, M. J. T., Broekhuijsen, M., Macro, J. A., Carrez, D., Contreras, R., van den Hondel, C. A. M. J. J., and Kinghorn, J. R., 1989b, The development of a homologous transformation system for *Aspergillus oryzae* based on the nitrate assimilation pathway: A convenient and general selection system for filamentous fungal transformation, *Mol. Gen. Genet.* **218:**99–104.

Upshall, A., 1986, Genetic and molecular characterisation of *argB*$^+$ transformants of *Aspergillus nidulans, Curr. Genet.* **10:**593–599.

Upshall, A., Kumar, A. A., Bailey, M. C., Parker, M. D., Favreau, M. A., Lewison, K. P., Joseph, M. L., Maraganore, J. M., and McKnight, G. L., 1987, Secretion of active human tissue plasminogen from the filamentous fungus *Aspergillus nidulans, Bio/Technology* **5:**1301–1304.

van Brunt, J., 1986, Fungi, the perfect hosts?, *Bio/Technology* **4:**1057–1062.

van den Hondel, C. A. M. J. J., van Gorcom, R. F. M., Goosen, T., van den Broek, H. W. J., Timberlake, W. E., and Pouwels, P. H., 1985, Development of a system for analysis of regulation signals in *Aspergillus,* in: *Molecular Genetics of Filamentous Fungi* (W. E. Timberlake, ed.), Alan R. Liss, New York, pp. 29–38.

van den Hondel, C. A. M. J. J., Punt, P. J., and van Gorcom, R. F. M., 1991, Heterologous gene expression in filamentous fungi, in: *More Gene Manipulations in Fungi* (J. W. Bennett and L. L. Lasure, eds.), Academic Press, San Diego, pp. 396–428.

van den Hondel, C. A. M. J. J., Punt, P. J., and van Gorcom, R. F. M., 1992, Production of extracellular proteins by the filamentous fungus *Aspergillus, Anton. van Leeuw.* **61:**153–160.

van Gorcom, R. M. F., Pouwels, P. H., Goosen, T., Visser, J., van den Broek, H. W. J., Hamer, J. E., Timberlake, W. E., and van den Hondel, C. A. M. J. J., 1985, Expression of an *Escherichia coli* β-galactosidase fusion gene in *Aspergillus nidulans, Gene* **40:**99–106.

van Gorcom, R. F. M., Punt, P. J., Pouwels, P. H., and van den Hondel, C. A. M. J. J., 1986, A system for the analysis of expression signals in *Aspergillus, Gene* **48:**211–217.

van Hartingsveldt, W., Mattern, I. E., van Zeijl, C. M. J., Pouwels, P. H., and van den Hondel, C. A. M. J. J., 1987, Development of a homologous transformation system for *Aspergillus niger* based on the *pyrG* gene, *Mol. Gen. Genet.* **206:**71–75.

Vollmer, S. J., and Yanofsky, C., 1986, Efficient cloning of genes of *Neurospora crassa, Proc. Natl. Acad. Sci. U.S.A.* **83:**4869–4873.

Ward, M., 1991, Chymosin production in *Aspergillus,* in: *Molecular Industrial Mycology: Systems and Applications for Filamentous Fungi* (S. A. Leong and R. M. Berka, eds.), Marcel Dekker, New York, pp. 83–106.

Ward, M., Wilkinson, B., and Turner, G., 1986, Transformation of *Aspergillus nidulans* with a cloned, oligomycin-resistant ATP synthase subunit 9 gene, *Mol. Gen. Genet.* **202:**265–270.

Ward, M., Wilson, L. J., Carmona, C. L., and Turner, G., 1988, The *oliC3* gene of *Aspergillus niger:* Isolation, sequence and use as a selectable marker for transformation, *Curr. Genet.* **14:**37.

Ward, M., Kodama, K. H., and Wilson, L. J., 1989, Transformation of *Aspergillus awamori* and *A. niger* by electroporation, *Exp. Mycol.* **13:**289–293.

Ward, M., Wilson, L. J., Kodama, K. H., Rey, M. W., and Berka, R. M., 1990, Improved production of chymosin in *Aspergillus* by expression as a glucoamylase–chymosin fusion, *Bio/Technology* **8:**435–440.

Ward, P. P., May, G. S., Headon, D. R., and Conneely, O. M., 1992a, An inducible expression system for the production of human lactoferrin in *Aspergillus nidulans, Gene* (in press).

Ward, P. P., Lo, J. Y., Duke, M., May, G. S., Headon, D. R., and Conneely, O. M., 1992b, Production of biologically active recombinant human lactoferrin in *Aspergillus oryzae, Bio/Technology* **10:**784–789.

Wernars, K., Goosen, T., Wennekes, L. M. J., Visser, J., Bos, C. J., van den Broek, H. W. J., van Gorcom, R. F. M., van den Hondel, C. A. M. J. J., and Pouwels, P. H., 1985, Gene amplification in *Aspergillus nidulans* by transformation with vectors containing the *amdS* gene, *Curr. Genet.* **9:**361–368.

Wernars, K., Goosen, T., Wennekes, B. M. J., Swart, K., van den Hondel, C. A. M. J. J., and van den Broek, H. W. J., 1987, Cotransformation of *Aspergillus nidulans:* A tool for replacing fungal genes, *Mol. Gen. Genet.* **209:**71–77.

Woloshuk, C. P., Seip, E. R., Payne, G. A., and Adkins, C. R., 1987, Genetic transformation system for the aflatoxin-producing fungus *Aspergillus flavus, Appl. Environ. Microbiol.* **55:**86–90.

Yelton, M. M., Hamer, J. E., and Timberlake, W. E., 1984, Transformation of *Aspergillus nidulans* by using a *trpC* plasmid, *Proc. Natl. Acad. Sci. U.S.A.* **76:**5259–5263.

Yelton, M. M., Timberlake, W. E., and van den Hondel, C. A. M. J. J., 1985, A cosmid for selecting genes by complementation in *Aspergillus nidulans:* Selection of the developmentally regulated *yA* locus, *Proc. Natl. Acad. Sci. U.S.A.* **82:**834–838.

Solid-State Fermentations of the Genus *Aspergillus*

5

P. GERVAIS and M. BENSOUSSAN

1. INTRODUCTION

For the majority of filamentous fungal species, solid-state media are the natural life media. Growth can occur on the surface or within the whole substrate, depending on the porosity. Industrial solid-state fermentations have been developed largely in traditional food industries such as cheese, Oriental fermentations, fermented vegetables, meat, and other products, and in modern biotechnological industries such as antibiotics and enzymes. At the industrial scale, the latter fermentations allow for a decrease in the drying costs of the final products.

The use of water-insoluble materials in microbial metabolism has been characterized in solid-state fermentation (Moo-Young *et al.*, 1983). Since substrate utilization could also occur in a suspension, the definition of solid-state fermentation requires the absence or near absence of free liquid or "free" water (Hesseltine, 1972; Lonsane *et al.*, 1985). Such an absence has been related to a moisture content that could vary between 40 and 80% (Cannel and Moo-Young, 1980a). The empty or void space must then be filled by a gaseous phase (Durand *et al.*, 1983).

It is well recognized that the growth and metabolism of microorganisms always or almost always occur in the liquid phase. Solute as well as dissolved gas transfers take place in the aqueous film surrounding the microorganisms. So, even in solid-state fermentations, the microorganisms are in a liquid medium. Solute diffusion in the substrate must

P. GERVAIS • ENSBANA Département de Génie des Procédés Alimentaires et Biotechnologiques, Université de Bourgogne, 21000 Dijon, France. M. BENSOUSSAN • ENSBANA, Département de Microbiologie-Biotechnologie, Université de Bourgogne, 21000 Dijon, France.

Aspergillus, edited by J. E. Smith. Plenum Press, New York, 1994.

occur in the liquid phase, and gaseous diffusion in the substrate can occur in the liquid as well as in the gaseous phase. Thus, the replacement of the water phase by the gaseous phase increases oxygen transfer but prevents solute diffusion in most of the substrates. Figure 1 demonstrates this duality in the case of aerobic fermentations.

For discontinuous media (e.g., agglomerated particules) of high porosity with a large gas phase, growth is limited by the nutrients, which can diffuse only in the thin liquid layers around the microorganism, and could also be limited by the presence of high concentrations of metabolites in such layers. Mostly, the three phases (gas, solid, and liquid) are present in the liquid as well as in the solid media, but not in the same proportions, and so kinetic, diffusional, rheological, and thermodynamic (a_w, osmotic pressure) properties are modified.

It is difficult to give an exact definition of solid-state fermentations based on the proportion of each phase because the biophysical properties of the mixture depend on the nature of the liquid and solid components. The major difference between liquid fermentations and solid-state fermentations is the mixing property. Liquid fermentations are generally perfectly mixed reactions, and so each part of the reactor

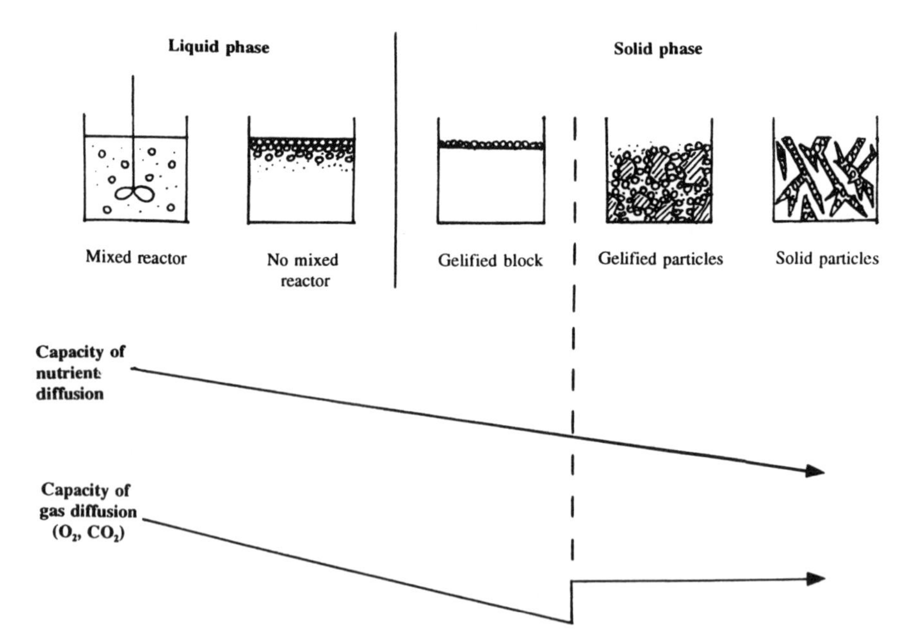

Figure 1. Effect of the replacement of the water phase by the air phase on nutrients and gas diffusion.

contains at any time the same quantity of microorganisms, of nutrients, and of metabolites.

Solid-state fermentations, which are characterized by a reduced liquid phase (approximately the water-holding capacity of the substrate in the case of the water-liquid phase), present a high viscosity. Thus, mixing such media leads to great shear forces that drastically injure the microbial cells. Solid-state media could then be defined by high viscosity and no continuous possibility of mixing; such media could be considered as heterogeneous media in terms of microbial population and solute concentration.

Perfect mixing of liquid fermentations prevents the formation of limit layers around the microorganisms. Such layers drastically limit the diffusion phenomena of nutrients and metabolites and prevent the removal of the heat of the fermentation process, which could represent quite high values, and so limit the metabolism of the microorganisms.

In solid-state fermentations, three alternatives could be used in order to improve the mixing:

1. Sequential mixing.
2. The use of mobile microorganisms, most often filamentous fungi; the forward growth of such microorganisms allows the occupation of new substrate with high levels of nutritive components.
3. Forced aeration through the solid substrate, which allows the control of water content and therefore water activity; the removal of volatile compounds and CO_2 together with heat produced during the fermentation; and supply of O_2 for aerobic metabolism.

2. FUNGAL GROWTH CHARACTERISTICS IN A SOLID-STATE MEDIUM

Relative to previous considerations on mixing, two different types of solid-state fermentation must be considered. The first is characterized by a point inoculation without any mixing and corresponds to the natural contamination process and to most gelosed surface cultivations. In such systems, the specific growth rate μ varies not only with time but also with space. Indeed, the specific growth rate relative to the apical or peripheral zone of the fungal colony is extremely different from the μ value relative to the rear zones. The second type of fermentation is characterized by a homogeneous inoculation; in such cultivations, the specific growth rate can be considered as the same for the whole colony at a given time.

In order to characterize growth in solid-state media, gelosed surface culture can be used as a model for the definition of specific parameters.

2.1. Radial Extension Rate and Hyphal Growth Rate

In the case of a point inoculation, the radial extension rate is widely used to evaluate the growth ability of a fungus. Indeed, after a few micrometers of exponential growth following the emergence of the germ tube [120 μm for *Aspergillus nidulans* described by Trinci (1969)], the radial extension occurred at a constant rate (Scott, 1957) and could be related to the hyphal growth rate in the peripheral zone, which was also constant for given conditions of biophysical parameters (Trinci, 1978). Hyphal contribution to radial growth is given by the projection of the vector hyphal growth rate on the radial axis of the culture.

Nevertheless, we have recently shown (Molin *et al.*, 1992) that the mean hyphal growth rate is greater than the radial extension rate of the colony due to the statistical distribution of the direction of the apices around the radial direction of the colony. Such a distribution, which could be easily bound to the mycelium density, has been shown to be dependent on external conditions.

Radial extension rate has often been used as a growth indicator (Brancato and Golding, 1953; Ryan *et al.*, 1943), but recent works (Inch and Trinci, 1987; Lemière, 1992; Molin *et al.*, 1992) have shown that it could not be satisfactorily linked to biomass production because of greater mycelial density variations depending on the physical conditions (a_w, nutrients) (Gervais *et al.*, 1988a).

2.2. Branching

Branching is another parameter of hyphal growth rate to complement biomass measurements. Nevertheless, branching frequency, which is proportional to the specific growth rate for a constant growth rate, is not constant for all the zones.

Branching can occur apically or laterally, or both, as for *A. nidulans*, in the peripheral zone just behind the apical zone (Fiddy and Trinci, 1976). Katz *et al.* (1972) have also proposed a model for branch initiation for *A. nidulans*. From the branching point for a zone of about 100 μm, growth is exponential. This zone corresponds to the apical zone described previously by Trinci (1971). Recent results developed by Lemière (1992) have shown the high value of the branching frequency in this zone ($Fr = 11\ hr^{-1}$), which confirms previous results (Ryan *et al.*, 1943). Behind this zone, the width of which depends on the strain used, branching is reduced, and growth is essentially due to hyphal extension. Moreover, this extension is limited by the mutual inhibition of hyphal proximity as shown for *A. nidulans* by Saunders and Trinci (1979). Branching frequency measurements are easy in the peripheral zone, where hyphal density is low, but much more difficult to perform in high-density zones.

2.3. Biomass

Biomass formation is the most informative parameter relative to the growth of the colony; it is the result of the combination of the hyphal growth rate and the hyphal branching frequency. Nevertheless, as shown previously, these two parameters vary with hyphal growth:

1. In the apical zone: In the case of only lateral branching, there is no branching and growth extension is exponential; biomass formation is exponential vs. time.
2. In the branching zone: Branching frequency and hyphal extension are constant, so biomass formation is still exponential vs. time.
3. In the rear zone: Hyphal extension occurs at a constant growth rate; biomass formation is quadratic vs. time.

Consequently, indirect measurement of biomass kinetics of a solid-state fermentation through branching frequency and radial or hyphal growth rate would be extremely difficult.

Direct methods of biomass measurement could be used, such as dry-matter weight or tracer-molecule analysis of wall components (e.g., chitin, glucosamine, ergosterol). Such methods have been reviewed (Matcham *et al.*, 1984). Nevertheless, substrate components could interfere, and such measurements do not distinguish between dead and live mycelium. Enzyme activity measurement with epifluorescence methods could be advantageously used toward this objective.

2.4. Differentiation and Metabolite Production

As in liquid media, specific metabolites are produced by particular specific physiological states of filamentous fungi. For example, heptanone (blue cheese flavor) was shown to be produced essentially by the spores of *Penicillium roqueforti* (Larroche *et al.*, 1988) or by the mycelium of *Trichoderma viride* (Gervais and Sarrette, 1990).

We have shown, in the case of the use of *T. viride* with central inoculation on a gelosed medium, that the concentric localization of mycelium of the same age may explain the observed variations in aroma production. Maximum heptanone production was found to occur in an annular zone aged between 3.4 and 6.5 days (Gervais and Sarrette, 1990). In the same way, we previously observed (Gervais *et al.*, 1988b) that the sporulation of *T. viride* was also cyclic and related to circadian rhythms.

Trinci and Banbury (1967) studied the growth of the tall conidiophores of *A. giganteus*.

In order to optimize such productions, it would be interesting to favor the induction of specific stages of filamentous fungi cultivated in solid-state media. The use of plug flow reactors with static mixing could be considered for this objective.

Growth of filamentous fungi has been shown to be dependent on physicochemical parameters such as thermodynamics (a_w, temperature) (Scott, 1957; Gervais *et al.*, 1988c) and chemicals (glucose, O_2 and CO_2, nutrient concentrations) (Trinci and Collinge, 1973; Trinci, 1969). Numerous workers have shown the evidence of such dependence, but there are many results that are sometimes difficult to synthesize. The major parameter involved in the difference between solid and submerged fermentation is water. Specific interest has been focused in this chapter on the properties of water that are involved in such solid-state media.

3. DYNAMIC AND THERMODYNAMIC PROPERTIES OF A SOLID-STATE MEDIUM

3.1. Water Properties

In a heterogeneous system such as a solid-state medium, numerous parameters can describe the state of water. Two types of parameters can be distinguished:

1. Thermodynamic parameters, relative to steady-state systems, describe the energy state of water, the mainspring of water transfer between two systems such as cells and substrates. Evolution of biological systems is very low compared with that of physical systems, so such systems can be considered at a steady state.
2. Dynamic or kinetic parameters define the resistance of the medium to the water transfer. Solute diffusion as well as cellular absorption occurs in aqueous medium. The diffusional properties of water interact with these transfers, and the dynamic properties of water molecules are modified when they are near other

molecules. The diffusion laws are deduced from the atomic theory of Brownian movement established by Einstein. The relation between the flow and the concentration gradient is given by Fick's law, and the coefficient of proportionality is called *diffusion coefficient* or *diffusivity;* its value is closely dependent on water concentration in the medium.

In fact, thermodynamic parameters and kinetic parameters, although fundamentally different, are always related to the water content of the medium and therefore dependent. Nevertheless, this degree of dependence is extremely variable, and the limiting factors for the transfer of water in a system can have a kinetic or thermodynamic origin.

3.1.1. Thermodynamic Parameters

The parameters generally used to characterize the thermodynamic state of water in a system may be defined from the two fundamental laws of thermodynamics.

3.1.1a. Chemical Water Potential. The chemical water potential of a system is equivalent to the partial free energy of water:

$$\left(\frac{\delta G}{\delta n_w}\right)_{T,P,n_j} = \mu_w = \overline{G}_w \tag{1}$$

where T is temperature, P is pressure, w is the water suffix, n is the number of moles, j is the solute suffix, G is free energy, $\overline{G}_w$ is the partial molar free energy of water, and μ_w is the water chemical potential in the solution.

The potential difference can be estimated in relation to a standard reference state:

For a solution:

$$\mu_w - \mu_w^o = \overline{V}_w(P - \pi) \tag{2}$$

with

$$\pi = RT\left(\frac{C}{M} + A_2C^2 + A_3C^3 + \ldots\right) \tag{3}$$

For a porous solid:

$$\mu_w - \mu_w^o = \overline{V}_w(P - \tau) \tag{4}$$

with $\tau = h\rho g$

Aspergillus, edited by J. E. Smith. Plenum Press, New York, 1993.
For a heterogenous system: as a solid-state medium (solutes and porus solids):

$$\mu_w - \mu_w^o = \overline{V}_w(P - \pi - \tau) \tag{5}$$

where μ_w^o is the chemical potential of pure water, π is the osmotic pressure, C is the solute concentration, M is the solute molar weight, A_i is the i[st] viriel osmotic coefficient, R is the perfect gas constant, $\overline{V}_w$ is the water molar volume, τ is the suction pressure,

3.1.1b. Water Potential of a System. The water potential (Ψ) of a system is defined by:

$$\Psi = \frac{\mu_w - \mu_w^o}{\overline{V}_w} \quad (Pa) \tag{6}$$

The water potential Ψ can be written as the sum of terms of different origins:

$$\Psi = \Psi_p + \Psi_s + \Psi_m \tag{7}$$

where $\Psi_p = P$, water potential pressure (hydrostatic pressure), $\Psi_s = -\pi$, osmotic water potential (solute effect: osmotic pressure), and $\Psi_m = -\tau$, matrix water potential (solid matrix effect: adsorption, capillarity).

Other parameters like temperature or gravity can affect the water energy state of a system. Nevertheless, in biological systems, most of the time, only two components Ψ_s and Ψ_m can describe the systems. The distinction between Ψ_s and Ψ_m is especially valid at the macroscopic level; at the molecular level, the same physicochemical phenomena such as adsorption and size effects are implicated.

3.1.1c. Water Activity. The water activity concept is used to describe the equilibrium between a heterogeneous system and the surrounding water vapor phase (considered as a perfect gas). The water activity of a solution is given by:

$$a_w = \frac{P_w}{P_w^o} = \gamma_w \cdot X_w \tag{8}$$

where P_w is the partial pressure of water vapor equilibrated with the solution, P_w^o is the partial pressure of water vapor above pure water in the same conditions, X_w is the water molecular fraction, and γ_w is the water activity coefficient (dependent on the concentration).

Guggenheim (1967) demonstrated the relation between the chemical potential and water activity:

$$\mu_w - \mu_w^o = RT \ln a_w \tag{9}$$

and so, according to equation (6):

$$\Psi = \frac{RT \ln a_w}{\overline{V}_w} \tag{10}$$

3.1.1d. Molecular Basis of the Thermodynamic Properties of Water

Case of Solutions. The thermodynamic water properties of a solution are related to three types of molecular interactions: solute–solute, solute–solvent, and solvent–solvent. The hydration phenomenon corresponds to water–solute interactions and can be studied alone at solute concentrations corresponding to an infinite dilution, where the solute–solute interactions therefore become negligible. For electrolyte solutions, this dilution threshold can be estimated using the Debye-Hückel law; for nonelectrolytes, an objective method to estimate this threshold does not yet exist. The thermodynamic reference for the hydration term is composed by the ideal solution, which presents no solute–water interaction. From this model, some deviations are observed that can be attributed to three types of phenomena:

1. Solute molecular size vs. solvent molecular size. Flory (1953), using the theory of the "liquid network," showed that the binary solution entropy and consequently its eventual deviation from an ideal solution depends on the respective size of the solvent and solute molecules.
2. Solute molecule hydration or solvation.
3. Other molecular interactions (solute–solute).

Case of Insoluble Substances. The previous types of interactions can occur depending on the chemical nature of the polymers, hydrophilic for cellulose and hydrophobic for EPTF (Teflon) and cellulose gels.

But, in addition, the energy state of water is closely dependent on the thermodynamic level of the geometry of the medium and especially on the dimensions of the pores. Fisher and Israelachvili (1979) showed that the thermodynamic properties of liquids trapped in microscopic pores can be described by Kelvin's equation, the basis of the nucleation theory, for pore sizes as small as a few nanometers.

The Kelvin equation can be written:

$$\ln \frac{P_w}{P_w^o} = \frac{M_w}{\rho_w RT}\left[-\frac{\sigma}{r_m} - (P_w - P_w^o)\right] \tag{11}$$

where r_m is the radius of the curve of the water meniscus in the pore, σ is the surface tension at temperature T, and ρ_w is the water volumetric weight.

$P_w - P_w^o$ is negligible for $\frac{-\sigma}{r_m}$, so:

$$\frac{RT \ln a_w}{M_w} = -\frac{\sigma}{\rho_w r_m} \tag{12}$$

and then:

$$\frac{RT \ln a_w}{\overline{V}_w} = -\frac{\sigma}{r_m} = \Psi \tag{13}$$

In this case, the energy state of water is dependent only on the curve radius of the meniscus and then on the particles or pore sizes of the material. The porosity of a heterogeneous system, such as a solid substrate, is modified during drying or hydration (bulking), as the curve radius of the water meniscus decreases. Nevertheless, for a definite water content, the matrix water potential provides an estimation of liquid–gas and solid–liquid interactions, generated by very thin pore size (capillarity), which can characterize biological structures such as membranes or macromolecules.

3.1.2. Dynamic Parameters

These parameters allow the characterization of the water resistance to transfer in a medium. The parameters that are used are the coefficients of the rotational and translational diffusion of water.

These coefficients are related to the friction coefficient by Einstein's equation:

$$D = \frac{kT}{\xi} \tag{14}$$

where D is the diffusion coefficient of a particle in a liquid, ξ is the friction coefficient, k is Boltzmann's constant, and T is the temperature.

The hydrodynamic model allows the calculation of ξ in the case of large molecules with a radius a for the translational diffusion $\xi = 6\pi a\eta$ (η = liquid viscosity) and for the rotational diffusion $\xi = 8\pi a^3\eta$.

Water content is a parameter that represents the diffusional properties of a slightly hydrated medium. In fact, in the case of the study of the rotational diffusivity of the solute molecules (Le Meste and Voilley, 1988), as in the case of the translational diffusivity of water molecules (Simatos and Karel, 1988), the relation proposed between diffusivity and water content is established for definite scales of water content.

The thermodynamic and kinetic parameters of a medium are linked by the water content. Nevertheless, this link, represented by the sorption isotherm of the medium, can be formalized by equations only for simple model systems. In most cases, this link can be established only experimentally.

Previous studies (Bruin and Luyben, 1980) have confirmed that the water diffusivity in polymer solutions is essentially dependent on the water content of the solution. Le Meste and Voilley (1988) even showed an antagonism between the kinetic and thermodynamic parameters.

A unique factor representing both the energy state of water and its solvent properties in a heterogeneous system does not exist. Nevertheless, it seems possible to draw general conclusions. For dilute solutions, the solute diffusion is not a limiting factor, and the solute nature has only a small influence. If the solvent and the solute molecules are of compatible sizes, there is little deviation from Raoult's law. In this case, thermodynamic factors allow the prediction of the water transfer in the system.

For more concentrated solutions, homogeneous or heterogeneous, as found in solid-state media, the previous hypotheses are no longer true; the diffusion phenomena can become limiting according to the sorption properties of the solutes or solids. The physicochemical nature of the molecules can then interfere.

For certain enzymatic activities, which can be developed over almost the total scale of water content (5–100% wt./wt.) (Bouanda, 1983) or of the water activity (Goldberg *et al.*, 1988), the previous considerations are particularly important to the understanding of the systems. For microbial activity, there may be a hierarchy in the action of kinetic or thermodynamic parameters on the cellular physiology. In submerged cultivations, limitation would essentially be of a thermodynamic origin, but in solid-state cultivation, limitation should be thermodynamic as well as kinetic in origin.

3.2. Heat and Mass Transfer Properties in a Solid-State Medium

3.2.1. Heat Transfer

Transfer of heat by conduction occurs in solid-state media, which are not mixed. In such systems, heat transfers follow Fourier's law:

$$\frac{q_x}{A} = -k\frac{dT}{dx} \tag{15}$$

where q_x is the heat transfer rate in the x direction in W, A is the cross section of area normal to the direction of flow of heat in m^2, T is the temperature in °K, x is the distance in m, and k is the thermal conductivity in W/m°.K.

Thermal conductivity, which is the kinetic parameter, is compared among water, gas, and cellulose in Table I.

The low thermal conductivity of air should be noticed in such conductive transfers. It ensures that heat is much more difficult to extract from a porous solid-state medium than from a liquid medium.

Natural convection heat transfer occurs when a solid surface is in contact with a gas or a liquid that is at a different temperature from the surface. Such phenomena occur necessarily when forced aeration is realized in a solid-state fermentation, and constitutes the major part of heat transfers.

In convective transfer, the rate of heat transfer from a solid to a convective fluid is represented by the following equation:

$$q = hA[T_w - T_g] \tag{16}$$

where q is the heat transfer rate in W, A is the area in m^2, T_w is the temperature of the solid surface in °K, T_g is the average temperature of the fluid flowing in °K, and h is the convective heat transfer coefficient in W/m^2.°K Different values of h are presented in Table II.

Table I. Thermal Conductivities at 273°K (W/m · °K)

Water	0.569
Air	0.0242
Cellulose	0.130

Table II. Approximate Magnitude of the Heat Transfer Coefficient

Mechanism	$W/m^2 \cdot °K$
Condensing steam	5,700–28,000
Moving water	50–30,000
Moving air	11–35
Still air	3–23

In the case of a convective heat transfer, the coefficient in moving water is about 10^3 times greater than in moving air. Thus, convection heat transfer phenomena are also favored in liquid media with regard to solid substrate.

3.2.2. Mass Transfer

The two previous modes of heat transfer, conduction and convection, are analogous to molecular diffusion and convective mass transfer.

3.2.2a. Molecular Diffusion. Diffusion of molecules is due to a concentration gradient; the equation for molecular diffusion of mass is Fick's law, which is written as follows for a constant total concentration in a fluid:

$$J_{Ax} = -D_{AB} \frac{dC_A}{dx} \tag{17}$$

where J_{Ax} is the molar flux of component A in the x direction due to molecular diffusion in kg/sec·m^2, D_{AB} is the molecular diffusivity of molecule A in B in m^2/sec, C_A is the concentration of A in kg/m^3, and x is the distance of diffusion in m.

It is evident that the rate of molecular diffusion in liquids is considerably slower than in gases. In fact, the diffusion coefficient in a gas will be of an order of magnitude about 10^5 times greater than in a liquid.

Thus, O_2 diffusion will be largely facilitated in a porous solid-state fermentation through forced aeration in comparison to submerged fermentation.

Diffusion of fluids or solutes in porous solids can follow Fick's law as previously seen, in the case where the diffusion does not depend on the structure of the solid. If the solid cannot be treated as a homogeneous material—for example, in the case of porous solids that have pores or interconnected voids—such porosity affects the diffusion.

An effective diffusion coefficient (D_{Aeff}) can then be used as:

$$D_{Aeff} = \frac{\varepsilon}{\tau} D_{AB} \text{ in m}^2/\text{sec} \quad (18)$$

where ε is the open void fraction and τ is the tortuosity factor.

For inert-type solids, τ can vary from 1.5 to 5. Such a coefficient must be used for the study of solutes or gas diffusion in solid-state fermentation media.

3.2.2b. Convective Mass Transfer. Such transfers occur when the fluid velocity is increased until turbulent mass transfers occur. Such a regime occurs in solid-state fermentations when forced aeration is applied for the transfer between the liquid phase and the air phase and, particularly in the evaporation of water, the exit of CO_2 and volatile components. Nevertheless, such convective mass transfer cannot occur with the majority of solutes, which are very weak volatiles.

4. INFLUENCE OF PHYSICOCHEMICAL PARAMETERS OF THE MEDIUM IN SOLID-STATE FERMENTATION

4.1. Influence of Water Properties of Solid Culture Medium on Microbial Physiology

The introduction of the water activity concept (and then of the water potential) in microbiology dates from 1953 (Scott, 1953). This parameter has been and still is widely used as a predictive criterion of the physiological functions of microorganisms and thus of food preservation (although physicochemical factors such as nonenzymatic oxidations also intervene). Ben-Amotz and Avron (1978) have used the influence of water activity or osmotic pressure to increase the production of glycerol by unicellular algae. In recent years, the systematic utilization of this parameter to qualify the water situation of a medium has been debated by specialists (Lilley, 1985; Franks, 1985). For these authors, the thermodynamic considerations are insufficient to explain the water effect in a medium. In their view, kinetic parameters should also be included. The respective influence of kinetic and thermodynamic parameters in fermentation media should be investigated.

4.1.1. Influence of Water Thermodynamic Parameters of the Culture Medium on Microorganism Activity

4.1.1a. Cellular Mechanisms. It is recognized that there is no active water transport in cells and that water moves freely according to the gradient of the water potential.

Thermodynamic models relating the energy equilibrium of the intra- and extracellular media have been developed by many authors (e.g., Griffin, 1981; Steudle *et al.*, 1983).

At equilibrium, the water potential of the protoplasm Ψ^P is equal to the water potential of the medium Ψ^E:

$$\Psi^E = \Psi^P \tag{19}$$

The variation scale of Ψ^P according to Ψ^E is dependent on the species. For xerotolerant species, Ψ^P can decrease to -40 MPa without damaging the cells.

The protoplasmic water potential Ψ^P can be restated in terms of its constituent elements as follows:

$$\Psi^P = \Psi^P_p + \Psi^P_s + \Psi^P_m \tag{20}$$

where $\Psi^P = P - \pi - \tau$, with P the hydrostatic pressure, π the osmotic pressure, and τ the suction pressure. Ψ^P_s and Ψ^P_m are the water osmotic and matrix potentials of the protoplasm, and Ψ^P_p the turgor potential of the protoplasm created by the plasmic membrane and the cellular walls. This latter potential corresponds to the pressure permitting the initiation of cellular plasmolysis. For most filamentous fungi, $\Psi^P_p \approx -1$ MPa, a value that can rise to -3 MPa for xerotolerant species.

An increase in the osmotic pressure (or decrease in the water activity for a heterogeneous medium) of the extracellular medium leads to a two-phase cell response (Zimmermann, 1978):

1. A very fast but passive exit of cell water, for a few seconds, with a corresponding decrease in cell volume for protoplasts or decrease in turgor pressure for cells with walls. For such cells, the variation of the turgor potential is related to cellular volume modifications through the elastic properties of the cellular walls, defined by the elastic volumetric coefficient ε according to the equation:

$$\Delta P = \epsilon \frac{\Delta V}{V} \tag{21}$$

This coefficient is related to turgor pressure and cellular volume; it can vary from 5 bars to 600 bars according to species and pressure conditions.

This first phase can lead, in the case of large gradients of osmotic pressure, to either an exit of most of the free water from the cell, and thus to a metabolically inert cell that can survive over long periods, or to cellular death essentially due to the alteration of membrane properties (Cudd and Steponkus, 1987).

2. A second compensation phase, during which the cell allows the input of permeable solutes or the biosynthesis of intracellular metabolites. This accumulation in response to a hydric stress has been identified in microorganisms and in plant and animal cells (Gilles, 1975; Hellebust, 1976; Luard, 1982) and allows the cell to recover its initial volume.

The turgor potential of the cell seems to interact with the growth mechanisms of microorganisms; a growing cell constitutes a system of nonequilibrium, and the intracellular medium is slightly hypertonic vs. the extracellular medium. During growth, the concentrations of intracellular solutes increase and then the Ψ_S^p decreases, leading to entrance of water into the cell, which reequilibrates the system, and then to an increase in volume corresponding to growth. During the stationary phase of the growth of the microorganisms, the turgor potential is constant and the system can be considered in equilibrium.

Numerous experiments have demonstrated the influence of the water activity of the culture medium on the survival, growth, and metabolism of microorganisms cultivated on different types of medium (Mossel, 1975; Troller, 1980; Richard-Molard *et al.*, 1982; Jakobsen, 1985). In all cases, optimum growth and metabolic production occurred at media a_w values of less than 1.

4.1.1b. Effects on Solid-State Fermentation of Filamentous Fungi. The characteristic shape of the curve relating water activity of the medium to the growth of a microorganism, which can be estimated by measuring the radial extension rate on agar-agar medium for filamentous fungi, or by measuring cell division number for yeast in liquid medium, has long been known (Scott, 1957) (Fig. 2).

Optimum growth occurs in a medium in which the water activity is clearly less than 1, corresponding to a water osmotic potential scale between -0.1 MPa and -10 MPa.

Esener and colleagues (Esener *et al.*, 1981) tried to relate growth kinetics to the water activity of the media in a model. From their data, it is evident that biosynthesis is less efficient at a reduced a_w level. They

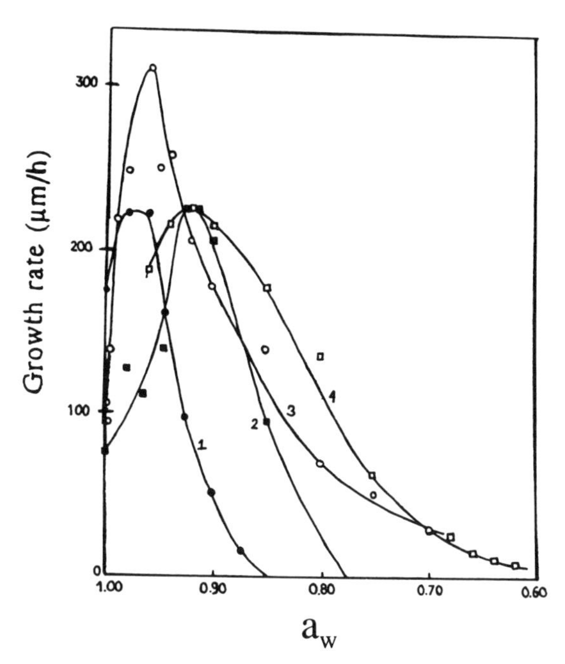

Figure 2. Influence of the a_w on the growth rate of filamentous fungi (1) *Aspergillus niger* at 20°C; (2) *A. glaucus* at 20°C; (3) *A. amstelodami* at 25°C; (4) *Xeromyces bisporus* at 25°C. From Scott (1957).

found an optimum a_w level lower than 1 for the yield of biomass relative to substrate. Such an optimum is confirmed by plotting the thermodynamic efficiency parameter η given previously by Roels (1980) vs. a_w.

Previous workers (Stouthamer and Bettenhaussen, 1973) showed that both parameters Y^{max}_{sx} and m_s of the Pirt (1965) model are a function of a_w:

$$q_s = [\mu/Y^{max}_{sx}] + m_s \tag{22}$$

where q_s is the specific substrate consumption rate, μ is the specific growth rate, Y^{max}_{sx} is the maximal yield for the substrate (mole/mole), and m_s is the maintenance coefficient (mole/mole·hr).

This assumption prevents the construction of a thermodynamic model. Results obtained on the growth of *Trichoderma viride* T.S. and of *Penicillium roqueforti* confirmed these observations (Gervais *et al.*, 1988c).

4.1.1c. Measurement and Control. Different indirect control methods of the water content have been proposed by several authors

(Lonsane *et al.*, 1985; Sato *et al.*, 1982). Up until now, no sensor was available to measure on line the water content of a solid substrate in the range necessary for microbial metabolism. Thus, all methods lacked feedback and consequently were not extremely accurate. To regulate the water activity on line, we have developed a sensor allowing the continuous measurement of this variable (French Patent No. 860572). The principle of this sensor (Fig. 3) is to equilibrate the relative humidity of the

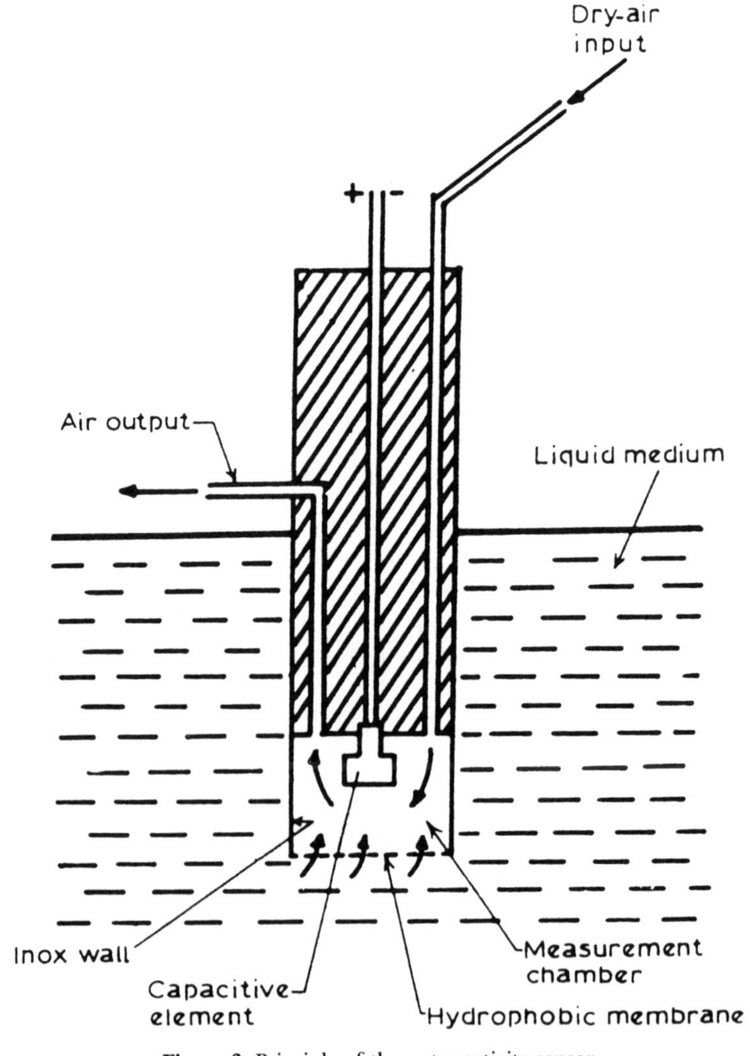

Figure 3. Principle of the water activity sensor.

gas in a small chamber (6 ml) with the water activity of the fermentation substrate.

A capacitive element specific to relative humidity variation (trademark CORECI) is placed in the chamber. This element is surrounded by an inox matrix, so the whole sensor is sterilizable. This sensor, developed by Gervais and colleagues (Gervais, 1987), is connected to a regulator and can control the water activity of a low hydrated liquid or solid fermentations.

In the case of insoluble media—such as cellulose, for which sorption isotherms are such that for very low water content, the a_w value is close to 1—the use of nonaqueous solvents with specific a_w-lowering properties, such as polyols (compatible solutes), is recommended in order to allow solute diffusion at reduced a_w values.

4.1.2. Water Dynamic Effects on Solid-State Fermentation

In the case of solid-state fermentations, the presence of an insoluble matrix generates a matrix potential (Ψ_m) through the intermediary of water-solid interactions. The presence of solutes in the liquid medium gives another kind of potential linked with solute–liquid interaction, the osmotic potential (Ψ_s). These two potentials make up the total water potential of a fermentation medium if temperature, gravity, and pressure are constant.

By measuring the respective effects of the matrix potential and the osmotic potential, Griffin (1981) has determined that these two potentials have a different effect on the growth of microorganisms at the same pressure. For the same value of a_w, the growth measured on media with only solid–liquid interactions, i.e., the matrix potential, is always less than the growth measured on media with only solute–liquid interactions, i.e., the osmotic potential. However, these two types of potential should always have the same effect on the growth of microorganisms at a thermodynamic level.

Some authors have suggested that this variation is due to the interference of diffusion phenomena with the growth of microorganisms (Adebayo and Harris, 1971; Griffin, 1972). In unpublished work, we have demonstrated that variations in kinetic properties of the culture medium in relation to hydration can explain the physiological modifications of microorganisms.

The hydration of the gelified substrate was related to the growth of the filamentous fungus *Trichoderma viride* (Gervais *et al.*, 1988d). The experimental design was intended to discriminate the respective influence on the fungal growth of thermodynamic and kinetic (or dynamic)

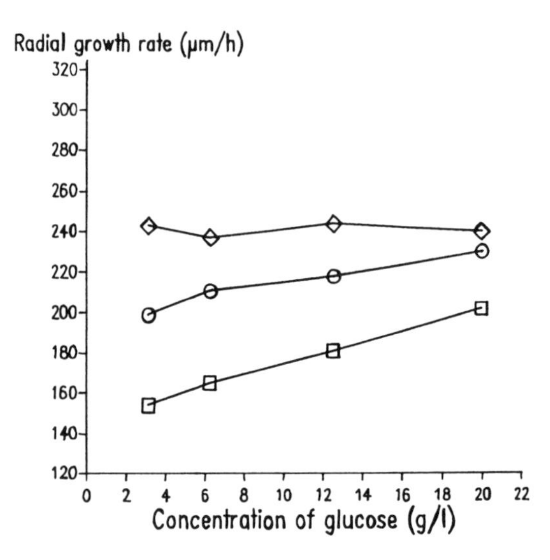

Figure 4. Radial growth rate of *Trichoderma viride* grown on gels with water activity 0.99. ◇ Medium A [63.4 g water/g insoluble dry matter (DM)]; ⊖ medium B (5.47 g water/g insoluble DM); ⊟ medium C (2.22 g water/g insoluble DM).

parameters related to water. For the same thermodynamic level of water (a_w) and by adding silica gel to the medium, it was found that the decrease in water content could generate diffusional limitation of the fungal growth.

The results presented in Fig. 4 show that for the same a_w value, the increase in glucose concentration significantly increases the radial extension rate for very low water content value, but has no effect for high water content values. In order to confirm the assumption of a diffusion limitation, the diffusion of glucose in previous media was evaluated. The translational diffusion coefficient proposed in Table III was determined following Lyrett's method. Such diffusivity was found to decrease sharp-

Table III. Coefficients and Ratio of Diffusion of Glucose in Different Media[a]

Parameter	Water content of the medium (g water/g insoluble DM)		
	Medium A (63.40)	Medium B (5.47)	Medium C (2.22)
$D_{trans} \times 10^{10}$ m²/sec	6.8	4.6	2.5
$D_{rot} \times 10^{-9}$ sec^{-1}	4.8	4.5	1.3
$D_{rot}/D_{trans} \times 10^{-19}$ m^{-2}	0.70	0.98	0.52

[a]Results are for the same a_w = 0.99.

ly when silica gel was added to the media. These results confirm the hypothesis that diffusion could be limiting when the water content of the fermentation decreases. The rotational diffusivity (D_{rot}) measurements determined by using electronic spin resonance are presented in Table III.

Similar results described previously showed a decrease in diffusivity of solutes and water molecules when there was a decrease in water content (Karel, 1976; Biquet and Guilbert, 1986). The values for the D_{rot}/D_{trans} ratio in Table III show that this ratio stayed almost constant whatever the water content, implying that the tortuosity of the media used had little or no influence on solute diffusion in these media. The limiting factor on the growth of microorganisms was the flow rate of solutes in gel and not the tortuosity of the medium. When there was a large amount of polymer, the water content was reduced and solute and water diffusion decreased for the same a_w. The higher the proportion of the matrix potential, the lower the water content became. In fact, analysis of the sorption curves showed that insoluble polymers required a smaller quantity of water in order to reach the same a_w as most solutes. MacDonald and Duniway (1978) carried out experiments on the same kind of soils as those used by Griffin (1981) and Griffin and Luard (1979). It can be deduced that in order to obtain a matrix potential level that corresponds to water activities 0.98 and 0.99 in their culture medium, extremely low water contents were required. Thus, diffusional factors combined with a low water content in the culture can limit the growth of microorganisms in such solid media.

4.2. Other Physicochemical Parameters

4.2.1. Aeration

In solid-state fermentations, aeration has essentially two functions (Chahal, 1983; Tabak and Cooke, 1968): (1) oxygen supply for aerobic metabolism and (2) removal of CO_2, heat, water vapor, and volatile components produced during the metabolism.

The aeration rate must also depend on the thickness of the layer and the porosity of the medium. The effects of pO_2 and pCO_2 in the gas environment are important variables and must be optimized for each medium, microorganism, and process (Bajracharya and Mudgett, 1985; Huang *et al.*, 1985; Levonen-Muñoz and Bone, 1985; Lonsane *et al.*, 1985; Raimbault and Alazard, 1980). Aeration also has very important effects on hydration properties and heat regulation in solid-state fermentations. As shown previously, the water potential could be monitored through the relative humidity of the air, and most of the heat evolved would be transferred to the air latent heat of water vaporization. Sato *et al.* (1982) estimated this latent heat to account for about 80% of the total heat evolved. The aeration in solid-state fermentations which controls air distribution in the medium can be characterized by means of residence time distribution determinations (Gervais *et al.*, 1986).

As defined by Moo-Young and colleagues (Moo-Young *et al.*, 1983), the most important O_2 transfer mechanism is interparticle mass transfer. The dew point of the air used for aeration must be defined as a function of the air inlet temperature in order to control the relative humidity of the solid substrate. Modification of the air status occurs when air is passing through the solid bed. The relative humidity of the air generally increases due to evaporation; such phenomena can be localized and lead to heterogeneity in the medium if the flow rate is too slow. Pressure drop could also vary as the chemical composition of the solid state evolved. An air flow rate control loop could prevent these variations.

4.2.2. Temperature

Temperature regulation is directly related to water potential and aeration. Temperature is a factor that strongly influences culture growth, and in aerobic fermentations, a large amount of heat is produced during microbial growth, as shown by Gonzalez-Blanco and colleagues (Gonzalez-Blanco *et al.*, 1990) for *A. terreus*.

A great limitation of solid-state fermentations is the difficulty of removing excess heat due essentially to the low thermal conductivity of the solid substrate. Saucedo-Castaneda and colleagues (Saucedo-Castaneda *et al.*, 1990) have estimated the dimensionless Biot and Peclet numbers corresponding to the heat transfer during a cassava solid-substrate fermentation by *A. niger.* It was shown that conduction through the 11-fixed bed was the main heat transfer resistance. To maintain the optimal culture conditions, it was necessary to immediately eliminate the heat generated. Convective heat transfer from the reactor wall was not effective due to low mixing and internal heat transfer limitations. The method of evaporative cooling that inserts partially saturated air at lower temperature is the most interesting, as proposed by Durand and colleagues (Durand *et al.*, 1983) and Grajek (1987).

In practice, only air is used for temperature control in solid-state fermentations. The heat capacity of the air at maximal water saturation is less than the heat capacity of water for cooling in submerged fermentations. For these reasons, solid-state fermentations need a large quantity of air, exceeding the amount necessary for microbial respiration. At a semiindustrial scale, the limitation in cooling technique in solid-state fermentations enhances the use of thermophilic strains, as realized by numerous workers (Chahal, 1983; Moo-Young *et al.*, 1977; 1983; Viesturs *et al.*, 1981).

4.2.3. pH

The hydrogen ion concentration is a very important factor in solid-state fermentations. It is well known that pH affects the physiology of fungi and thereby strongly influences the growth of microorganisms.

The control of pH in solid-state fermentations is very difficult because of the problems of achieving an equal distribution of acid or base solution throughout the solid medium. One of the practical solutions is the use of media in which the fungi can maintain ionic balance at the same level during growth. Use of NH_4^+ ions leads to an acidification of the medium; in contrast, nitrate salts increase the pH values.

In practice, mixed nitrogen sources, which are simultaneously consumed, such as ammonium sulfate and nitrate forms of nitrogen sources, and amino acids and ionic salts are used to control pH.

In solid-state fermentations, there is also the possibility of introducing gaseous ammonia, but it is also necessary to protect the microorganisms against toxic ammonium concentrations. The use of acidophilic or acidotolerant microorganisms in solid-state fermentations is another approach to avoid pH control problems.

From the previous considerations, the physicochemical requirements of solid-state fermentations can be summarized as follows:

1. In order to optimize O_2 requirements without limiting the liquid diffusion of nutrients, it is generally beneficial to keep the level of water content just below the water holding capacity of the solid substrate.
2. In order to optimize heat, CO_2 volatile evacuation, and O_2 requirements, it is important to adjust the forced aeration with air with controlled relative humidity. Since respiration produces water, endothermic evaporation phenomena will be the major energetic factor of heat evacuation.
3. In order to prevent rapid water saturation of forced air and condensation in the upper part of the bed, it is necessary to take into account the depth of the fermentation bed and to calculate the air temperature and the air flow rate. Indeed, residence time of air must be sufficiently short in order to not be saturated at the exit of the bed.

4. In order to optimize water activity requirements, which are of major importance for growth, it is necessary to take into account the evolution of the water sorption properties of the solid substrate during the fermentation. Indeed, the division of high-molecular-weight polymers by extracellular enzymes such as cellulases could lead to drastic decreases in water activity for the same water content, as shown previously. Therefore, the relative humidity of the forced air must be adjusted during the fermentation.

5. *Aspergillus* PRODUCTS

5.1. Fermented Foods and Feeds

Oriental fermented foods and their production processes have been extensively discussed (Hesseltine, 1965; Steinkraus, 1983a).

Historically, solid-state fermentations made an early appearance in the Far East. Some of these fermentations are totally solid-state, such as ontjom and tempeh, whereas others (such as soy sauce) include only a part of the entire process as solid-state. These traditional fermentations were used to improve nutritional and organoleptic qualities of various agricultural products, such as rice or the soya bean (Serrano-Carreon *et al.*, 1992).

5.1.1. Koji Process

The koji process may be considered as the prototype of solid-substrate fermentations (Moo-Young *et al.*, 1983). The term koji has two connotations (Huang *et al.*, 1985), viz., the inoculum (the koji starter) and the final product (the koji), a cake including specific mixtures of amylases, proteases, lipases, and macerating enzymes. These enzymes catalyzed the degradation of solid raw materials to soluble products, providing fermentable substrates for yeasts and bacteria in a possible subsequent fermentation stage (Lotong, 1985).

While the first function of the koji is the production of amylase and protease enzymes by *A. oryzae* or *A. soyae*, the koji remains an intermediary product for other fermentations for human consumption (Steinkraus, 1983b).

Depending on the growth substrate, the fermented products are named kome koji (for rice) or muji koji (for barley); according to the colors given by the spore pigments, ki-koji (yellow), kuro-koji (black), and beni-koji (red) are related to sporulation of, respectively, *A. oryzae, A. awamori,* and *Monascus purpureus.* The name of the end product can also be identified, for instance, miso koji, sake koji, or shoyu (soy sauce) koji; these preparations are discussed below.

5.1.1a. Koji Strain Characteristics. Koji molds should be selected by the following criteria (Yokotsuka, 1981, 1983): (1) color and flavor of the final product; (2) ability to produce sufficient spores for the starter inoculum; (3) strong and rapid growth in solid culture; (4) low consumption of starchy substrates, but easy liberation of dextrinization products; (5) high enzymatic activities; (6) genetic stability; and (7) absence of toxin production.

More than 300 *Aspergillus* strains have been used in Japan for the preparation of shoyu (soy sauce). In practice, the koji starter is prepared from an *A. oryzae* or an *A. soyae* culture on steam cooked rice (Japan) or on a mixture of soybeans and wheat bran (China).

Only the *A. oryzae* cultures are able to give a fermented cake with low pH, high α-amylase, acid protease, and carboxypeptidase enzyme activities, and low levels of free sugars; in the same conditions, *A. soyae* strains resulted in a product with lower acidity and lower viscosity related to, respectively, the low amount of citric acid and the high level of free sugar (Yokotsuka, 1983).

In fact, commercial inocula of *A. oryzae* comprise several strains. Protoplast fusion experiments to produce a stable hybrid having an optimum enzymatic balance were unsuccessful (Ushijima and Nakadai, 1984).

5.1.1b. Preparation of Soy Sauce. The main substrates of Japanese soy sauce are soybeans and wheat. Soybeans are soaked up to 18 hr with regular changing of the water to avoid bacterial buildup. The wheat is roasted to provide the brown reaction products that later contribute to the aroma, flavor, and appearance of the soy sauce. Then, after being crushed, the fine particles of roasted wheat, free of contaminants, are adsorbed onto the soybean surface and reduce excessive moisture. The overall process involves two stages and a pasteurization (Sugiyama, 1984) (Fig. 5).

First Stage. The solid medium inoculated with the starter (0.1–0.2%, mole/mole) is incubated aerobically between 25 and 35°C for 3 days in small boxes or on perforated trays. At the beginning of the fermentation, the extension of the mycelium needs higher humidity (43%) than during the sporulation phase (30%). Further, to cool the

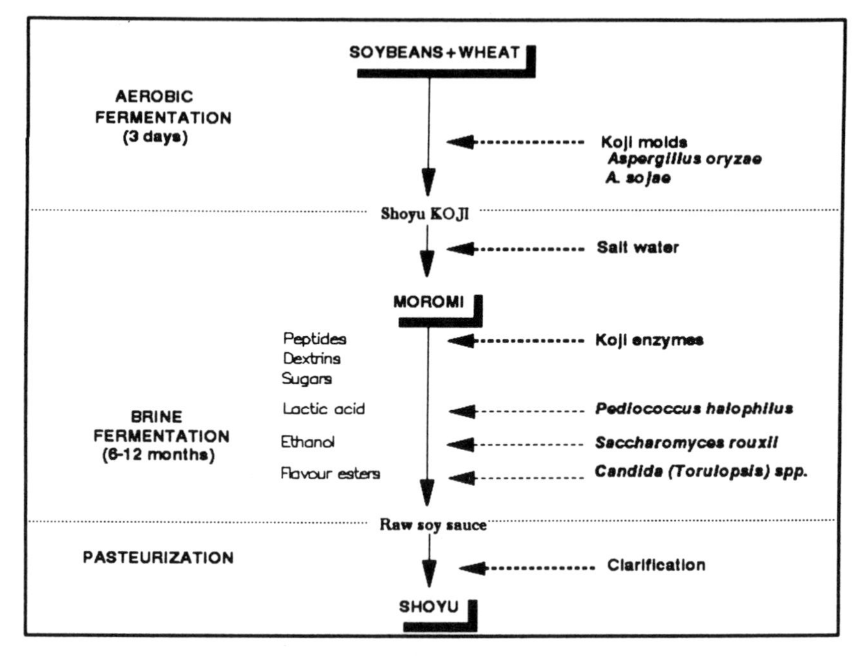

Figure 5. Successive stages in the preparation of Japanese soy sauce (Sugiyama, 1984).

fermented substrate, stirring is usually done twice at about 20–40 hr (Lotong, 1985).

Second Stage. When the koji has reached maturity, it is mixed with an equal amount of saline water (22–23% salt) to form the moromi. The moromi mash is held for several months. The pH value of this shoyu mash starts generally at 6.5–7.0, depending on the extent of contamination of the koji, but only salt-tolerant microorganisms such as lactic acid bacteria or yeast are active. *Micrococcus* or *Bacillus* is inhibited by the low water activity level, which will reach 0.80 in a well-aged mash. In this anaerobic stage, the koji enzymes hydrolyze the proteins, while amylases convert starch into sugars, which are then fermented to lactic acid, alcohol, and flavor substances. The major lactic acid bacterium in shoyu mash belongs to the genus *Pediococcus.* Its optimum pH is 5.5–9.9; it can grow in 24% salt solution and tolerates an a_w level of 0.81 or more.

The lactic fermentation and the degradation of the proteins lead to a rapid decrease of the pH value of the shoyu mash. At a pH value of 5.5 or less, the yeast *Saccharomyces rouxii* takes the place of the lactobacilli and becomes the predominant yeast in the mash.

Other yeasts of the genus *Candida (Torulopsis),* which support 24–

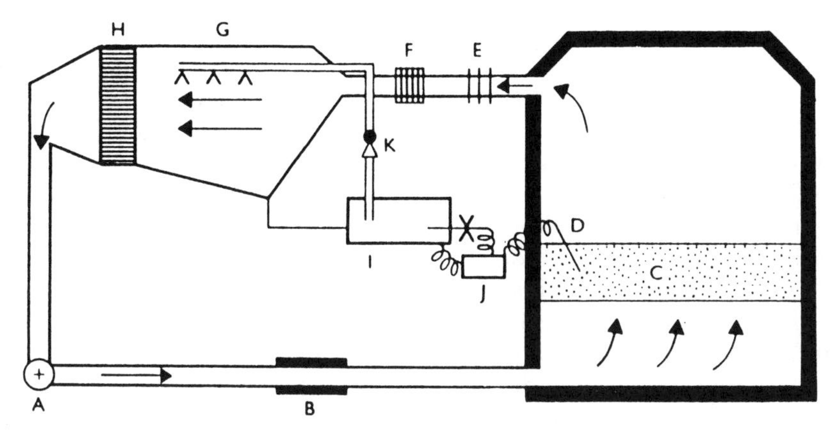

Figure 6. Schematic representations of a koji bioreactor (Cannel and Moo-Young, 1980b). (A) Fan; (B) heater; (C) fermenting solid mass; (D) thermometer; (E) air filter; (F) damper; (G) shower; (H) drain eliminator; (I) water bath; (J) temperature regulator; (K) pump.

26% salt concentration for growth, appear in the latter steps of the moromi transformation and add flavor and aroma to the mash (Yokotsuka, 1983).

Final Stage. The moromi is filtered and the microflora is inactivated by adjustment of salt concentration (17%), addition of sodium benzoate, and pasteurization (70–80°C for 1 hr). After clarification by sedimentation, the brown liquid containing 3% of ethanol can be bottled.

Several solutions have been proposed for automation of the koji process (Cannel and Moo-Young, 1980a,b), one of which is presented in Fig. 6. In the production of koji for soy sauce brewing, the process comprises a continuous cooker and large shallow vats with automatic inoculation and mixing. Evaluating the more effective control of the contamination of the koji provided by this automatic process, Fukushima (1978) reported that the soy sauce obtained by the process has a better aroma than that prepared by the traditional method.

5.1.2. Protein-Enriched Fermented Feeds

In connection with their enzymatic capabilities, members of the genus *Aspergillus* exhibit best growth on starchy materials and have been used for protein enrichment of raw materials by fermentation for animal feeding and some human consumption. Starchy substrates (cassava, potatoes) are of great interest because of their high yield per hectare and the good rate of bioconversion that can be achieved by many fast-growing microorganisms.

To be economically competitive, protein enrichment of starchy materials should be adapted to the rural level, in a system combining the availability of the raw material, the fermentation process, and the local utilization of the fermented product (Senez, 1983); for these reasons, fermentation processes developed for animal feeding have been principally solid-state fermentations.

At laboratory and pilot-scale levels (Karanth and Lonsane, 1988), various substrates have been inoculated with *A. niger:* cassava (Raimbault and Alazard, 1980), banana wastes (Sethi and Grainger, 1978; Baldensberger *et al.,* 1985; de Leon, 1988), mango stones (Sethi and Grainger, 1981), citrus peel (Rodriguez *et al.,* 1985), and coffee pump (Penaloza *et al.,* 1985; de Leon, 1988). However, there have been many problems associated with the scale-up of these processes to the commercial level (Lonsane *et al.,* 1992).

5.2. Metabolites

Among the few organic acids that have been investigated by solid-state fermentation with *Aspergillus* species, citric acid production has found a dominant place. However, gallic acid production has also been described (Miall, 1975).

5.2.1. Citric Acid

Citric acid was first identified as a metabolite from fungal origin in 1893 by Wehmer (1919), who was investigating the production of oxalic acid by *A. niger.* Fermentation processes for citric acid production by this species were developed at the beginning of the century. Currie (1917) opened the way for successful production of citric acid on a commercial scale with granulated sugar as the raw material. Molliard (1919) further showed that culture media of *A. niger* with deficiencies in some nutrients could more readily accumulate citric acid. A solid-state fermentation process involving *A. niger* growing on a re-usable support (bagasse) impregnated by a sugar solution was described by Cahn (1935). Modern microbial production of citric acid has been reviewed by Miall (1978) and Kapoor and colleagues (Kapoor *et al.,* 1982) and involves three process technologies, viz., solid-state fermentation, liquid surface culture, and submerged culture (see Chapter 6).

In Japan, up to 20% of total citric acid production is by solid-state fermentation. Hang and colleagues (Hang *et al.,* 1987) justified this mode of production by the nature of the reduced dimensions of the equipment because of the low quantity of free water present, by an energetic balance that permitted a decrease of the production costs, and by higher yields than in liquid media.

While many *Aspergillus* species can produce some citric acid (*A. awamori, A. flavus, A. fonsecaeus, A. fumaricus, A. lanosus, A. luchensis, A. niger, A. phoenicus, A. saitoi, A. usami,* and *A. wentii*), all the industrial solid-state strains belong to the *A. niger* species.

Although sweet potato starch was used in earlier fermentations, it was soon replaced by wheat bran. In addition, other solid substrates have been investigated, e.g., apple pomace (Hang and Woodams, 1984, 1986a), grape pomace (Hang and Woodams, 1985, 1986b), kiwi fruit peel (Hang *et al.*, 1987), orange peel (Kumagai *et al.*, 1981), and a mixture of cornstarch and extruded wheat bran (Japan Patent, 1983).

Substrate moisture is of central importance in the process: It must be greater than 65% (Hang and Woodams, 1987) and is normally between 70 and 80% (Lockwood, 1979). The initial pH of the culture must be adjusted between 4 and 5. Normally, the process is inoculated with a precultured koji or with a suspension of conidiospores. After a week at 30°C, the solid substrate is transferred to percolators, and citric acid is extracted by water.

Some substances are known to stimulate citric acid production. Methanol, the precise role of which remains undetermined (Kapoor *et al.*, 1982), is currently used at low concentrations (2–4%) (Hang and Woodams, 1984, 1986a,b). Dithiocarbamates such as tetramethylthiuram disulfide improved the production by 74% (Khanna and Gupta, 1986). Citric acid production was also shown to be dependent on the level of divalent trace elements such as Fe^{2+}, Mn^{2+}, Mg^{2+}, Cu^{2+}, and Zn^{2+}.

The yield of citric acid from solid-state fermentations based on sugar consumption were about 55–65% from orange peel and citrus molasses (Kumagai *et al.*, 1981), more than 60% from kiwi fruit peel (Hang *et al.*, 1987) or grape pomace (Hang and Woodams, 1985, 1986b), and 77–87% from apple pomace (Hang and Woodams, 1984, 1986a).

5.2.2. Mycotoxins

The aflatoxins are a major group of mycotoxins produced by *A. flavus* and *A. parasiticus* and are discussed together with other *Aspergillus* mycotoxins in Chapters 2 and 8. In almost all cases in nature, mycotoxins are formed when the producer organism is growing on solid particles such as cereal grains and oil seeds (Smith and Henderson, 1991). Aflatoxins are produced in much higher concentrations under solid-state fermentation conditions than by liquid fermentation (Hesseltine, 1977).

In a similar way, ochratoxin A production by *A. ochraceus* was shown to be a thousand times more efficient by solid-state than by liquid fermentation (Hesseltine, 1972).

5.3. Enzymes

The production of enzymes by solid-state fermentation started in the United States at the turn of the century with glucoamylase production (Takamine, 1894, 1914). At present, several countries, particularly Japan, operate solid-state fermentation processes for enzyme production.

The composition of fungal enzymatic mixtures produced by solid-state fermentations has been shown to be significantly different from those obtained by submerged fermentation systems (Alazard and Raimbault, 1981; Raimbault, 1988).

5.3.1. Amylases

The "moldy bran process" was the first industrial solid-state fermentation enzyme process and utilized *A. oryzae,* the species long used in making sake in the Far East. The process was largely derived from the koji process and used rice and wheat bran, previously treated with steam and acid. This material is deposited on trays, inoculated with spores of *A. oryzae,* and maintained in a place with controlled temperature and humidity. After some days, the highly colonized substrate can be harvested, dried, and crushed. It can be used as is or, after extraction, as a commercial mixture of concentrated amylases. Takamine (1914) proposed the use of rotating drums in place of the static tray fermenters. This modification was to be later adopted for other solid-state fermentation systems (Silman, 1980; Pandey, 1991).

After Takamine's investigations, few studies on fungal amylase production on solid media were described during two decades. They were reactivated for the production of industrial alcohol (Underkofler *et al.,* 1939).

More recently, solid-state fermentation processes have been studied for the production of α-amylases and glucoamylases. Bajracharya and Mudgett (1985) have examined the influence of O_2 and CO_2 partial pressures on enzyme production, while Narahara and colleagues (Narahara *et al.,* 1982) studied the effects of moisture and other parameters on α-amylase production by *A. oryzae.*

The α-amylase produced by *A. oryzae* hydrolyzes the glucosidic bonds in the inner region of the starch molecule and liberates dextrins of varying lengths. The resulting hydrolysis products have a lower viscosity than the starch substrate and are water-soluble. In the early stages, dextrinization leads to maltose and a low concentration of glucose. The levels of glucose may be increased with further hydrolysis.

At present, this fungal α-amylase is used in the production of maltose syrups as an economical substitute for β-amylase and is widely employed in baking as a flour supplement to ensure adequate maltose for yeast fermentation. Its dextrinizing capabilities are used in the clarification of beer, wine, and fruit juices, in which the presence of starch causes haze or filtration problems.

Attention has also been focused on glucoamylase activity in order to find enzymes able to hydrolyze raw starch. Glucoamylsae is an exo-acting glucan hydrolase that cleaves both α-1,4- and α-1,6-glucosidic linkages. Mitsue and colleagues (Mitsue *et al.*, 1979) and Ueda (1981) have observed three different forms of the enzyme. Oriol and colleagues (Oriol *et al.*, 1988) have extensively examined the glucoamylase activity of *A. niger.*

Glucoamylase hydrolysis has now replaced acid hydrolysis for the conversion of starch into glucose. The fungal enzyme has only limited action on starch in its native form. In some cases, acid treatment or thermostable α-amylase is used prior to saccharification with glucoamylase. Glucoamylase is widely used for dextrose production (see Chapter 6 for further details).

5.3.2. Ligninases

Lignocellulosic wastes can be readily fermented by many microorganisms. Oxygen pressure above atmospheric levels has been reported to stimulate lignin conversion during submerged cultures (Kirk, 1980) and solid-state fermentation (Reid, 1983). However, some studies have suggested that lignin degradation by a strain of *Aspergillus* was depressed by elevated carbon dioxide levels in late fermentation (Drew and Kadam, 1979).

5.3.3. Proteases

Very few data have been published about *Aspergillus* proteases produced by solid-state cultivation. Nevertheless, fungal extracellular proteases represent a basic feature of many Oriental solid-substrate fermentations (Raimbault, 1988) and are involved in the meat flavor of fermented soybean products (Flegel, 1988).

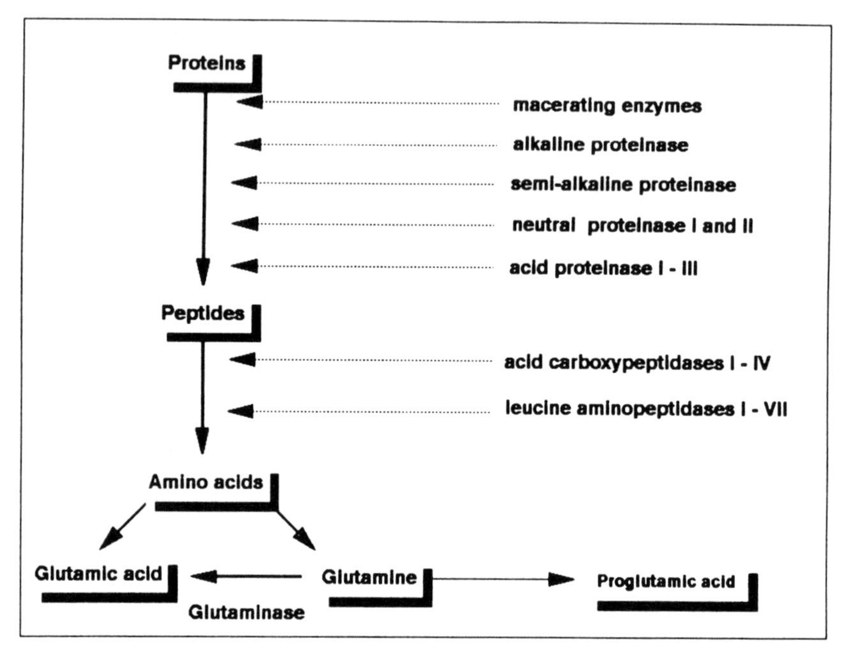

Figure 7. Proteases and peptidases involved in koji protein hydrolysis (Fukushima, 1982).

For the soy sauce koji process, the proteolytic system has been extensively described. There are at least seven major proteases with four different pH optima. The proteases of importance are neutral and alkaline proteases and are found together with several exo-peptidases (Fig. 7).

The activity of these proteases may be strongly reduced (up to 80%) by the level of salt in the moromi stage. Flegel (1988) has suggested that the moromi fermentation stage be begun by addition of small amounts of fresh water.

Fukushima (1982) observed that many proteases and peptidases excreted in solid-state fermentation by *A. oryzae* were not similarly produced in submerged cultures. Narahara and colleagues (Narahara *et al.*, 1982) reported that a temperature of 30°C enhanced the production of proteases in solid-state fermentation.

Aspergillus proteases are preferred to other microbial proteases for bread-making because of their rapid inactivation during baking (Ward, 1983).

5.3.4. Pectinases

Pectinase preparations have been extracted from various *Aspergillus* species, particularly from *A. niger, A. oryzae, A. ventii* (Raimbault, 1988),

A. awamori (Mushikova *et al.*, 1978), or *A. carbonarius* (Ghildyal *et al.*, 1981) growing under solid-state fermentation conditions.

All preparations contained polygalacturionidases, pectinesterases, and pectinlyases. Fungal cultures were conducted on complex media, however; in the crude enzymatic preparations, there are also other kinds of enzymatic activities such as cellulases, hemicellulases, glucosidases, oxidoreductases, and esterases (Sheppards, 1986).

With the use of separation techniques, preparations comprising only one activity have been developed (Aunstrup, 1983), e.g., pectinesterases produced by *A. japonicus* (Ishii *et al.*, 1979) and *A. niger* (Baron *et al.*, 1980), and employed to demethylate high methoxypectin.

At the industrial scale, three processes are used: solid-state fermentation, submerged culture, and a two-stage submerged process. For efficiency reasons, the solid-state fermentation is the most frequently used (Rombouts and Pilnik, 1980; Fogarty and Kelly, 1983).

Two different procedures have been used for pectinase enzyme production by solid-state fermentation: (1) cultures distributed in thin layers on perforated trays and (2) cultures in horizontal rotating drums half filled with solid medium and oxygenated by the flow of air through the system.

Medium composition is an important factor for pectinase production. The highest enzymatic yields were obtained with carbohydrate mixtures such as glucose, molasses, starch, and wheat bran. Recent studies have shown that wheat bran could be substituted by cassava fibrous waste residue (Budiatman and Lonsane, 1987), by pretreated lemon peel (Maldonado *et al.*, 1986), or by henequen pulp (Huitron *et al.*, 1984). Since pectinases are inducible enzymes, it is necessary to supply the culture medium with a pectin source such as citrus peel, apple pomace, or sugar-bette cassettes.

The enzymes are usually produced when the sucrose concentration is extensively decreased. Initial pH is around 4.5 and drops gradually to 3.5, at which point pectinase production normally stops. The inoculum is normally a spore suspension. The fermentation is carried out over a period of 3–7 days at 30°C. At the end of the fermentation, the enzymes are extracted from the culture and concentrated by organic solvents or inorganic salts.

5.3.5. Lipase

A United States patent describes a process involving the cultivation of *A. oryzae* on a bran medium fortified with vegetable oil sludge, mineral salts, and H_2O_2 or peroxide salts. The lipase-rich product is simply dried at 30°C, finely ground, and added to condiment (Grandel, 1959).

5.4. Perspectives

Initially utilized in the Far East and Africa for the production of indigenous fermented foods, many solid-state fermentation processes have been conducted up to the pilot scale. In some cases, despite problems associated with scale-up, solid-state fermentation processes have been shown to compete advantageously with submerged fermentations. Nevertheless, a basic characteristic of solid-state cultivation remains that in comparison to submerged culture, metabolites or enzymes with different specificities have been produced in solid-state fermentations.

REFERENCES

Adebayo, A. A., and Harris, R. F., 1971, Fungal growth responses to osmotic as compared to matrix water potential, *Soil Sci. Soc. Am. Proc.* **35:**465–469.

Alazard, D., and Raimbault, M., 1981, Comparative study of amylolytic enzymes production by *Aspergillus niger* in liquid and solid state cultivation, *Eur. J. Appl. Microbiol. Biotechnol.* **12:**113–117.

Aunstrup, K., 1983, Enzymes of industrial interest—Traditional products, in: *Annual Reports on Fermentation Processes,* Vol. 6, (G. T. Tsao, ed.), Academic Press, New York, pp. 175–201.

Bajracharya, R., and Mudgett, R., 1985, Effects of controlled gas environments in solid substrate fermentations of rice, *Biotechnol. Bioeng.* **22:**2219–2235.

Baldensberger, J., Le Mer, J., Hannibal, L., and Quinto, P. J., 1985, Solid-state fermentation of banana wastes, *Biotechnol. Lett.* **7:**743–748.

Baron, A., Rombouts, F., Drilleau, J. F., and Pilnick, W., 1980, Purification properties of the pectinesterase produced by *Aspergillus niger, Lebensm. Wiss. Technol.* **13:**330–333.

Ben-Amotz, A., and Avron, M., 1978. The mechanism of osmoregulation in *Dunaliella,* in: *Developments in Halophilic Microorganisms,* Vol. 1, *Energetics and Structure of Halophilic Microorganisms* (S. R. Caplan and M. Ginzburg, eds.), Elsevier–North-Holland, Amsterdam, pp. 529–536.

Biquet, B., and Guilbert, S., 1986, Diffusivités relatives de l'eau dans les aliments modèles à humidité intermédiaire, *Lebensm. Wiss. Technol.* **19:**208–214.

Bouanda, R., 1983, Contribution à l'étude de l'activité enzymatique en milieu faiblement hydraté: Influence de la mobilité du substrat, Thèse de doctorat, Université de Bourgogne, Dijon, France.

Brancato, F. P., and Golding, N. S., 1953, The diameter of the mold colony as a reliable measure of growth, *Mycologia* **45:**848–864.

Bruin, S., and Luyben, K. C. A., 1980, Drying of food materials: A review of recent developments, in: *Advances in Drying,* Vol. 1 (A. Mujumdar, ed.), Hemisphere, London, pp. 155–215.

Budiatman, S., and Lonsane, B. K., 1987, Cassava fibrous waste residue: A substitute to wheat bran in solid state fermentation, *Biotechnol. Lett.* **9:**597–600.

Cahn, F., 1935, Citric acid fermentation on solid materials, *Ind. Eng. Chem.* **27:**881–883.

Cannel, E., and Moo-Young, M., 1980a, Solid state fermentation systems, *Process Biochem.* **15:**2–7.

Cannel, E., and Moo-Young, M., 1980b, Solid state fermentation systems, *Process Biochem.* **15:**24–28.

Chahal, D. S., 1983, Growth characteristics of microorganisms in solid state fermentation for upgrading of protein values of lignocelluloses and cellulases production, *Am. Chem. Soc. Symp. Ser. 207:*421–442.

Cudd, A., and Steponkus, P. L., 1987, Osmotic dehydration-induced lamellar to hexagonal-II phase transitions in liposomes of rye plasma membrane lipids, *Cryo 87:24th Annual meeting,* Society for Cryobiology, Edmonton Canada, pp. 33–35.

Currie, J. N., 1917, The citric acid fermentation of *Aspergillus niger, J. Biol. Chem.* **31:**15–37.

de Leon, R., 1988, Brief description of SSF processes at AICATI, in: *Solid State Fermentation in Bioconversion of Agro-Industrial Raw Material* (M. Raimbault, ed.), O.R.S.T.O.M., Paris, pp. 139–143.

Drew, S. W., and Kadam, K. L., 1979, Lignin metabolism by *Aspergillus fumigatus* and white-root fungi, *Dev. Ind. Microbiol.* **20:**153–161.

Durand, A., Arnoux, P., Teilhard de Chardin, O., Chereau, D., Boquien, C. Y., and Larios de Anda, G., 1983, Protein enrichment of sugar-beet pulp by solid state fermentation, in: *Production and Feeding of Single-Cell Protein* (M. P. Ferranti and A. Fietcher, eds.), Applied Science Publishers, London and New York, pp. 120–123.

Esener, A. A., Bol, G., Kossen, N. W. F., and Roels, J. A., 1981, Effect of water activity on microbial growth, *Advances in Biotechnology Scientific and Engineering Principle Intern. Fermentation,* Vol. 1, Symp. No. 6, pp. 339–344.

Fiddy, C., and Trinci, A. P. J., 1976, Mitosis, septation, branching and duplication cycle in *Aspergillus nidulans, J. Gen. Microbiol.* **97:**169–184.

Fisher, L. R., and Israelachvili, J. N., 1979, Direct experimental verification of the Kelvin equation for capillary condensation, *Nature (London)* **277:**548–549.

Flegel, T. W., 1988, Yellow-green *Aspergillus* strains used in Asian soybean fermentations, *ASEAN Food J.* **4:**14–30.

Flory, P. J., 1953, Statistical thermodynamics of polymer solutions, in: *Principles of Polymer Chemistry* (P. J. Flory, ed.), Cornell University Press, Ithaca, New York, pp. 495–540.

Fogarty, W. M., and Kelly, C. T., 1983, Pectic enzymes, in: *Microbial Enzymes and Biotechnology* (W. M. Fogarty, ed.), Applied Science Publishers, London, pp. 131–182.

Franks, F., 1985, Water and aqueous solutions: Recent advances, in: *Water Properties of Food* (D. Simatos and J. L. Multon, eds.), Martinus Nijhoff, Dordrecht, The Netherlands, pp. 1–23.

Fukushima, D., 1978, Soy sauce manufacturing techniques, in: *Report of the ASEAN Workshop on Soy Manufacturing Techniques,* 22–24 January, Singapore.

Fukushima, D., 1982, Koji as an important source in the Orient and its unique composite system of proteinases and peptidases, in: *Use of Enzymes in Food Technology* (P. Dupuy, ed.), Lavoisier, Paris, pp. 381–388.

Gervais, P., 1987, Development of a sensor allowing the continuous measurement of the water activity value of a liquid medium, *Biotechnol. Techniques* **1:**15–18.

Gervais, P., and Sarrette, M., 1990, Influence of age of mycelium and water activity of the medium on aroma production by *Trichoderma viride* grown on solid substrate, *J. Ferment. Bioeng.* **69**(1)**:**46–50.

Gervais, P., Bazelin, C., and Voilley, A., 1986, Patterns of aeration in a solid substrate fermentor through the study of the residence time distribution of a volatile tracer, *Biotechnol. Bioeng.* **28:**1540–1543.

Gervais, P., Bensoussan, M., and Grajek, W., 1988a, Water activity and water content: Comparative effects on the growth of *Penicillium roqueforti* on solid substrate, *Appl. Microbiol. Biotechnol.* **27:**389–392.

Gervais, P., Fasquel, J. P., and Molin, P., 1988b, Water relations of fungal spore germination, *Appl. Microbiol. Biotechnol.* **29:**586–592.

Gervais, P., Molin, P., Grajek, W., and Bensoussan, M., 1988c, Influence of the water activity of a solid substrate on the growth rate and sporogenesis of filamentous fungi, *Biotechnol. Bioeng.* **31:**457–463.

Gervais, P., Belin, J. M., Grajek, W., and Sarrette, M., 1988d, Influence of water activity on aroma production by *Trichoderma viride* TS growing on a solid substrate, *J. Ferment. Technol.* **66:**403–407.

Ghildyal, N. P., Ramakrishna, S. V., Nirmala Devi, P., Lonsane, B. K., and Asthana, H., 1981, Large scale production of pectinolytic enzymes by solid state fermentation, *J. Food Sci. Technol.* **18:**244–257.

Gilles, R., 1975, Mechanisms of ion and osmoregulation, in: *Marine Ecology,* Vol. 2 (O. Kinne, ed.), Wiley-Interscience, Chichester, England, Part 1, pp. 259–347.

Goldberg, M., Parvaresh, F., Thomas, D., and Legoy, M. D., 1988, Enzymatic ester synthesis with continuous measurement of water activity, *Biochim. Biophys. Acta* **957:**359–362.

Gonzalez-Blanco, P., Saucedo-Castaneda, G., and Viniegra-Gonzalez, G., 1990, Protein enrichment of sugar cane by-products using solid state cultures of *Aspergillus terreus, J. Ferment. Bioeng.* **70**(5)**:**351–354.

Grajek, W., 1987, Cooling aspects of solid state fermentation and thermophilic fungi, *J. Ferment. Technol.* **66**(6)**:**675–679.

Grandel, F., 1959, United States Patent No. 2,888,385.

Griffin, D. M., 1972, General ecology of soil fungi, in: *Ecology of Soil Fungi,* Vol. 1, Syracuse University Press, London, pp. 2–67.

Griffin, D. M., 1981, Water and microbial stress, *Adv. Microb. Ecol.* **5:**91–136.

Griffin, D. M., and Luard, E. J., 1979, Water stress and microbial ecology, in: *Strategies of Microbial Life in Extreme Environments* (M. Shilo, ed.), Verlag Chemie-Weinheim, pp. 49–63.

Guggenheim, E. A., 1967, *Thermodynamics: An Advanced Treatment for Chemists and Physicists,* 5th ed., North-Holland, Amsterdam.

Hang, Y. D., and Woodams, E. E., 1984, Apple pomace: A potential substrate for citric acid production by *Aspergillus niger, Biotechnol. Lett.* **6:**763–764.

Hang, Y. D., and Woodams, E. E., 1985, Grape pomace: A novel substrate for microbial production of citric acid, *Biotechnol. Lett.* **7:**253–254.

Hang, Y. D., and Woodams, E. E., 1986a, Solid state fermentation of apple pomace for citric acid production, *J. Appl. Microbiol. Biotechnol.* **2:**283–287.

Hang, Y. D., and Woodams, E. E., 1986b, Utilization of grape pomace for citric acid production by solid state fermentation, *Am. J. Enol. Vitic.* **37:**141–142.

Hang, Y. D., and Woodams, E. E., 1987, Effect of substrate moisture content on fungal production of citric acid in a solid state fermentation system, *Biotechnol. Lett.* **9:**183–186.

Hang, Y. D., Luh, B. S., and Woodams, E. E., 1987, Microbial production of citric acid by solid state fermentation of kiwi peel, *J. Food Sci.* **52:**226–227.

Hellebust, J. A., 1976, Osmoregulation, *Annu. Rev. Plant Physiol.* **27:**485–505.

Hesseltine, C. W., 1965, A millenium of fungi, food and fermentation, *Mycologia* **57:**149–157.

Hesseltine, C. W., 1972, Solid state fermentation, Part 1, *Biotechnol. Bioeng.* **14:**517–532.

Hesseltine, C. W., 1977, Solid state fermentation, *Process Biochem.* **9:**29–32.

Huang, S. Y., Wei, C. H., Malaney, G. W., and Tanner, R. D., 1985, Kinetic responses of the koji solid state fermentation process, in: *Topics in Enzyme and Fermentation: Biotechnology,* Vol. 10 (A. Wiseman, ed.), John Wiley & Sons, New York, pp. 88–108.

Huitron, C., Saval, S., and Acuna, M. E., 1984, Production of microbial enzymes from agroindustrial by-products, *Ann. N.Y. Acad. Sci.* **434:**110–114.
Inch, J. M. M., and Trinci, A. P. J., 1987, Effects of water activity on growth and sporulation of *Paecilomyces farinosus* in liquid and solid media, *J. Gen. Microbiol.* **133:**247–252.
Ishii, S., Kino, K., Sugiyama, S., and Sugimoto, H., 1979, Low-methoxypectin prepared by pectinesterase from *Aspergillus japonicus*, *J. Food Sci.* **44:**611–614.
Jakobsen, M., 1985, Effect of a_w on growth and survival of Bacilliaceae, in: *Properties of Water in Foods* (D. Simatos and J. L. Multon, eds.), Martinus Nijhoff, Dordrecht, The Netherlands, pp. 259–272.
Japan Patent, 1983, Citric acid fermentation on solid media, Kyowa Hakko Kogyo Co., Ltd., J.P. 5800 879.
Kapoor, K. K., Chaudhary, K., and Tauro, P., 1982, Citric acid, in: *Industrial Microbiology* (G. Reed, ed.), Avi Publishing Co., Westport, Connecticut, pp. 709–747.
Karanth, N. G., and Lonsane, B. K., 1988, Laboratory and pilot scale production of enzymes and biochemicals by solid state fermentation at C.F.T.R.I., Mysore, in: *Solid State Fermentation in Bioconversion of Agro-Industrial Raw Materials* (M. Raimbault, ed.), O.R.S.T.O.M., Paris, pp. 113–120.
Karel, M., 1976, Technology and application of new intermediate moisture foods, in: *Intermediate Moisture Foods* (R. Davies, G. G. Birch, and K. J. Parker, eds.), Applied Science Publishers, London, pp. 1321–1354.
Katz, D., Goldstein, D., and Rosenberger, R. F., 1972, Model for branch initiation in *Aspergillus nidulans* based on measurements of growth parameters, *J. Bacteriol.* **109:**1097–1100.
Khanna, V., and Gupta, K. G., 1986, Effect of dithiocarbamates on citric acid production by *Aspergillus niger*, *Folia Microbiol.* **31:**288–292.
Kirk, T. K., 1980, Physiology of lignin metabolism by white-rot fungi, in: *Lignin Biodegradation: Microbiology, Chemistry and Applications*, Vol. 2 (T. K. Kirk, T. Higushi, and H. Chang, eds.), CRC Press, Boca Raton, Florida, pp. 51–63.
Kumagai, K., Usami, S., and Hattori, S., 1981, Citric acid production from mandarin orange wastes by solid culture of *Aspergillus niger*, *Hakko Kogaku Kaishi* **59:**454–461.
Larroche, C., Tallu, B., and Gros, J. B., 1988, Aroma production by spores of *Penicillium roqueforti* on a synthetic medium, *J. Ind. Microbiol.* **3:**1–8.
Le Meste, M., and Voilley, A., 1988, Influence of hydration on rotational diffusivity of solutes in model systems, *J. Phys. Chem.* **92:**1612–1616.
Lemière, J. P., 1992, Hydratation du milieu et développement des champignons filamenteux, Thèse de Doctorat, Université de Bourgogne, Dijon, France.
Levonen-Muñoz, E., and Bone, D. H., 1985, Effect of different gas environments on bench-scale solid state fermentation of oat straw by white-rot fungi, *Biotechnol. Bioeng.* **27:**382–387.
Lilley, T. H., 1985, Water activity as a thermodynamic device: Definitions and assumptions, in: *Water Activity: A Credible Measure of Technological Performance and Physiological Viability*, Discussion conference of the Royal Society of Chemistry, Cambridge, pp. 1–6.
Lockwood, L. B., 1979, Production of organic acids by fermentation, in: *Microbial Technology*, Vol. 1 (H. G. Peppler and D. Perlman, eds.), Academic Press, New York, pp. 355–367.
Lonsane, B. K., Ghildyal, N. P., Butiatman, S., and Ramakrishna, S. V., 1985, Engineering aspects of solid state fermentation, *Enzyme Microb. Technol.* **7:**258–265.
Lonsane, B. K., Saucedo-Castaneda, G., Raimbault, M., Roussos, S., Viniegra-Gonzalez, G., Ghildyal, N. P., Ramakrishna, M., and Krishnaiah, M. M., 1992, Scale-up strategies for solid state fermentation systems, *Process Biochem.* **27:**259–273.

Lotong, N., 1985, Koji, in: *Microbiology of Fermented Foods,* Vol. 2 (B. J. B. Wood, ed.), Elsevier, London, pp. 237–270.

Luard, E. J., 1982, Growth and accumulation of solutes by *Phytophthora cinnamomi* and other lower fungi in response to changes in external osmotic potential, *J. Gen. Microbiol.* **128:**2583–2590.

MacDonald, J. D., and Duniway, J. M., 1978, Influence of the matrix and osmotic components of water potential on zoospore discharge in *Phytophthora, Phytopathology* **68:**751–757.

Maldonado, M. C., Navarro, A., and Callieri, D. A. S., 1986, Production of pectinases by *Aspergillus* sp. using differently pretreated lemon peel as the carbon source, *Biotechnol. Lett.* **8:**501–504.

Matcham, S. E., Jordan, B. R., and Wood, D. A., 1984, Methods for assessment of fungal growth on solid substrates, *Soc. Appl. Bacteriol. Technic. Ser.* **19:**5–18.

Miall, L. M., 1975, Historical development of the fungal fermentation industry, in: *The Filamentous Fungi,* Vol. 1, *Industrial Mycology* (J. E. Smith and D. R. Berry, eds.), E. Arnold, London, pp. 104–107.

Miall, L. M., 1978, Organic acids, in: *Economic Microbiology,* Vol. 2 (A. H. Rose, eds.), Academic Press, London, pp. 47–119.

Mitsue, T., Saha, B. C., and Ueda, S., 1979, Glucoamylase of *Aspergillus oryzae* cultured on steamed rice, *J. Appl. Biochem.* **1:**410–422.

Molin, P., Gervais, P., Lemière, J. P., and Davet, T., 1992, Direction of hyphal growth: A relevant parameter of the development of filamentous fungi, *Res. Microbiol.* **143:**777–784.

Molliard, M., 1919, Production d'acide citrique par le *S. nigra, C. R. Acad. Sci.* **168:**360–363.

Moo-Young, M., Chahal, D. S., Swan, J. E., and Robinson, C. W., 1977, SCP production by *Chaetomium cellulolyticum:* A new cellulolytic fungus, *Biotechnol. Bioeng.* **19:**527–538.

Moo-Young, M., Moriera, A. R., and Tengerdy, R. P., 1983, Principles of solid state fermentation, in: *The Filamentous Fungi,* Vol. 4, *Fungal Technology* (J. E. Smith, D. R. Berry, and B. Kristiansen, eds.), Edward Arnold Publishers, London, pp. 117–144.

Mossel, D. A. A., 1975, Water and micro-organisms in foods: A synthesis, in: *Water Relation in Foods* (R. B. Duckworth, ed.), Academic Press, New York, pp. 347–361.

Mushikova, L. H., Lociakova, L. A., and Jarovenko, V. L., 1978, Pectinase production by *Aspergillus awamori*-16 by solid substrate fermentation, *Microbial Synthesis (Moscow)* **1:**25–28.

Narahara, H., Koyama, Y., Yoshida, T., Pichangkura, S., Ueda, R., and Taguchi, H., 1982, Growth and enzyme production in solid-state culture of *Aspergillus oryzae, J. Ferment. Technol.* **60:**311–319.

Oriol, E., Schettino, B., Viniegra-Gonzalez, G., and Raimbault, M., 1988, Solid-state culture of *Aspergillus niger* on support, *J. Ferment. Technol.* **66:**56–62.

Pandey, A., 1991, Aspects of fermenter design for solid-state fermentation, *Process Biochem.* **26:**355–361.

Penaloza, W., Molina, H. R., Gomez Brenes, R., and Bressani, R., 1985, Solid-state fermentation: An alternative to improve the nutritive value of coffee pulp, *Appl. Environ. Microbiol.* **49:**388–393.

Pirt, S. J., 1965, The maintenance of energy of bacteria in growing cultures, *Proc. R. Soc. B* **163:**224–231.

Raimbault, M. (ed.), 1988, *Solid State Fermentation in Bioconversion of Agro-Industrial Raw Materials,* Documents ORSTOM, Paris, 147 pp.

Raimbault, M., and Alazard, D., 1980, Culture method to study fungal growth in solid fermentation, *Eur. J. Appl. Microbiol.* **9:**199–209.

Reid, I. D., 1983, Effects of nitrogen supplements on degradation of aspen wood lignin and carbohydrate components by *Phanerochaete chrysosporium, Appl. Environ. Microbiol.* **45:**830–837.

Richard-Molard, D., Bizot, H., and Multon, J. L., 1982, Water activity: Essential factor of the microbiological evolution of foods, *Sci. Aliments.* **2:**3–17.

Rodriguez, J. A., Echevarria, J., Rodriguez, F. J., Sierra, N., Daniel, A., and Martinez, O., 1985, Solid state fermentation of dried citrus peel by *Aspergillus niger, Biotechnol. Lett.* **7:**577–580.

Roels, J. A., 1980, Application of macroscopic principles to microbial metabolism, *Biotechnol. Bioeng.* **22:**2457–2514.

Rombouts, F. M., and Pilnik, W., 1980, Pectic enzymes, in: *Economic Microbiology,* Vol. 5, *Microbial Enzymes and Bioconversions* (A. H. Rose ed.), Academic Press, London, pp. 227–282.

Ryan, F. J., Beadle, G. W., and Tatum, E. L., 1943, The tube method of measuring the growth rate of *Neurospora, Am. J. Bot.* **30:**784–799.

Sato, K., Nagatami, M., and Sato, S., 1982, A method of supplying moisture to the medium in a solid state culture with forced aeration, *J. Ferment. Technol.* **60:**607–610.

Saucedo-Castaneda, G, Gutierrez-Rojas, M., Bacquet, G., Raimbault, M., and Viniegra-Gonzales, G., 1990, Heat transfer simulation in solid substrate fermentation, *Biotechnol. Bioeng.* **35:**802–808.

Saunders, P. T., and Trinci, A. P. J., 1979, Determination of tip shape in fungal hyphae, *J. Gen. Microbiol.* **110:**469–473.

Scott, W. J., 1953, Water relations of *Staphylococcus aureus* at 30°C, *J. Biol. Sci.* **6:**549–564.

Scott, W. J., 1957, Water relations of food spoilage microorganisms, *Adv. Food Res.* **7:**83–127.

Senez, J. C., 1983, New developments in the field of protein enrichment of foods and feeds (P.E.F.F.), *Acta Biotechnol.* **3:**299–308.

Serrano-Carreon, L., Baudron, S., Bensoussan, M., and Belin, J. M., 1992, A simple protocol for extraction and detection of pyrazines produced by fungi in solid state fermentation, *Biotechnol. Techniques* **6:**19–22.

Sethi, R. P., and Grainger, J. M., 1978, Solid fermentation of banana to produce animal feed, *J. Appl. Bacteriol.* **45:**12–15.

Sethi, R. P., and Grainger, J. M., 1981, Conversion of mangostone into SCP for animal feed by *Aspergillus niger* using solid substrate fermentation, in: *Advances in Biotechnology,* Vol. 2 (M. Moo-Young, ed.), Pergamon Press, London, pp. 319–325.

Sheppards, G., 1986, The production and uses of microbial enzymes in food processing, in: *Progress in Industrial Microbiology,* Vol. 23, *Microorganisms in the Production of Food* (M. R. Adams ed.), Elsevier, Amsterdam, pp. 237–283.

Silman, R. W., 1980, Enzyme formation during solid-substrate fermentation in rotating vessels, *Biotechnol. Bioeng.* **22:**411–420.

Simatos, D., and Karel, M., 1988, Characterization of the condition of water in foods: Physico-chemical aspects, in: *Preservation of Foods in Tropical Regions by Control of the Internal Aqueous Environment* (C. G. Seow, ed.), Elsevier, London, pp. 1–41.

Smith, J. E., and Henderson, R. S. (eds.), 1991, *Mycotoxins and Animal Foods,* CRC Press, Boca Raton, Florida, 876 pp.

Steinkraus, K. H. (ed.), 1983a, *Handbook of Indigenous Fermented Foods: Microbiology Series,* Vol. 9, Marcel Dekker, New York.

Steinkraus, K. H., 1983b, Industrial applications of Oriental fungal fermentations, in: *The Filamentous Fungi,* Vol. 4, *Fungal Technology* (J. E. Smith, D. R. Berry, and B. Kristiansen, eds.), Edward Arnold, London, pp. 171–189.

Steudle, E., Tyerman, S. D., and Wendler, S., 1983, Water relations of plant cells, in: *Effects of Stress on Photosynthesis* (R. Marcelle, H. Clijsters, and M. Van Poucke, eds.), Martinus Nijhoff, The Hague, pp. 95–109.

Stouthamer, A. H., and Bettenhaussen, C., 1973, Utilization of energy for growth and maintenance in continuous and batch cultures of microorganisms, *Biochim. Biophys. Acta* **301:**54–69.

Sugiyama, S., 1984, Selection of microorganisms for the use in the fermentation of soy sauce, *Food Microbiol.* **1:**339–347.

Tabak, H. H., and Cooke, W. M., 1968, The effects of gaseous environment on the growth and metabolism of fungi, *Bot. Rev.* **34:**126–252.

Takamine, J., 1894, Process of making diastatic enzyme, United States Patent No. 525 823.

Takamine, J., 1914, Enzymes of *Aspergillus oryzae* and the application of its amyloclastic enzyme to the fermentation industry, *Ind. Eng. Chem.* **6:**824–828.

Trinci, A. P. J., 1969, A kinetic study of the growth of *Aspergillus nidulans* and other fungi, *J. Gen. Microbiol.* **57:**11–24.

Trinci, A. P. J., 1971, Influence of the peripheral growth rate of fungal colonies, *J. Gen. Microbiol.* **67:**325–334.

Trinci, A. P. J., 1978, Wall and hyphal growth, *Sci. Prog.* **61:**75–99.

Trinci, A. P. J., and Banbury, G. H., 1967, A study of the growth of the tall conidiophores of *Aspergillus giganteus, Trans. Br. Soc.* **50**(4)**:**525–538.

Trinci, A. P. J., and Collinge, A., 1973, Infuence of L-sorbose on the growth and morphology of *Neurospora crassa, J. Gen. Microbiol.* **78:**179–192.

Troller, J. A., 1980, Influence of water activity on microorganisms in foods, *Food Technol.* **5**(34)**:**76–82.

Ueda, S., 1981, Fungal glucoamylases and raw starch digestion, *Trends Biochem. Sci.* March:89–90.

Underkofler, L. A., Fulmer, E. I., and Schoene, L., 1939, Saccharification of starchy grain mashes for the alcoholic fermentation industry: Use of mold amylase, *Ind. Eng. Chem.* **31:**734–738.

Ushijima, S., and Nakadai, T., 1984, *Aspergillus sojae* strain improvement by protoplast fusion, *Annu. Meet. Agric. Chem. Soc.,* Osaka, Japan, April.

Viesturs, U. E., Aspite, A. F., Laukevics, J. J., Ose, V. P., and Bekers, M. J., 1981, Solid state fermentation of wheat straw with *Chaetomium cellulolyticum* and *Trichoderma lignorum, Biotechnol. Bioeng. Symp.* **11:**359–369.

Ward, O. P., 1983, Proteinases, in: *Microbial Enzymes and Biotechnology* (W. M. Fogarty, ed.), Applied Science Publishers, London, pp. 251–317.

Wehmer, C., 1919, Verlust des Oxalsäure-Bildungsvermögens bei einem degenerierten *Aspergillus niger, Zentralbl. Bacteriol.* **49:**145–148.

Yokotsuka, T., 1981, Recent advances in shoyu research, in: *The Quality of Beverages,* Vol. 2 (G. Charalambous and G. Inglett, eds.), Academic Press, New York, pp. 171–196.

Yokotsuka, T., 1983, Fermented protein foods in the Orient, with emphasis on shoyu and miso in Japan, in: *Microbiology of Fermented Foods,* Vol. 1 (B. J. B. Wood, ed.), Elsevier, London, pp. 197–247.

Zimmermann, U., 1978, Physics of turgor and osmoregulation, *Annu. Rev. Plant Physiol.* **29:**121–148.

Liquid Fermentation Systems and Product Recovery of *Aspergillus* 6

L. M. HARVEY and B. McNEIL

1. INTRODUCTION

The genus *Aspergillus* has an important commercial role within the fermentation industry. In the developed countries, the term "fermentation industry" usually implies an emphasis on submerged liquid cultivation systems.

In some ways, it is surprising that a genus with morphology that fits it so well to the colonization of surfaces should be a success in the comparatively "unnatural" environment found in fermentation vessels. Table I illustrates some of the features commonly attributed to each system. The first point that should be made is that in most cases, these differences are only a matter of degree. Thus, for example, an *Aspergillus* species growing over a nutrient material will rapidly run into O_2 limitation as the thickness of the biofilm increases and diffusional resistance mounts. A similar limitation will be quickly encountered by the same *Aspergillus* species if growing rapidly in a fermenter vessel. In simple terms, it is unlikely that the organism can differentiate among the differing causes of limitation of a given nutrient or gas, and the consequences for the organism are likely to be the same if the nutrient or gas is considered in isolation.

The "success" of *Aspergillus* is due largely to the metabolic versatility of members of this genus, which over the years have provided a rich source of organic acids, enzymes, and food additives and in the future hold some promise as host systems for the production of foreign pro-

L. M. HARVEY and B. McNEIL • Department of Bioscience and Biotechnology, University of Strathclyde, Glasgow G1 1XW, Scotland.

Aspergillus, edited by J. E. Smith. Plenum Press, New York, 1994.

Table I. Some Features of Surface and Submerged Cultivation of the Aspergilli

Feature	Growth on solid surfaces	Growth within liquid nutrient media
Shear forces	Low	Often high
O_2 transfer	Gas phase to cell	Gas to liquid to cell
Nutrient source	Substratum	Bulk liquid
a_W	Low	High
Degree of differentiation	Often pronounced and usually sporulating	Usually mycelial or pellet form and rarely sporulating

teins. It thus seems likely that the industrial future of *Aspergillus* will rest not only on the production of "mature" products, such as the organic acids, but also on the development and commercialization of new products, some of them arrived at by means of the techniques of modern molecular biology.

2. SUBMERGED CULTIVATION

The development of the bioreactor has been an essential factor in the successful commercialization of many products produced by the growth of suitable microorganisms on relatively simple substrates in order to optimize the synthesis of useful products, e.g., the development of the antibiotics industry, production of food additives, and enzymes. It is necessary for the bioreactor to provide the optimum conditions to allow the formation of the product to proceed at the desired rate. It is not always possible for a bioreactor to do this, however, as there are various problems that can be associated with particular bioprocesses, problems associated with broth rheology, such as heat and mass transfer difficulties, and problems with mixing and therefore process control. Many of these difficulties are particularly pronounced when the fermentation process involves a filamentous microorganism, such as an *Aspergillus* species. In assessing the particular problems associated with each bioprocess, the morphological character of the microorganism is a vitally important consideration, since it can affect so many aspects of the process.

2.1. Morphology

The morphological form of the organism has a significant effect on the rheological nature of a fermentation broth. In fermentations involving *Aspergillus* species, one encounters different morphological forms that result in different types of broth rheology; e.g., strains used for the production of citric acid tend to form small pellets that form Newtonian-type broths, whereas strains involved in the production of enzymes or heterologous proteins generally grow in a filamentous manner and consequently exhibit non-Newtonian behavior. The effects of broth rheology on mass, momentum, and heat transfer within a bioreactor have been well documented (Moo Young *et al.*, 1987; Charles, 1985; Funahashi *et al.*, 1988). In aerobic fermentations, in particular, an increase in the viscosity of the fermentation broth has a marked effect on mixing and on mass and heat transfer, which have a combined effect on the organism's response to its environment, often resulting in a decrease in productivity.

In low-viscosity systems following Newtonian behavior, as would be found in a fermentation in which the *Aspergillus* species formed pellets, the fermentation broth is often assumed to be well mixed, and problems associated with poor bulk mixing and mass transfer from gas to liquid phase are assumed to be slight in smaller, lab-scale fermenters (Oosterhuis and Kossen, 1984). However, at production scale, where overall oxygen transfer in the reactor is influenced by bulk mixing, long mixing times may result in the formation of concentration gradients, particularly oxygen. Such a gradient can stress the organism when it enters a zone of limitation, thus causing a decrease in productivity. Phenomena of this kind are often referred to in the literature as "scale-up" problems.

Fermentations in which the *Aspergillus* species grows in the form of dispersed filaments usually exhibit a rather more complex rheological character than those discussed above. At moderate to high biomass levels, these broths display shear thinning or pseudoplasticity, while some may even display a marked yield stress. These effects can have a number of generally undesirable consequences, including poor mass transfer rates outside the well-mixed high-shear impeller zone, and may even lead to stagnant, nonaerated areas in the reactor. In such systems, the problems of poor mixing and mass and heat transfer are much more severe, with poor bulk mixing being exhibited even at the laboratory scale of operation. The resultant stagnant zones often lead to a decreased level of metabolic activity in the culture, thus decreasing the overall productivity of the fermenter vessel.

Mixing times are greatly increased with non-Newtonian broths of this kind, being of the order of minutes in some fermentations (Bylinkina *et al.*, 1973). Thus, the limitations imposed on the organism are considerable, and product formation can be inhibited to a relative degree. Additional considerations that require attention with fermentation broths of this type, associated with the poor bulk mixing within the bioreactor, are problems of process monitoring and control, particularly dissolved oxygen concentration (Metz *et al.*, 1979). Concentration gradients set up within the bioreactor mean that the location of the probes and the titrant addition points is of paramount importance if the fermentation is to be controlled with any degree of accuracy. Unfortunately, there are few guidelines that can be referred to in this case, and experience with the specific fermentation is the only directive that can be used. Power consumption is significantly higher with viscous fermentations. Increased impeller diameter, number, and speed and increased air flow rates are often employed to try and minimize the mixing problems. Consequently, process economics are considerably affected.

2.2. Shear Effects

Shear effects are an important consideration in bioreactor design and usage. Shear is required in a bioreactor in order to facilitate mass, heat, and momentum transfer (Merchuk, 1991), but it is necessary to control the shear effects within a bioreactor in order that the cells are not damaged to the extent where productivity and cell growth are affected. Shear effects can be controlled to some extent by impeller speed, but the design of the impeller is also important, e.g., the use of a marine propeller for the cultivation of fragile mammalian cells.

The aspergilli are often regarded as robust organisms, well able to withstand the effects of hydromechanical stress. Although at most impeller speeds the aspergilli show no signs of gross morphological damage caused by shearing effects, stirrer speed may influence the final product concentration in other ways. The productivity of the citric acid fermentation, for example, is strongly influenced by the impeller speed and, consequently, the shear forces exerted (Ujcova *et al.*, 1980).

3. CURRENT TECHNOLOGY

3.1. Stirred Tank Reactors

At present, approximately 90% of the bioreactors currently used in industry are of the stirred tank configuration (see Fig. 1), even though it is widely recognized that the stirred tank reactor (STR) has certain limitations, e.g., high power consumption, high shear stresses, and poor mixing, mass and heat transfer in highly viscous broths. However, one of the main reasons for the apparent unwillingness of manufacturers to change is the substantial capital investment already made, to-

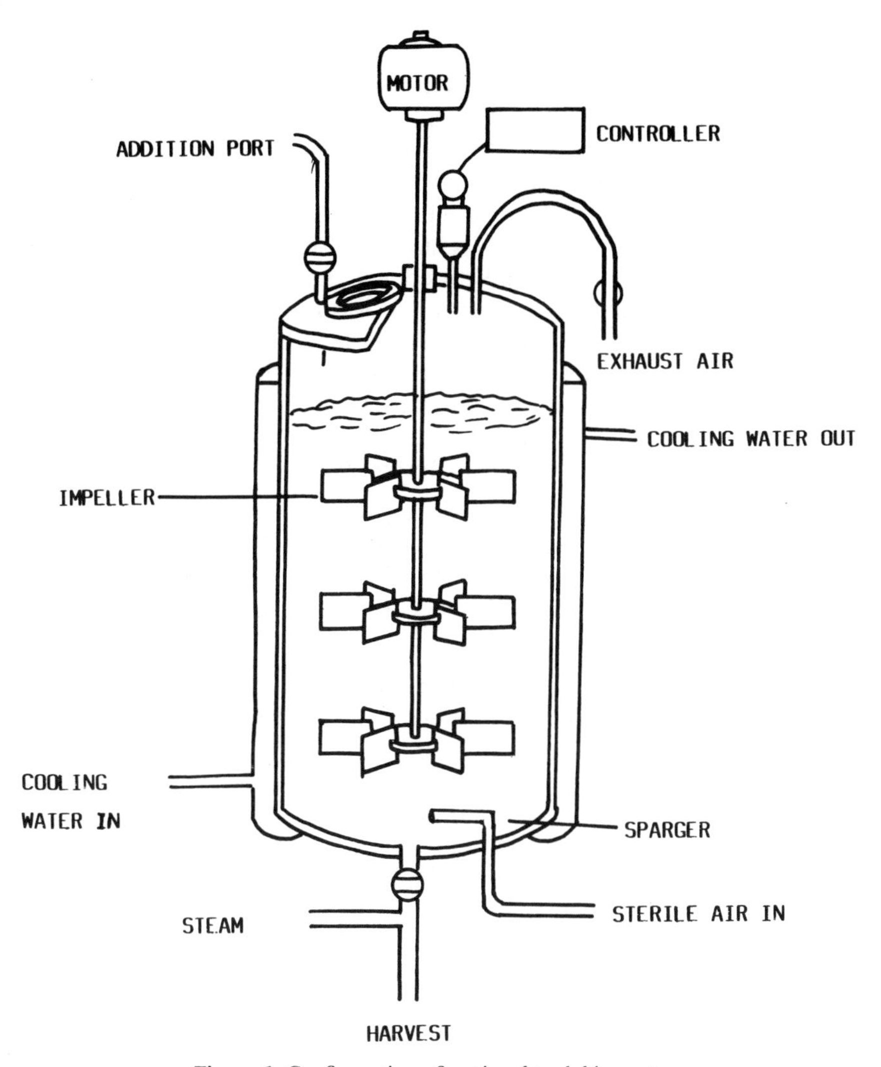

Figure 1. Configuration of a stirred tank bioreactor.

gether with the fact that the STR is thought to be a well-understood system that can be used, with few changes, for many fermentation processes. Attempts have been made to improve the performance of these bioreactors by retrofitting new impeller types, with some limited success. The bulk of fermentations involving the aspergilli employ STRs, used in the batch or, more commonly, the fed-batch mode.

3.2. Airlift Bioreactors

Airlift bioreactors are simple in design and construction (see Fig. 2), having no form of mechanical agitation and requiring relatively low power inputs per unit volume, thus making them an economically attractive proposition. Their application at the industrial level depends entirely on their ability to meet the desired heat, mass, and momentum transfer rates for a particular fermentation (Chisti, 1989). The airlift fermenter has been successfully used for citric acid production and many other fermentation processes and has been found to be particularly useful in waste treatment (Moresi, 1981). Industrial use of airlift fermenters is likely to develop in areas where very low shear stresses are required, good oxygen transfer rates have been achieved at a shear stress of 0.05 Nm^{-2} (Wood and Thompson, 1986).

Two broad groupings of airlift fermenters are available:

1. Internal loop airlifts with an internal baffle splitting the vessel and creating a riser and a downcomer section.
2. External loop airlifts with the riser and the downcomer being separate sections connected by horizontal sections near the top and base of the bioreactor.

The design of these bioreactors is somewhat flexible, being altered to fit specific fermentation processes (Chisti, 1989; Kristiansen and McNeil, 1987).

The mixing in such bioreactor types is a consequence of the different gas hold-ups in the aerated and unaerated zones, thus resulting in differing bulk densities, which cause the fermentation broth to flow. Thus, mixing is not as efficient as that found in the STR, and oxygen transfer rates are consequently lower, which may result in lower specific productivities being achieved. Process control is also less effective, as concentration gradients are more pronounced in this type of bioreactor. Such systems are therefore usually considered to be unsuited to processes involving highly viscous fermentation fluids.

3.3. Modes of Operation of Fermentation Processes

There are few descriptions in the literature regarding the methodologies applied to submerged fermentation processes, due in part to the diverse range of fermentation processes and also to the closely guarded technologies employed by industry. By far the greatest number of industrial processes, including those developed for the aspergilli, employ either batch or fed-batch modes of operation. Continuous culture, although described in detail in many research publications, has not yet

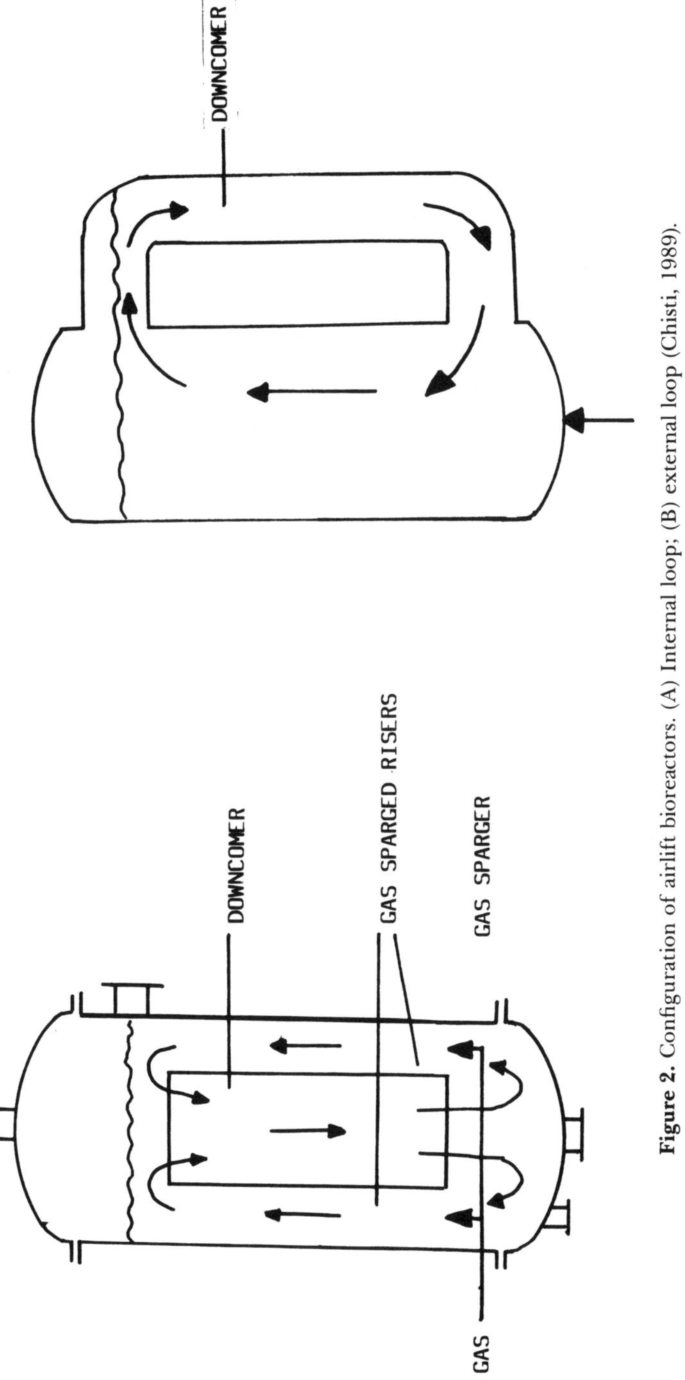

Figure 2. Configuration of airlift bioreactors. (A) Internal loop; (B) external loop (Chisti, 1989).

been adopted by industry to any great extent, although it does hold considerable promise in the production of growth-associated products, such as cell mass (Trinci *et al.*, 1990).

3.3.1. Batch Cultivation

In this mode of operation, the organism is inoculated into the complete medium required for the entire fermentation process. The only additions to the system are of titrant for the control of culture pH and, if necessary, of an antifoaming agent to suppress foam production. This latter feature is particularly necessary where proteinaceous materials are present in the culture broth, as such systems tend to form very stable foams when aerated and agitated.

Batch cultivations have the disadvantage of being unsuitable for fermentations involving organisms that are subject to inhibition either by the substrate or by the product produced by the organism itself. Batch fermentations can also be unsuited to the cultivation of species that form non-growth-associated products, produced when the organism enters the stationary phase of growth. Once the cells enter this phase of growth and before the organism enters the declining or death phase, there may not be sufficient time or nutrients available to allow formation of product at an economically viable rate. These organisms are more suited to a fed-batch process.

The production of citric acid by *A. niger* is an example of a fermentation carried out in the batch mode. Fortunately, this organism is not strongly subject to inhibition by the substrate, molasses, and so sufficient sugars can be present in the medium at the end of the growth phase to allow production of the organic acid to proceed. Many enzymes produced by the *Aspergillus* group are also produced by simple batch cultivation.

3.3.2. Fed-Batch Mode

A fed-batch cultivation is essentially one in which a batch culture is supplied either with additional nutrients, on a continuous or intermittent basis, with the aim of avoiding inhibition by the substrate, or O_2 limitation, or with additives required for product formation, e.g., precursors. Fed-batch cultivation techniques have led to higher final-product concentrations, particularly in fermentations in which the product is growth-associated, and to more efficient use of the substrate constituents.

Fed-batch culture can also be useful for reducing broth viscosity, thus allowing improved mass and heat transfer, particularly in fermentations in which the viscosity of the broth is due to the filamentous nature of the microorganism, as is the case in many fermentations involving the aspergilli.

Fed-batch-culture control techniques have been developed based mainly on those that operate on a feedback control loop, e.g., direct feedback control based on substrate concentration and indirect feedback control based on fermentation parameters such as pH, respiratory quotient, and dissolved oxygen levels. There are a number of other feeding techniques that do not operate in this fashion, e.g., incremental feeding, constant rate additions, and exponential rate additions (Brown, 1990).

Disadvantages of this method are increased capital costs due to the additional instruments required for feedback control; deviations in fermentation profile where incremental feeding, for example, is employed, resulting in reduced control of the system; and the requirement for skilled operators to ensure that the system runs smoothly.

Fed-batch techniques have been successfully employed in the industrial production of organic acids and enzymes.

3.3.3. Continuous Culture

Continuous culture techniques have been successfully employed on an industrial scale for many years, particularly for the treatment of waste, e.g., trickling filters and activated sludge processes. Industry has been somewhat reluctant, however, to use the technique for the manufacture of products by submerged liquid fermentation.

The most common type of continuous culture system is the chemostat. This technique involves feeding the microorganisms with fresh nutrients while removing spent medium from the system containing cells and products of metabolism, and at the same time keeping several parameters constant, such as culture volume, cell concentration, product concentration, pH, and dissolved oxygen.

The main advantages of this type of system are that biomass and product formation proceed under optimal conditions and fermenter down time is significantly reduced, thus making processes more economical, from both a financial and a labor-saving point of view. The major disadvantages are the increased risk of contamination, increased capital expenditure, and the practical difficulties of selecting and installing a suitable system for controlling the continuous feed and withdrawal of spent culture broth. There may also be problems with genetic instability, especially where plasmids have been introduced into a production strain. Currently, no continuous cultivation of the aspergilli occurs on an industrial level.

4. CITRIC ACID PRODUCTION

4.1. Introduction

Citric acid is widely used in industry for various purposes, the principal areas being food, confectionery, and beverages (75%), pharmaceutical (10%), and industrial (15%). The estimated world production of citrate has been reported as 500,000 tons per year (Bu'Lock, 1990), and it is therefore one of the most important fermentation products currently manufactured. It is expected that an upward trend in usage and applications will continue.

4.2. Organisms

A large number of fungi have the ability to produce citric acid. For example, *A. niger, A. clavatus, A. fumigatus, A. japonicus, A. wentii, Penicillium luteum, P. citrinum, Paecilomyces divaricatum, Mucor piriformis, Ustulina vulgaris, Trichoderma viride, Candida lipolytica,* and other *Candida* species have all been shown to be able to produce significant quantities of citrate. The organism used most frequently for industrial production in the West, usually in submerged culture systems, is *A. niger.* Other species reported to have some industrial potential as sources of citrate include *A. fumigatus, A. japonicus, A. wentii,* and the yeast *C. lipolytica.*

4.3. Production Processes

Citric acid is produced by a number of different processes, with submerged liquid fermentation being the most important in terms of the tonnage produced. A very small amount of the world total, less than 1%, is still produce from citrus fruits, e.g., in South America, where citrus fruits are available in quantity at a low price. The three main fermentation processes used are: (1) the koji process, (2) the liquid surface process, and (3) submerged liquid fermentation.

Due to the relatively low profitability, the economics of operation are very constricted (Bu'Lock, 1990), and hence the bulk of citric acid is increasingly produced by submerged liquid fermentation operated, of necessity, on a very large scale.

4.3.1. Koji Process

The koji process accounts for approximately one fifth of the annual citrate produced in Japan. Wheat bran or sweet potato waste is steam sterilized, the substrate picking up sufficient moisture during the sterilization process to give a final water concentration of 70–80%. The resulting "mash" is then inoculated with spores of *A. niger* and spread out in trays or on the floor of a "clean" room. Air is circulated using large fans, which helps to maintain a temperature of around 28°C. The fermentation process runs for a period of 5–8 days, after which the mash is collected and the citrate is extracted using water (Lockwood, 1975) (see also Chapter 5).

4.3.2. Liquid Surface Process

The liquid surface process was developed by Pfizer for large-scale production of citrate (Rohr *et al.*, 1983). The fermentation is carried out in shallow pans, traditionally made of either stainless steel or very high quality aluminium, filled with an appropriate medium. The pans are located in a "clean" room that has been disinfected. Several substrates can be employed, e.g., linear paraffins and carbohydrates; the current trend is to use molasses as the carbohydrate source. Beet molasses is the preferred substrate. The molasses is diluted with water to a suitable sugar concentration, about 150 $kg \cdot m^{-3}$, and brought to a suitable pH (5–7); additional nutrients can be added at this stage, before the medium is sterilized. The medium is then run into the pans, to a depth of 0.05–0.2 m.

Spores of the appropriate strain of *Aspergillus*, selected to minimize the production of by-products such as oxalate, are then sprayed onto the surface of the medium, and the organism is left to grow for a period of 6–12 days at 30°C. A continuous flow of sterile air passes over the pans, supplying oxygen to the growing mycelial mat and helping to reduce the heat produced by the metabolic functions of the organism. The rate of flow of air is therefore controlled in order to facilitate these functions. Since oxygen transfer into the cells is by this surface aeration method only, the fermentation times are longer than those achieved using submerged fermentation techniques. Overall productivities are also lower.

At the end of the fermentation period, a dense mycelial mat has formed, which can be easily separated from the liquid. The liquid, containing the citric acid, is then sent for downstream processing.

This method is still employed by major manufacturers, but since it is very labor-intensive, it tends to be used in countries where labor is cheap and technology limited. The major advantages of this process are that the method is simple and requires unsophisticated equipment, separation of the process liquor from the biomass is easy, and the power requirements are significantly less than those for submerged fermentation processes.

4.3.3. Submerged Liquid Fermentations

Submerged liquid fermentation is the main method employed for the production of citrate (Bu'Lock, 1990). A suitable substrate, usually beet or cane molasses (or a mixture of both), glucose syrup, or starch hydrolysates, is sterilized and inoculated with a strain of *Aspergillus*, usually *A. niger.* A typical production medium is shown in Table II. It is important that the levels of heavy metals often found in the complex substrates employed be kept below certain critical levels, since otherwise the yield of citrate is affected and the loss of productivity results in an uneconomical fermentation. Ion exchange can be used to reduce the heavy-metal loading in molasses, but since this technique can be expensive, addition of potassium or sodium ferrocyanide will effectively remove the heavy-metal ions by chelation (Clarke *et al.*, 1965; Hustede and Rudy, 1976a).

Production of citric acid is generally carried out in a batch or fed-batch mode, continuous cultivation being restricted to research purposes only (Kristiansen and Sinclair, 1979; Kristiansen and Charley, 1981).

The inoculum can be introduced either by direct inoculation of a spore suspension into the fermentation vessel or by use of a vegetative inoculum, the spores having been germinated in an inoculum medium prior to introduction into the main fermentation vessel. The inoculum

Table II. Typical Production Medium for Citric Acid[a]

Component	Concentration ($kg \cdot m^{-3}$)
Sugar (molasses)	140
Ammonium nitrate	2.5
Potassium dihydrogen phosphate	2.5
Magnesium sulfate · $7H_2O$	0.25
Cu^{2+}	0.00006
Zn^{2+}	0.00025
Fe^{2+}	0.0013
Mn^{2+}	0.001

[a]Adapted from Atkinson and Mavituna (1991).

stage, whether vegetative or spore, is vital for the success of the subsequent fermentation. In particular, the morphology of the organism has been shown to be of great importance, with small, smooth pellets, 1.2–2.5 mm in diameter, being the optimal morphological form for good citrate yields (Clark, 1962b; Hustede and Rudy, 1976). The pelleted form gives rise to a Newtonian-type broth that has improved bulk mixing and lower oxygen consumption than its filamentous counterpart (Gomez *et al.*, 1988). Consequently, the economics of the process are greatly improved. Control of morphological form is not always easy, as a number of different factors influence an organism's ability to form pellets or aggregates. From a microbiological point of view, the factors that affect aggregation/pellet formation are cell wall composition, inoculum size, growth rate, nutrition, carbon/nitrogen ratio and the genetic predisposition to pellet. From a physicochemical aspect, factors that affect pellet formation are shear forces, presence of suractive agents, pH, temperature, ionic strength of the medium, suspended solids, and the concentration of calcium ions (Braun and Vecht-Lifshitz, 1991). Control of morphological form has been achieved in the citric acid fermentation by controlling either nutritional or environmental parameters (Clark, 1964; Fried and Sandza, 1959; Clark *et al.*, 1966; Berry *et al.*, 1977). Further advantages of using a pelleted system are reduction in problems such as wall growth and pipe blockage and ease of separation of the biomass from the citrate-containing fermentation broth, which reduces the load on the downstream processing aspect of the process.

The medium is either aerated, if an airlift fermenter is employed, or aerated and agitated if a stirred tank fermenter is used.

A period of rapid growth, during which little or no citric acid is formed, is followed by the production phase, where cell growth ceases and most of the substrate available is directed toward product formation. The fermentation time (2–5 days) is much faster than that achieved in the koji or surface processes described above.

Temperature and pH are more accurately controlled in submerged fermentation processes than in the liquid surface processes; this control, together with the increased oxygen transfer rates, leads to improved productivity.

The fermentation is normally carried out at temperatures between 25 and 30°C, depending on the strain used. Reports on optimal pH values for good citrate production vary considerably. However, initial pH values between 5 and 7 appear to be the industrial preference for processes using molasses. The pH of the culture will decrease as the organism grows and produces citrate, with final pH values falling to around pH 2. It is important, for the prevention of oxalate and gluconate production, that the pH be maintained at 2 or below during the production phase (Kubicek and Rohr, 1986; Marison, 1988).

The citric acid fermentation is highly aerobic and it has sometimes been found necessary, when airlift fermenters are used, to supplement process air with pure oxygen (Martin and Waters, 1952). Critical dissolved oxygen levels have been found to be 9–10% of air saturation for growth of the organism and 12–13% of air saturation for production of citrate (Dawson *et al.*, 1986). The high requirement for oxygen explains why the bulk of the citrate produced is made by submerged liquid fermentations in stirred tank fermenters.

4.4. Conclusions

The citric acid process is probably one of the best understood and intensively researched fermentations, having been in existence for the best part of a century. New applications and increased usage for citrate, and its derivatives, will no doubt lead to a continued increase in demand and production. Manufacturers are constantly trying to improve the process, with current developments centering around strain improvement to increase productivity, decrease fermentation time, and also increase the tolerance of the organism to heavy metals (an important consideration when molasses is the chosen carbon source). Improved understanding of the genetics of this group of fungi will benefit production processes in time; in the meantime, traditional strain-improvement techniques will continue to play a major role.

It is perhaps instructive to note that the number of companies involved in citrate production in the West has been declining, while the scale of production has generally been increasing in those companies that remain in the market. In fermentation terms, citrate is thus clearly a "mature" product, characterized by intense competition among surviving producers.

5. ENZYME PRODUCTION

5.1. Introduction

The success of commercially produced enzymes is due to the wide variety of applications they have in modern industry. Applications include roles in the textile, dairy, brewing, baking, food processing, and detergent industries. Current trends are directed toward the therapeutic use of enzymes, which is likely to open up a new area of application.

Table III. Worldwide Enzyme Market[a]

Enzyme	Market (in millions of $U.S.)
Alkaline protease	$150
Neutral protease	70
Rennins	60
Other proteases	50
Isomerase	45
Amylase	100
Pectinase	40
Carbohydrase	10
Lipase	20
Other	55
TOTAL	600

[a]From Arbige and Pitcher (1989).

The world market for industrial enzyme production is approximately $600 million (U.S.), of which enzymes produced by *Aspergillus* species contribute a significant proportion (Arbige and Pitcher, 1989). The worldwide enzyme market is shown in Table III.

The large-scale commercial production of enzymes by fermentation is becoming increasingly attractive, due in part to the availability of powerful expression systems. Organisms for producing enzymes were originally isolated from their natural environment, with enzyme production being enhanced by medium manipulation and mutagenesis of the appropriate species. Currently, genes for the desired enzyme can be cloned into suitable hosts to facilitate large improvements in yield and functionality. The production of enzymes using rDNA technology is discussed in Section 6 and in Chapter 4. Enzymes can be "programmed" for production either intra- or extracellularly (Kouser *et al.*, 1987), but most are produced extracellularly, involving less complicated extraction procedures and reducing downstream processing costs.

The past decade has seen a shift in emphasis in the market share of different enzymes, with a significant rise in the use and application of high-specificity enzymes, e.g., enzymes required in starch processing. There has been a concommitant decrease in the relative importance of low-specificity enzymes, such as those used in the detergent industry. Both types of enzymes do play, however, a significant role in the multimillion-dollar market of the present-day enzyme industry.

The aspergilli as a group are able to produce a wide range of en-

Table IV. Industrial Enzymes Produced by the Aspergilli

Enzyme	Organism
α-Amylase	*A. niger, A. oryzae*
Glucoamylase	*A. niger, A. oryzae*
Glucose oxidase	*A. niger, A. oryzae*
Lipase	*A. malleus, A. oryzae*
Neutral protease	*A. ochraceous, A. oryzae*
Catalase	*A. niger*
Hemicellulase	*A. niger*
Lactase	*A. niger, A. oryzae*
Endo-polygalacturonase	*A. niger*
Endo-polymethylgalacturonase	*A. foetidis, A. niger*
Exo-polygalacturonase	*A. niger, A. foetidis*
Oligo-1,6-glucosidase	*A. awamori*
Pectin transeliminase	*Aspergillus* species
α-Glucosyltransferase	*Aspergillus* species
β-glucosyltransferase	*Aspergillus* species
β-1,4-xylanase	*A. niger*
Cyclohexaglucanase	*A. oryzae*
Cycloheptaglucanase	*A. oryzae*
β-1,2-glucanhydrolase	*A. fumigatus*
Anthocyanase	*A. niger*

zymes that fall into specific commercial groupings. Table IV illustrates the variety of enzymes, mostly extracellular, produced by the genus.

5.2. Production Processes

Processes used for enzyme production from *Aspergillus* species are very similar to those employed for citric acid production, viz., the traditional koji process, no longer favored in the West, and production by submerged liquid fermentation. Submerged liquid fermentations are the most common and are generally batch or fed-batch processes. Although, in general, continuous processes have been slow to catch on for reasons discussed in Section 3.3.3, with the increased energy savings from the use of continuous sterilizers and continuous fermentations, together with reduced "down" time for cleaning and preparation of the culture vessel, the trend toward continuous culture for the production of enzymes is marked (Arbige and Pitcher, 1989).

In batch and fed-batch processes, the fermentation is allowed to proceed for the appropriate length of time, with the specific enzyme titer being closely monitored as the main criterion determining harvest point.

5.2.1. Media

Media for the production of bulk enzymes have to be relatively cheap, chemically consistent, and available in bulk. Consequently, the most popular substrates are relatively crude, such as molasses, hydrolyzed starches, whey, and various cereals (the latter being particularly important for the koji processes). Care must be taken to ensure that enzyme production is not repressed by the presence of certain carbon sources; e.g., glucoamylase production can be subject to catabolite repression in some species, due to the presence of excessive concentrations of starch, glycerol, or glucose. Likewise, the presence of inducers of enzyme formation can often speed up the production rate; e.g., the presence of maltose or isomaltose acts as an inducer to glucoamylase production by *A. niger* when starch is used as the substrate. Common sources of nitrogenous material are soy and fish meals and cottonseed.

It can be seen from the types of substrates employed that typical industrial media contain a high concentration of solids, which can lead to problems with mixing and mass transfer and also can present problems with downstream processing. There is, however, a correlation between dry matter content, cell growth, and enzyme yield (Aunstrup, 1977).

5.2.2. Process Conditions

Process parameters such as temperature, pH, aeration, and agitation levels are very process-specific.

The majority of fermentations are carried out in the batch mode of operation, usually in stirred tank reactors, although in the case of species that are susceptible to shear damage, airlift bioreactors provide a good alternative.

5.3. Glucoamylase

Glucoamylase, or amyloglucosidase as it is also known, is probably the most important carbohydrase produced by the aspergilli. The enzyme has important applications in the starch industry, being responsible for splitting starch into glucose molecules. Glucoamylase has a molecular weight ranging from 20,000 to over 100,000, with a pH optimum between 4.5 and 5.0. The enzyme is less thermostable than the α-amylases, with denaturation occurring around 70°C.

Aspergillus niger is the main industrial strain used for glucoamylase production. The product is non-growth-associated, with the enzyme being expressed extracellularly after growth has ceased. While it has been shown that glucoamylase production may in fact benefit from oxygen-limited conditions (Aunstrup, 1977; Rousset and Schlich, 1989), oxygen should not be allowed to become limiting during the growth phase of the fermentation, if final productivities are to remain high. Typical fermentation yields are 1–1.5% (wt./vol.) biomass and 0.5–1.0% (wt./vol.) enzyme protein (Aunstrup, 1977).

5.4. Conclusions

It is probable that a gradual and steady increase in the use of industrial enzymes will continue in the coming decades, fueled by the "synthesis" of new enzymes and the development of new fields of application to which enzymes can be "tailor-made" using the latest genetic technologies. With the extensive number of enzymes produced by the *Aspergillus* group, it is therefore likely that this group of fungi will remain at the forefront of enzyme production and engineering. The success of any new industrial enzyme will largely be in areas in which chemical technologies have been unsuccessful and in which the specific properties of the enzyme can provide a viable and economical process. Another area in which enzymes produced by fungi may have some potential is that of enzyme therapy.

6. HETEROLOGOUS PROTEIN PRODUCTION

6.1. Introduction

Although considerable success has been achieved in the production of heterologous proteins using both bacteria and lower eukaryotes, such as yeasts, there is currently great interest in the use of the filamentous fungi for such production. A number of species of aspergilli have been successfully used (Table V); because of the extensive understanding of its genetics, *A. nidulans* has received most attention and has been used as host to a number of foreign genes. In heterologous protein production,

Table V. Some Heterologous Proteins Secreted from Aspergilli[a]

Host	Promoter/source	Product	Amount
A. niger	*god*/*A. niger*	Fungal glucose oxidase	$3.5\ U \cdot mg^{-1}$
A. awamori	*glaA*/*A. niger*	Calf chymosin	$1.3\ kg \cdot m^{-3}$
A. nidulans	*adhA*/*A. niger*	Human tissue plasminogen activator (ETPA)	$1 \times 10^{-3}\ kg \cdot m^{-3}$
A. oryzae	*amyA*/*A. oryzae*	Fungal protease	$3.3\ kg \cdot m^{-3}$

[a]From Saunders *et al.* (1989).

the aspergilli have been demonstrated to have both advantages and disadvantages as host organisms.

6.1.1. Potential Advantages of the Use of *Aspergillus*

A number of features make these species attractive systems for the production of heterologous proteins (Saunders *et al.,* 1987). These include the following:

1. Some of these organisms are prodigious exporters of homologous proteins. For example, *A. niger* can produce certain enzymes in quantities of kilograms per cubic meter under the right conditions.
2. Organisms such as *A. niger* and *A. oryzae* have a long history of usage within the fermentation industry and are generally regarded as safe (GRAS). This often facilitates the path toward regulatory approval of the production system.
3. Following on from the foregoing, the fermentation industry is, in particular, very familiar with the conditions required to maximize production of homologous proteins in *Aspergillus.* In this way, it provides a good starting point for the identification of those physicochemical influences that are likely to be of greatest importance to heterologous protein production using a similar strain.
4. *Aspergillus* is capable of carrying out efficient posttranslational modifications of products, e.g., glycosylation. This is especially important for some proteins derived from higher eukaryotes, e.g., chymosin.
5. *Aspergillus* species are effective secretors of heterologous proteins, often in an authentic, correctly folded form. They tend not to accumulate large quantities of the protein intracellularly, in the form of inclusion bodies, as do some bacteria and yeasts.

6. *Aspergillus* is a particularly useful production system for heterologous proteins derived from other filamentous fungi.
7. Transformant stability is relatively high, so the threat of revertants is less pronounced.

6.1.2. Potential Drawbacks to the Use of *Aspergillus*

Conversely, a number of points can be made against *Aspergillus* as a potential system for heterologous protein production. These might include:

1. Low frequency of transformation, especially when compared to transformation in bacteria or yeasts. It has been proposed that the factor limiting the frequency of transformation is the degree of success in generating protoplasts.
2. Although capable of pellet formation, *Aspergillus* normally grows in liquid culture as dispersed filaments. This has the usual consequences with respect to mass and heat transfer limitations, mixing, and controlling/monitoring the process as the concentration of the biomass increases.
3. Some aspergilli secrete significant quantities of other products, e.g., organic acids, which may reduce the pH of the medium. This represents a diversion of the carbon energy source from the desired activity, which cannot always be overcome by merely increasing the initial amount of the carbon source, due to the possibility of inhibition/repression at higher sugar concentrations. A second possible difficulty lies in the effect on the heterologous protein of being secreted into a low-pH environment. This may cause modifications to, or even denaturation of, the protein.

   A number of aspergilli are well known for their ability to produce extracellular proteases (often most active at low pHs). Potentially, such proteases could damage the product directly, but will also contaminate the final product, unless steps are taken to purify it. This, of course, adds to the recovery/clean-up costs.

   Interestingly, some authors have claimed that inclusion-body formation in bacteria is actually advantageous, since it gives some protection from protease attack in species such as *Escherichia coli* (Enfors *et al.*, 1990).
4. The relationship between copy number and gene expression is not a simple one in some of the systems studied. Thus, an increase in the number of copies of the gene encoding for the

protein within the host might not always result in increased expression or levels of the gene product.

6.2. Production of Calf Chymosin in *Aspergillus*

6.2.1. Background

Calf chymosin is an aspartyl protease that hydrolyzes the peptide bond between phenylalanine and leucine in K-casein, promoting curd formation under acidic conditions (Law and Mulholland, 1991).

Due to limitations in the availability of the natural product, there has been a long-standing interest in the development of substitutes, such as the aspartyl protease from *Mucor* (Saunders *et al.,* 1989) or, in chymosin, produced by means of recombinant DNA technology (Law and Mulholland, 1991; Jeenes *et al.,* 1991).

6.2.2. Other Systems

Escherichia coli was the first microorganism used to produce chymosin (or more correctly preprochymosin), but the amount of chymosin formed was relatively low (<5% of total protein), and it was usually found in the form of inclusion bodies (Emtage *et al.,* 1983). The next system used involved *Saccharomyces cerevisiae,* but once again, though no inclusion bodies were apparent, most of the enzyme was intracellular (Goff *et al.,* 1984; Mellor *et al.,* 1983). Calf chymosin has also been successfully produced in the yeast system *Kluyveromyces lactis* by Gist Brocades in the Netherlands, and the development of this process has been discussed by Seizen and Geurts (1991).

6.2.3. *Aspergillus*

The general sequence of secretion of chymosin can be seen in Fig. 3 (see also Chapter 4).

6.3. Possible Limitations to Heterologous Protein Production

6.3.1. Transformation in *Aspergillus*

Transformation frequency in *Aspergillus* is low (Fincham, 1989), but conversely, mitotic stability of transformants is high (Cullen *et al.,* 1987). Thus, once transformants have been isolated, the loss of the transforming DNA is unlikely, though problems may arise due to persistence and

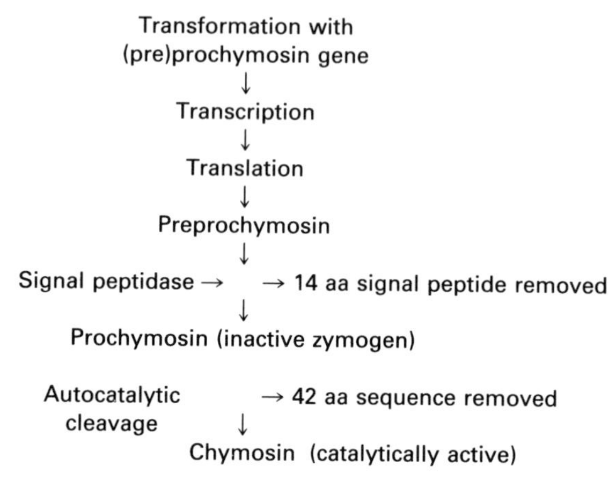

Figure 3. Chymosin secretion from *Aspergillus*.

replication of untransformed nuclei if precautions have not been taken (Buxton and Radford, 1984).

There appears to be no simple relationship between copy number of the transforming gene and protein production in some species. This ambiguity may be a consequence of the often rather random nature of the integration process; also, the specific area(s) of the genome into which the foreign DNA is introduced may affect expression level (Saunders *et al.*, 1989).

6.3.2. Transcription

A range of promoters is available in the aspergilli. These include both strong inducible promoters, such as the glucoamylase promoter of *A. niger,* and constitutive promoters, such as the α-amylase promoter from *A. oryzae.* Jeenes *et al.*, (1991) and Saunders *et al.*, (1989) have described some of the promoters that have been used.

For a fermentation process involving heterologous protein production, an inducible promoter that could be switched on at a given point in the fermentation would probably be favored. Jeenes *et al.*, (1991) have suggested that the preheterologous protein secretion process is not restricted at the transcriptional level.

6.3.3. Translation

Kozak (1986) has demonstrated the influence of a specific sequence upstream of the initiation code that may influence translation. It is as yet unclear whether protein production is markedly influenced in the filamentous fungi at the stage of translation (Jeenes *et al.*, 1991).

6.3.4. Posttranslational Modifications

In order to be efficiently secreted in eukaryotes, proteins usually require a signal peptide at the amino terminus. These signal peptides have a number of features in common, and those of filamentous fungi are generally similar to those of higher eukaryotes (Gierasch, 1989). The signal peptides are removed before secretion of the protein. High levels of chymosin production were achieved in *A. nidulans* when the glucoamylase promoter and signal peptide sequence were fused to the prochymosin gene (Saunders *et al.*, 1989).

Many eukaryotic proteins are glycosylated. Evans and co-workers (Evans *et al.*, 1990) proposed that glycosylation may promote correct protein folding, thus ensuring an active molecule. Glycosylation may also aid the protein in its passage through the secretory system. Chymosin is not normally glycosylated, nor is it glycosylated when produced in *A. nidulans, A. oryzae,* or *A. awamori* (Saunders *et al.*, 1989), although Seizen and Geurts (1991) indicated that glycosylation of chymosin may occur in some species of *Aspergillus.*

6.4. Physiological Control

Once a suitable producer of a heterologous protein is constructed, it must still be cultivated in a fermenter vessel. Both Saunders and co-workers (Saunders *et al.*, 1989) and Jeenes and colleagues (Jeenes *et al.*, 1991) have emphasized the importance of optimizing conditions within the bioreactor. The optimization of heterologous production is best approached in a logical fashion by considering what is known about the technology to be employed, the microbial strain, and the likely interactions of the two.

6.4.1. Fermenter Type

In all probability, the stirred tank will be the dominant bioreactor type to be used for production. Although airlift reactors are attractive from a number of viewpoints (simplicity of construction and maintenance, low average shear rates, and lower running costs), they are not suited to all processes. In particular, organisms with high O_2 requirements would not be well suited for cultivation in such reactors.

One point worthy of note is that since most protein synthesis occurs at the apical tip (Nielsen, 1992), the generally low shear regime in airlift type reactors is unlikely to maximize this synthesis. Instead, since *Aspergillus* is relatively insensitive to shear damage, the higher shear regime in stirred tank reactors, which maximizes the number of hyphal tips, might be favored.

6.4.2. Inoculum

Inoculum age and condition have been shown to have an impact in a number of *Aspergillus* processes (Charley, 1981). Similarly, with spore inocula, concentration of spores can significantly influence the amount of heterologous protein produced from *A. niger* (Archer *et al.*, 1990).

Vegetative inocula give a much shorter lag phase than spores and may be favored for this reason. Parton and Willis (1990) indicated that the use of an autolyzing vegetative seed gave improved biomass formation in an *Aspergillus* process compared with a more "traditional" exponential-phase inoculum. They also advocated the use of minifermenters rather than shake flasks to control the inoculum characteristics more closely and noted marked improvements in subsequent pilot- or production-scale fermentations.

6.4.3. Media

Most industrial processes involving *Aspergillus* species utilize fed-batch techniques to maximize production. In general, the media used would be those appropriate to the strain before genetic modification but taking into account the nature of the system introduced. For example, where the glucoamylase promoter is to be used, addition of starch or maltose at an appropriate time during the fermentation could be used to induce protein synthesis. Christensen and colleagues (Christensen *et al.*, 1988) emphasized the use of fed-batch techniques to maximize cell formation and thus heterologous protein production from an *A. oryzae* strain secreting an aspartyl protease derived from *Rhizomucor miehei.* This implied relationship between increased biomass concentration and improved concentrations of the heterologous protein would seem to indicate that the product-formation kinetics are of the non-growth-associated type.

The factors influencing the choice of medium type used, whether complex or synthetic, would include the nutritional requirements of the strain and also the nature of the downstream processing steps. Media containing large quantities of solids and proteinaceous materials, such as corn, steep liquor, and soybean meal are often encountered in industrial fermentation media and might be expected to result in more prolonged and costlier clean-up stages. Conversely, chemically defined media, if they support good growth, should result in an easier recovery process, but at industrial scale their costs might be prohibitive.

6.4.4. pH

The necessity to closely control culture pH, especially in fermentations involving strains that secrete organic acids, has already been referred to. Unfortunately, due to the rheological character of certain *Aspergillus* broths, this control may be practically rather difficult (Olsvik, 1992).

A number of authors have shown that *Aspergillus* cultures, when grown in the dispersed filamentous form, display pronounced non-Newtonian characteristics (Allan and Robinson, 1987; Kemblowski *et al.*, 1990). Such filamentous broths show marked pseudoplasticity, and their shear thinning nature may cause restriction in mass heat transfer and mixing. As a consequence, chemical gradients may exist within the fermenter (Nienow, 1990). Thus, in poorly mixed or stagnant zones, pH may be very different from that in the well-mixed zone impeller zone. The rate of protein secretion might well be reduced in these low-pH areas, while, in addition, protease activation, or denaturation, may cause a reduction in final product concentration. The overall result of non-ideal pH control regime is likely to be a reduced bioreactor productivity.

Scale-up of heterologous protein production involving filamentous organisms like the aspergilli clearly requires some additional physiological factors to be considered.

6.4.5. Temperature

Many of the difficulties discussed in relation to pH control may also apply to temperature control, but it should be noted that growth and protein secretion often display different temperature optima. The obvious approach to this problem is to carry out a "bi-staged" process such that the initial temperature is around the growth optimum, while in the later stages of the process, the temperature is reduced closer to the optimum for protein secretion (Uusitalo *et al.*, 1991). Jeenes and co-workers (Jeenes *et al.*, 1991) have indicated that the reduced temperature optimum for protein secretion relative to the growth optimum may be caused, at least in part, by a reduced level of protease activity.

6.5. Conclusions

Aspergillus looks set for some success as a system for expressing and secreting heterologous proteins in reasonable (i.e., industrially worthwhile) quantities. This is especially the case when the genes transferred have been derived from other filamentous fungi.

Some companies have indicated that scale-up of a recombinant microorganism is no more complex than that involving similar unmodified microorganisms. For *Aspergillus,* certainly, and probably for most other genetically modified microorganisms, this relative simplicity does not always hold true.

In scaling up a process involving such an organism, one should not view the introduced genes as being inert. There will be a number of interactions between the new genetic material and the physiology of the microorganism, and these in turn may be affected by the conditions within the fermenter. If the process is to be successful, the best starting point is a thorough understanding of the physiology of the unmodified organism, of how the new genetic material is likely to change that, and of what conditions are required to give the best results from the system as a whole.

7. DOWNSTREAM PROCESSING

The number and complexity of the steps involved in arriving at a stable marketable product depend on the nature of the product. At its simplest, the product-recovery stage may consist simply of de-watering and drying when, for example, the biomass is intended as a feed or food product. As might be expected, the number of stages in the recovery process for a product that is a biologically active protein will be greater.

It has been suggested that the renewed interest in product-recovery or separation technology in recent years is due in part to the development of a range of protein or glycoprotein products produced by biotechnology (Roe, 1987). Certainly, there is a greater appreciation of the need to emphasize not only the fermentation step but also the recovery/purification stages (Lambert, 1983). This impetus has led to a movement toward "integrated bioprocessing," which is usually taken to imply simultaneous production and recovery rather than sequential operation, as is the current norm.

The more complex and lengthy the recovery and purification steps, the greater the impact of this stage on the overall process economics. Thus, hard-won improvements in organism productivity and bioreactor

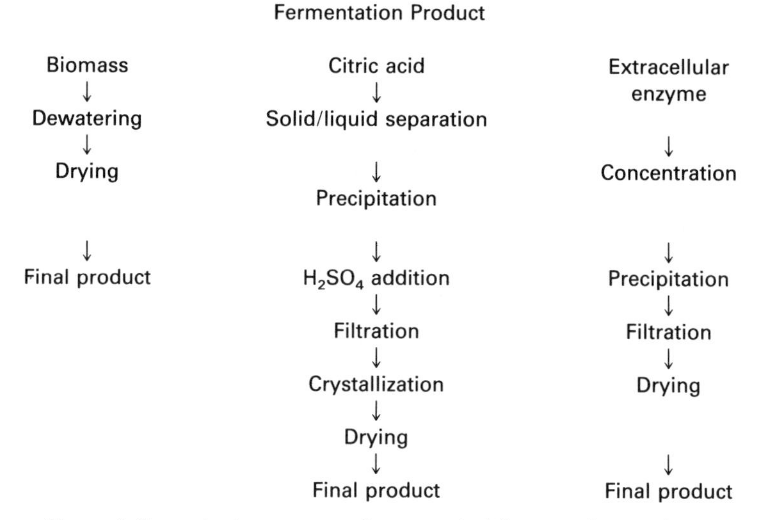

Figure 4. Stages in the recovery of some typical fermentation products.

control might be squandered if they are unmatched by similar developments downstream.

Atkinson and Mavituna (1991) give a concise summary of the recovery processes for a range of microbial metabolites. As can be seen in Fig. 4, most recovery processes involve an initial stage of solid/liquid separation. This stage is usually still a filtration stage, often employing a rotary vacuum filter, which is well suited to filtration of filamentous microbial biomass, such as that of *Aspergillus* cultures. Other alternative methods of achieving separation, such as centrifugation and foam separation, are even less common (Wang, 1987). Some recent developments in the area of cell separation are discussed by Bowden and colleagues (Bowden *et al.,* 1987).

7.1. Citric Acid

The conventional recovery process for citric acid commences with removal of solids by rotary vacuum filter. Calcium hydroxide is then added to give insoluble calcium citrate, which is recovered by filtration. Sulfuric acid is added to redissolve the citric acid, generating $CaSO_4$, which is removed by filtration and the liquid evaporated off, leading to crystallization. Additional steps may include washing, drying, ion exchange, and active carbon treatments.

Since there is such a large demand for citric acid, it is not surprising that there have been developments in both production and recovery. Many of these developments have had the aim of reducing the number of steps in the recovery procedure, thus minimizing product loss.

Liquid/liquid extraction has been proposed by Kertes and King (1985) as an alternative to the conventional precipitation method. They suggested that it is the relatively inefficient recovery stages for fermentation chemicals that have played a major part in restricting their penetration into the organic chemical market. The use of appropriate tertiary amines for citric acid extraction from cell-free fermentation fluids followed by back extraction into high-temperature water has been patented (Baniel, 1982).

7.1.1. Liquid Membranes

A development that offers considerable technical promise is the use of liquid membrane extraction for the recovery of metabolites from fermentation broths. Boey and co-workers (Boey *et al.*, 1987) examined the recovery of citric acid from *A. niger* broths and found that membrane-based extraction systems were suitable for citric acid recovery.

7.1.2. Liquid Membrane Emulsions

Both liquid membrane emulsion (LME) and supported liquid emulsion (SLM) have been investigated for citric acid extraction. Chaudhuri and Pyle (1987) have discussed the relative merits of each type. Briefly, the problems with LMEs relate to the use of an organic phase and the necessity to use a surfactant, which may cause difficulties with some processes. Another potential drawback lies in the need to break the emulsion and effect phase separation after extraction.

Conversely, LMEs have extremely high interfacial areas, giving high transfer rates, and, if carefully designed, may offer the ability to extract and concentrate direct from the broth. Continuous operation is also feasible. This opens up the possibility of continuous removal of citric acid, thus relieving product inhibition.

Boey and colleagues (Boey *et al.*, 1987) used an LME involving a tertiary amine (Alamine 336) as a carrier, Shellsol A as a solvent, and Span80 as a surfactant. Solutions of sodium carbonate were used as the internal phase. The transfer mechanism proposed is shown in Fig. 5.

The citric acid complexes reversibly with the carrier and across the membrane. At the interface with the internal phase (a solution of sodi-

External phase (aqueous)	Membrane (organic phase + surfactant)	Internal, stripping phase (aqueous)
$2C_6H_8O_2$	$6R_3N$	
	$2(R_3NH)_3C_6H_5O_7$	$3Na_2CO_3$
	$3(R_3NH)_2CO_3$	$2C_6H_5O_7Na$
$3H_2O + 3CO_2$	$6R_3N$	

Figure 5. Proposed transfer mechanism for citric acid using a liquid membrane emulsion. (R_3N) Tertiary amine (Alamine), carrier; $2(C_6H_8O_7)Na$ citric acid; $3(Na_2CO_3)$ sodium carbonate, stripping agent.

um carbonate), it reacts irreversibly with NA_2CO_3 to give the product, sodium citrate.

7.1.3. Supported Liquid Membranes

SLMs are conceptually very similar to LMEs, but in this case the membrane phase is entrapped within a thin porous support. These have been used for citric acid extraction (Sirman *et al.*, 1990). These systems potentially offer effective separation against a concentration gradient, low running costs, ease of scale-up, and the possibility of continuous product recovery.

One problem with SLMs relates to retention of the membrane on the solid support. Sirman and colleagues (Sirman *et al.*, 1990) again used Alamine 336 as a carrier in either heptane or xylene. They found that when heptane was used as a diluent, the system was unstable, but xylene was effective. The concentration of the carrier, the stripping agent (NA_2CO_3), and the thickness of the support all affected transfer rate. Diffusion across the membrane seemed to be the rate-limiting step.

7.1.4. Conclusions

These new technologies hold considerable technical promise for a single-step preliminary extraction and concentration of organic acids, such as citric acid. As yet, they have been operated only at laboratory or pilot scale, so their promise remains unrealized.

7.2. Enzymes

The genus *Aspergillus* is a versatile producer of enzymes. The range of enzymes produced from familiar species of *Aspergillus* is being steadily extended by the use of recombinant DNA technology. By comparison with the chemically stable primary metabolites, the recovery and puri-

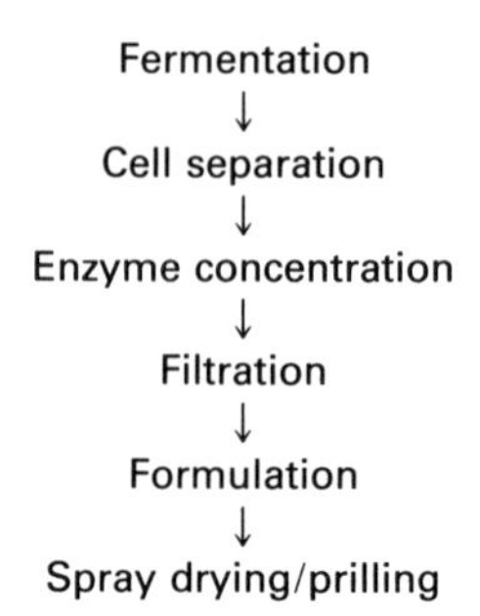

Figure 6. General scheme for the downstream processing of enzymes. From Arbige and Pitcher (1989).

fication of proteins from *Aspergillus* broths is generally more complex and prolonged (Fig. 6). In addition, it must be kept in mind at each step that the aim of the process is not only to produce the maximum amount of the product but also to give a product of high biological activity. This means, in general, that milder treatments are the preferred option.

Recovery and purification methods in current use are covered by Asenjo and Hong (1986) and Atkinson and Mavituna (1991). The general philosophy or thinking behind how protein recovery and purification are put together is discussed by Asenjo (1990). He divides the downstream processing of enzyme-containing broths into two steps:

1. Recovery/isolation, including:
   a. Solid/liquid separation.
   b. Cell disruption (when an intracellular enzyme is involved.
   c. Concentration, giving protein concentrations of up to 60–70 $kg \cdot m^{-3}$.
2. Purification, including:
   a. Primary isolation.
   b. High-resolution purification.
   c. Product polishing.

He also reemphasizes some basic "rule of thumb" guidelines for process-design strategy.

Perhaps one of the greatest potential advances in the approach to recovery and purification comes from the ability to genetically alter a microorganism in such a way as to render product purification easier. Thus, the techniques in genetic manipulation that lead to the stimulation of interest in recovery of protein from fermentation broths may themselves contribute to improvements in that area (Roe, 1987).

The use of fusion proteins as a means of achieving or improving the secretion of heterologous proteins from *Aspergillus* has already been mentioned. A very similar approach can be adopted with respect to endowing a protein product with an additional peptide or protein "tag" to aid in downstream processing (Sassenfeld, 1990). These purification "tags" have been used to improve purification of protein products using such techniques as affinity, ion exchange, hydrophobic, covalent, and metal–chelator interactions. Some of the features of these tags are shown in Table VI.

Protein engineering techniques may also be used to alter other characteristics of enzymes produced in fermentation systems to ease downstream processing. An area of particular interest is that of protein surface charge, alterations in which might affect ease of separation by means of liquid/liquid extraction or membrane techniques (Arbige and Pitcher, 1989).

Due to the frequent presence of proteases leading to degradation of the product enzyme, e.g., degradation of chymosin by native *A. niger* var. *awamori* aspergillopepsin, considerable care has to be taken to minimize the action of these "undesirable" proteases. Some steps taken may occur at the fermentation stage, for example, maintenance of a culture pH unsuitable for protease activation, but in many instances, the use of low temperatures as specific protease inhibitors may be necessary during subsequent processing.

A more elegant method of reducing degradative activity is to delete this activity altogether. Production of chymosin was increased threefold when the gene secreting aspergillopepsin A was deleted from *A. niger* var. *awamori* (Berka, 1992). This general line of approach is also advanced by Enfors *et al.*, (1990).

Yet another method that may reduce degradation by native proteases is the use of a fusion protein, e.g., the fusion of chymosin and glucoamylase (Berka, 1992). The other aspects of this approach have already been discussed. Two recent reviews by Weijers and van't Riet (1992a,b) cover general aspects of enzyme stability and how it is quantified during downstream processing.

Table VI. Potential Advantages of and Drawbacks to the Use of Purification Tags for Protein Purification

Potential drawbacks	Potential advantages
Necessity to remove "tag"	Improved specificity of separation
Possibility of interference with product functionality	Improved product stability
	Improved yield
Initial investment of resources	Higher activity
Increased load on synthetic mechanism of host	Possibility of product-specific assays

7.3. Integrated Fermentation and Recovery

The conventional approach to fermentative production of a biological material physically separates each individual operation in the production sequence. This reflects the process engineer's mental view of production as being the sum of the individual unit processes. Recently, however, there has been a marked trend to bring fermentation and recovery closer together, reflected in the coining of such terms as "integrated bioprocessing" and the more useful term "extractive bioconversions" (Mattiasson and Holst, 1991).

Much of the early work concentrated on simultaneous fermentation and recovery of relatively simple metabolites such as lactic acid and ethanol (Ramalingham and Finn, 1977), though the capacity to recover more complex secondary metabolites is developing. Mattiasson and Holst (1991) list some of the potential advantages, including relief from product inhibition, toxic effects of products and repression, and improved product stability, productivity, and yield. These are ideals, but one or more of these aims should be achievable for most microbial processes.

In many cases, membrane systems have been used to recycle cells or recover product or both. Chang and Furusaki (1991) have discussed developments in the use of membrane-linked bioreactors for extractive bioprocessing. One bioreactor that appeared promising was the dual hollow fiber bioreactor. It was used to produce citric acid using entrapped *A. niger* cells (Chung and Chang, 1988), allowing the separation of microorganism and liquid and gas phases. In addition to achieving very high productivities, this arrangement would ease subsequent processing, since the product stream would be effectively cell-free. Potential drawbacks to the system were shown to be overgrowth of the fungal cells, which led to fiber expansion and flow restriction.

Bioreactors such as this do hold considerable promise, allowing immobilization of biomass, which permits high cell densities and biomass reutilization; in addition, product recovery is simplified. Scale-up of such systems would be by duplication, and might thus be expected to be conceptually simpler.

8. OVERALL CONCLUSIONS

The importance of *Aspergillus* in the fermentation industry appears to be growing. In part, this is due to the exciting developments in our ability to genetically manipulate industrially important species such as *A. niger.* This development has led to improvements in the yield of existing enzymes and to production of a whole new range of biologically active protein products from *Aspergillus* species. In turn, this has led to a renewed interest and activity in the sphere of downstream processing, especially in relation to proteins, but also embracing the more traditional products such as citric acid. Integrated bioprocessing appears to hold much promise, but if the promise is to be realized, it must involve closer and continued communication among the geneticist, the microbiologist, and the biochemical engineer.

REFERENCES

Allan, D. G., and Robinson, C. W., 1987, The influence of slip on rheological measurements of a mycelial broth of *Aspergillus niger, Ann. N. Y. Acad. Sci.* **506:**589–593.

Arbige, M. V., and Pitcher, W. H., 1989, Industrial enzymology: A look towards the future, *Trends Biotechnol.* **7:**330–335.

Archer, D. B., Jeenes, D. J., MacKenzie, D. A., Brightwell, M., Lambert, N., Lowe, G., Radford, S. E., and Dobson, C. M., 1990, Hen egg white lysozyme expressed in, and secreted from, *Aspergillus niger* is correctly processed and folded, *Bio/Technology* **8:**741–745.

Asenjo, J. A., 1986, The rational design of large scale protein separation processes, in: *Separations for Biotechnology 2* (D. L. Pyle, ed.), Elsevier, New York, pp. 519–528.

Asenjo, J. A., and Hong, J., (eds.), 1986, Separation, purification, recovery and purification in biotechnology, *Am. Chem. Soc. Symp.,* Series 314, Washington.

Atkinson, B., and Mavituna, F., 1991, *Biochemical Engineering and Biotechnology Handbook,* 2nd ed., Stockton Press, New York.

Aunstrup, K., 1977, Production of industrial enzymes, in: *Biotechnology and Fungal Differentiation* (J. Meyrath and J. D. Bu'Lock, eds.), FEMS Symposium No. 4, Academic Press, New York, pp. 151–171.

Baniel, A. M., 1982, European Patent No. EP 49,429.

Berka, R. M., 1992, Systems for heterologous gene expression in filamentous fungi, presented at the 9th International Biotechnology Symposium, Crystal City, Virginia, August.

Berry, D. R., Chmiel, A., and Al Obaidi, Z., 1977, Citric acid production by *A. niger,* in: *Genetics and Physiology of Aspergillus* (J. E. Smith and J. A. Pateman, eds.), Academic Press, London, pp. 405–426.

Boey, S. C., Garcia del Cerro, M. C., and Pyle, D. L., 1987, Extraction of citric acid by liquid membrane extraction, *Chem. Eng. Res. Des.* **65:**218–223.

Bowden, C. P., Leaver, G., Melling, J., Norton, M. G., and Whittington, P. N., 1987, Recent and novel developments in the recovery of cells from fermentation broths, in: *Separations for Biotechnology* (M. S. Verrall and M. J. Hudson, eds.), Ellis Horwood, Chichester, England, pp. 49–61.

Braun, S., and Vecht-Lifshitz, S. E., 1991, Mycelial morphology and metabolite production, *Trends Biotechnol.* **9:**63–68.

Brown, A., 1990, Fed batch and continuous culture, in: *Fermentation—a Practical Approach* (B. McNeil and L. M. Harvey, eds), IRL Press, Oxford, pp. 113–130.

Bu'Lock, J. D., 1990, Swings and roundabouts for citric acid producers, *Biotechnol. Insight* **84:**5–6.

Buxton, F. P., and Radford, A., 1984, The transformation of mycelial spheroplasts of *N. crassa* and attempted isolation of an autonomous replicator, *Mol. Gen. Genet.* **196:**337–344.

Bylinkina, E. S., Ruban, E. A., and Nikitina, T. S., 1973, Studies of agitation intensity in antibiotic fermentation broths using isotopic tracers, *Biotechnol. Bioeng. Symp.* **4:**331–335.

Chang, H. N., and Furusaki, S., 1991, Membrane bioreactors: Present and prospects, *Adv. Biochem. Eng. Biotechnol.* **44:**28–64.

Charles, M., 1985, Fermentation scale-up: Problems and possibilities, *Trends Biotechnol.* **3:**134–149.

Charley, R. C., 1981, Production of citric acid in stirred tank reactors, PhD thesis, University of Strathclyde, Glasgow, Scotland.

Chaudhuri, J., and Pyle, D. L., 1987, Liquid membrane extraction, in: *Separations for Biotechnology* (M. S. Verrall and M. J. Hudson, eds.), Ellis Horwood, London, pp. 271–259.

Chisti, M. Y., 1989, *Airlift Bioreactors,* Elsevier, New York.

Christensen, T., Woedlike, H., Boel, E., Mortensen, S. B., Hjortshoej, K., Thin, L., and Hansen, M. T., 1988, High level expression of recombinant genes in *Aspergillus oryzae, Bio/Technology* **6:**1419–1422.

Chung, B. H., and Chang, H. N., 1988, Aerobic fungal cell immobilisation in a dual hollow-fibre: Continuous production of citric acid, *Biotechnol. Bioeng.* **32:**205.

Clark, D. S., 1962, Submerged citric acid fermentation of ferrocyanide treated beet molasses: Morphology of pellets of *A. niger, Can. J. Microbiol.* **8:**133–136.

Clark, D. S., 1964, Citric acid production, United States Patent No. 3,118,821.

Clark, D. S., Ito, K., and Tymchuk, P., 1965, Effect of potassium ferrocyanide addition on the chemical composition of molasses mash used in the citric acid fermentation, *Biotechnol. Bioeng.* **7:**269–278.

Clark, D. S., Ito, K., and Horitsu, H., 1966, Effect of manganese and other heavy metals on submerged citric acid fermentation of molasses, *Biotechnol. Bioeng.* **8:**465–471.

Cullen, D., Gray, G. L., Wilson, L. J., Hayenga, K. J., Larisa, M. H., Rey, M. W., Norton, S., and Berka, R. M., 1987, Controlled expression of bovine chymosin in *Aspergillus nidulans, Bio/Technology* **5:**369–376.

Dawson, M. W., Maddox, I. S., and Brooks, J. D., 1986, Effects of interruptions to the air supply on citric acid production by *Aspergillus niger, Enzyme Microb. Technol.* **8:**37–42.

Emtage, J. S., Angal, S., Doel, M. T., Harris, T. J. R., Jenkins, B., Lilley, G., and Lowe, P. A., 1983, Synthesis of calf prochymosin (prorennin) in *Escherichia coli, Proc. Natl. Acad. Sci. U.S.A.* **8:**3671–3765.

Enfors, S. O., Hellebust, H., Kohler, K., Straudberg, L., and Veide, A., 1990, Impact of genetic engineering on downstream processing of proteins produced in *E. coli, Adv. Appl. Biochem. Eng. Biotechnol.* **43:**31–42.

Evans, R., Ford, C., Siersk, M., Nikolor, Z., and Svensson, B., 1990, Activity and thermal stability of genetically truncated forms of *Aspergillus* glucoamylase, *Gene* **31:**131–134.

Fincham, J. R. S., 1989, Transformation in fungi, *Microbiol. Rev.* **53:**148–170.

Fried, J. H., and Sandza, J. G., 1959, Production of citric acid, United States Patent No. 2,910,409.

Funahashi, H., Hirai, K., Yoshida, T., and Taguchi, H., 1988, Mixing state of xanthan gum solution in an aerated and agitated fermenter, *J. Ferment. Technol.* **66:**103–109.

Gierasch, L. M., 1989, Signal sequences, *Biochemistry* **28:**923–930.

Goff, C. G., Moir, D. T., Kohno, T., Gravins, T. C., Smith, R. A., Yamasaki, E., and Taunton-Rigby, A., 1984, Expression of calf prochymosin in *Saccharomyces cerevisiae, Gene* **27:**35–46.

Gomez, R., Schnabel, I., and Garrido, J., 1988, Pellet growth and citric acid yield of *A. niger* 110, *Enz. Microb. Technol.* **10:**188–191.

Hustede, H., and Rudy, H., 1976a, Manufacture of citric acid by submerged fermentation, United States Patent No. 3,941,656.

Hustede, H., and Rudy, H., 1976b, Manufacture of citric acid by submerged fermentation, United States Patent No. 3,940,315.

Jeenes, P. J., MacKenzie, D. A., Roberts, I. N., and Archer, D. B., 1991, Heterologous protein production by filamentous fungi, in: *Biotechnol. Gen. Eng. Rev.* **9:**327–367.

Kemblowski, Z., Budzynski, P., and Owcarz, P., 1990, On line measurements of the rheological properties of fermentation broth, *Rheol. Acta* **29:**588–593.

Kertes, A. S., and King, C. J., 1985, Extraction chemistry of fermentation produced organic acids, *Biotechnol. Bioeng.* **28:**269–282.

Kouser, C. A., Preuss, D., Grisafi, P., and Botstein, D., 1987, Many random sequences functionally replace the secretion signal frequence of yeast invertase, *Science* **235:**3122–3127.

Kozak, M., 1986, Point mutations define a sequence flanking the AUG initiator codon that modulates translation by eukaryotic ribosomes, *Cell* **44:**283–292.

Kristiansen, B., and Charley, R., 1981, Continuous process for the production of citric acid, *Adv. Biotechnol.* **1:**221–227.

Kristiansen, B., and McNeil, B., 1987, The design of a tubular loop reactor for scale-up and scale-down of fermentation processes, in: *Bioreactors and Biotransformations* (G. W. Moody and P. B. Baker, eds.), Elsevier, London, pp. 321–334.

Kristiansen, B., and Sinclair, C. G., 1979, Production of citric acid in continuous culture, *Biotechnol. Bioeng.* **21:**297–315.

Kubicek, C. P., and Rohr, M., 1986, Citric acid fermentation, *Crit. Rev. Biotechnol.* **3**(4)**:**331.

Lambert, P. W., 1983, Industrial enzyme production and recovery from filamentous fungi, in: *The Filamentous Fungi,* Vol. 4 (J. E. Smith, D. R. Berry, and B. Kristiansen, eds.), Edward Arnold, London, pp. 210–237.

Law, B. A., and Mulholland, F., 1991, The influence of biotechnological developments on cheese manufacture, in: *Biotechnol. Gen. Eng. Rev.* **9:**369–409.

Lockwood, L. B., 1975, Organic acid production, in: *The Filamentous Fungi,* Vol. 1 (J. E. Smith and D. R. Berry, eds.), Edward Arnold, London, pp. 140–157.

Marison, I. W., 1988, Citric acid production, in: *Biotechnology for Engineers: Biological Systems in Technological Processes* (A. H. Scragg, ed.), Ellis Horwood, Chichester, England, pp. 323–326.

Martin, S. M., and Waters, W. R., 1952, Production of citric acid by submerged fermentation, *Ind. Eng. Chem.* **44:**2229–2240.

Mattiasson, B., and Holst, B., 1991, Extractive bioconversions—efforts to integrate fermentation and downstream processing, in: *Proceedings of the 3rd Conference on Upstream and Downstream Processing in Biotechnology* (A. Huyghebaert and E. van Damme, eds.), KVI, Antwerp, Belgium, pp. 1–7.

Mellor, J., Dobson, M. J., and Roberts, N. J., 1983, Efficient synthesis of enzymatically active calf chymosin in *Saccharomyces cerevisiae, Gene* **24:**1–14.

Merchuk, J. C., 1991, Shear effects on suspended cells, *Adv. Biochem. Eng. Biotechnol.* **44:**65–96.

Metz, B., Kossen, N. W. F., and van Suijdam, J. C., 1979, The rheology of mould suspensions, *Adv. Biochem. Eng.* **11:**103–156.

Moo-Young, M., Haland, B., Grant Allen, D., Burrell, R., and Kawase, Y., 1987, Oxygen transfer to mycelial fermentation broths in an air-lift fermenter, *Biotechnol. Bioeng.* **30:**746–753.

Moresi, M., 1981, Optimal design of airlift fermenters, *Biotechnol. Bioeng.* **23:**2537–2560.

Nielsen, J., 1992, Modelling the growth of filamentous fungi, *Adv. Biochem. Eng. Biotechnol.* **46:**187–223.

Nienow, A. W., 1990, Agitators for mycelial fermentations, *Trends Biotechnol.* **8:**224–233.

Olsvik, E. S., 1992, Rheological properties of a filamentous fermentation broth, Ph.D. thesis, University of Strathclyde, Glasgow, Scotland.

Oosterhuis, N. M. G., and Kossen, N. W. F., 1984, Dissolved oxygen concentration profiles in a production scale bioreactor, *Biotechnol. Bioeng.* **26:**546–554.

Parton, C., and Willis, P., 1990, Strain preservation, inoculum preparation and development, in: *Fermentation: A Practical Approach* (B. McNeil and L. M. Harvey, eds.), IRL Press, Oxford, pp. 39–64.

Ramalingham, A., and Finn, R. K., 1977, The Vacuferm process: A new approach to fermentation of alcohol, *Biotechnol. Bioeng.* **19:**583–589.

Roe, S. D., 1987, Whole broth extraction of enzymes from fermentation broths using commercially available adsorbents, in: *Separations for Biotechnology* (M. S. Verall and M. J. Hudson, eds.), Ellis Horwood, Chichester, England, pp. 210–216.

Rohr, M., Zehentgruber, O., and Kubicek, C. P., 1981, Kinetics of citric acid production by *A. niger* pilot plant scale, *Biotechnol. Bioeng.* **23:**2433–2455.

Rousset, S., and Schlich, P., 1989, Amylase production in submerged culture using principal component analysis, *J. Ferm. Bioeng.* **68:**339.

Sassenfeld, H. M., 1990, Engineering proteins for purification, *Trends Biotechnol.* **8:**88–93.

Saunders, G., Picknett, T. M., Tuite, M. F., and Ward, M., 1989, Heterologous gene expression in filamentous fungi, *Trends Biotechnol.* **7:**283–287.

Seizen, R. J., and Geurts, T. G. E., 1991, Chymosin: Production from genetically engineered microorganisms, in: *Biotechnological Innovations in Food Processing* (M. C. E. van Dam-Meiras and C. K. Leach, eds.), Butterworth-Heinemann, Oxford, pp. 197–201.

Sirman, T., Pyle, D. L., and Grandison, A. S., 1990, Extraction of citric acid using a supported liquid membrane, in: *Separations for Biotechnology 2* (D. L. Pyle, ed.), Elsevier, London, pp. 245–254.

Trinci, A. P. J., Robson, G. D., Wiebe, M. G., Cunliffe, B., and Naylor, T. W., 1990, Growth and morphology of *Fusarium graminearum* and other fungi in batch and continuous culture, in: *Microbial Growth Dynamics* (R. K. Poole, M. J. Bazin, and C. W. Keevil, eds.), IRL Press, Oxford, pp. 17–38.

Ujcova, E., Fencl, Z., Musilkova, M., and Seichert, L., 1980, Dependence of release of nucleotides from fungi on fermenter turbine speed, *Biotechnol. Bioeng.* **22**(1)**:**237–241.

Uusitalo, J. M., Nevalainen, K. M. H., Harkki, A. M., Knowles, J. K. C., and Penttila, M. E., 1991, Enzyme production by recombinant *Trichoderma reesei* strains, *J. Biotechnol.* **17:**35–50.

Wang, D. I. C., 1987, Separations for biotechnology, in: *Separations for Biotechnology* (M. S. Verrall and M. J. Hudson, eds.), Ellis Horwood, Chichester, England, pp. 17–46.

Weijers, S. R., and van't Riet K., 1992a, Enzyme stability in downstream processing: 1. Enzyme inactivation, stability and stabilisation, *Biotechnol. Adv.* **10:**237–249.

Weijers, S. R., and van't Riet K., 1992b, Enzyme stability in downstream processing: 2. Quantification of inactivation, *Biotechnol. Adv.* **10:**251–273.

Wood, L. A., and Thompson, P. W., 1986, Application of airlift fermenter, *Proceedings of the International Conference on Bioreactor Fluid Dynamics,* BHRA, Cambridge, England, pp. 157–172.

Enzymes of the Genus *Aspergillus* 7

W. M. FOGARTY

1. INTRODUCTION

Enzymes have been used throughout civilization either as animal or vegetable extracts or in the form of microorganisms. The Greek epic poems *The Odyssey* and *The Iliad,* dating around 700 B.C., both refer to the use of enzymes in cheese-making. The first microbial enzyme for industrial use was prepared by Takamine (1894), who developed a process for the manufacture of amylase using a strain of *Aspergillus oryzae.* The product, called "Taka-diastase," is still used commercially. The addition of enzymes (proteases) to detergents in the 1960s marked the beginning of an era of large-scale production and application of microbial enzymes. In addition to the detergent industry, enzymes have found many uses in a number of other industrial sectors. One of the most significant has been in the area of starch processing and starch technology, where different amylolytic enzymes have been used to supplant acid hydrolysis of starch. The outcome was increased efficiency, increased yields, decreased costs, and overall a much cleaner technology. Two of the success stories in this area were α-amylase and amyloglucosidase produced by *Aspergillus* species. The extension of these developments led to the introduction of glucose isomerase, which catalyzes isomerization of glucose to fructose and formed the basis of the high fructose syrups industry. Other developments included the introduction of proteases, pectinases, β-galactosidase, and other enzymes (Table I).

W. M. FOGARTY • Department of Industrial Microbiology, University College, Dublin 4, Ireland.

Aspergillus, edited by J. E. Smith. Plenum Press, New York, 1994.

Table I. Some Important Enzymes of *Aspergillus* Species

α-Amylase	*A. oryzae*
Amyloglucosidase	*A. niger*
Glucose oxidase	*A. awamori*
Glucose dehydrogenase	*Aspergillus* species
β-Glucosidase	*A. niger*
β-Glucanase	*Aspergillus* species
β-Galactosidase	*A. niger*
Lipase	*A. niger*
Pectinase	*A. niger*
Metalloproteinase	*A. niger*
Catalase	*A. niger*
Glycerol oxidase	*A. japonicus*

2. PRODUCTION AND RECOVERY OF ENZYMES

The two main methods used for production of commercial enzymes are submerged and solid-state fermentations. Although the bulk of enzymes are produced by the former method, certain fungal enzymes continue to be produced by solid-state fermentation (Table II). Solid-state fermentation involves the growth of microorganisms on water-insoluble material with no or almost no free liquid phase. The lower limit of the moisture content is about 12%, and it may be as high as 80% (Moo-Young *et al.*, 1983; Ramesh and Lonsane, 1990). The ancient koji process has been adapted for the production of commercial enzymes by solid-state fermentation. Originally, the koji process involved bamboo baskets in which the substrate was distributed, inoculated, and allowed to ferment. The baskets were stacked in layers so as to facilitate air circulation and incubated at 25–30°C in conditions of high humidity. Adaptations of this setup to improve control of the process and reduce labor

Table II. Enzymes Produced in Solid-State Fermentation

Microorganism	Enzyme	Reference
A. oryzae	Amyloglucosidase	Ono *et al.* (1988a)
	α-Amylase	Meyrath (1966)
A. hennebergi	Amyloglucosidase	Alazard and Baldensperger (1982)
A. phoenicis	β-Glucosidase	Deschamps and Huet (1984a,b)
A. awamori	β-Galactosidase	Silman (1980)
	Invertase	

include rotating drum and tray and bed-type systems (Cannel and Moo-Young, 1980; Hesseltine, 1972; Lonsane *et al.*, 1985). The medium used is always simple, and the most commonly used substrate in solid-state fermentations is wheat bran (Boing, 1981). The advantages and disadvantages of this method of fermentation for the production of enzymes have been outlined in other reviews (Lambert, 1983; Frost and Moss, 1987).

One of the main reasons for the use of submerged fermentation to produce the bulk of commercial enzymes is that it makes greater process control possible. The fermenters used for enzyme production are similar to those used in the antibiotic industry and for the production of other microbial metabolites. They are usually stainless steel stirred tank systems utilizing mechanical agitators to achieve mixing. Generally, batch or fed-batch processes are used (Doyle *et al.*, 1989). Various parameters are closely monitored and insofar as possible controlled during the various stages of the fermentation. The whole purpose of monitoring and controlling is to attempt to maintain optimum environmental conditions for maximum productivity.

Media formulation and development is a very important aspect of the success of submerged fermentation. The ingredients used in large-scale production of enzymes are limited to a relatively small number of mainly complex compounds (Table III). The choice and format of materials as well as the rate and time at which they are added to the fermenter must take into account such things as induction and catabolite repression as well as pH. Protocols for optimizing productivity include judicious selection of carbon and nitrogen source and concentration, as well as the carbon/nitrogen ratio. The effect of initial pH as well as pH control throughout the fermentation must also be established for each process. Inoculum age and size can also have a very important role to play in optimizing productivity. Dissolved oxygen concentration is measured throughout, but fermenter design may limit gaseous availability. Factors that affect production of microbial enzymes in submerged fermentation and regulation of enzyme synthesis have been reviewed by a number of workers (Aunstrup, 1977; Frost and Moss, 1987; Demain, 1990).

Table III. Medium Components for Submerged Fermentation

Starch, starch hydrolysates, molasses, sucrose, lactose, cereal products
Corn steep liquor, yeatex, soyabean, distillers' solubles, ammonium salts

3. DOWNSTREAM PROCESSING AND RECOVERY

Production of enzymes by submerged or solid-state fermentation and the subsequent recovery of these proteins are integral parts of a process geared toward obtaining the maximum amount of the desired product in the required state of purity at an acceptable cost. Recovery of enzymes from solid-state fermentation involves extraction using water by countercurrent techniques of percolation or simply drying and grinding the bran and compounding and selling the preparation as an enzyme bran.

Most but not all industrial enzymes are extracellular. A prerequisite to the recovery of an intracellular enzyme is its extraction and separation from the cells or cellular components. There are a number of methods by which extraction and separation can be achieved, and these methods have been detailed and assessed by a number of workers (Wimpenny, 1967; Atkinson *et al.*, 1987), although data on the application of these methods to the filamentous fungi relative to data on bacteria and yeasts are scarce.

The separation of an extracellular enzyme from the cells is usually achieved by either centrifugation (Axelsson, 1985) or filtration, e.g., using a rotary drum vacuum filter (Belter *et al.*, 1988). The recovery of an intracellular enzyme separated from cell debris and an extracellular enzyme separated from the cells involves similar steps. However, in the latter case, large volumes are involved in the processing, whereas in the former, the volumes are smaller and contain nucleic acids. The recovery steps used depend on the particular enzyme and its intended use as well as the availability of equipment (Table IV). For a liquid enzyme prepara-

Table IV. Methods for Large-Scale Recovery of Enzymes of *Aspergillus* Species

Precipitation and concentration	
Salt fractionation	Ultrafiltration/reverse osmosis
Ammonium sulfate, sodium sulfate	Rotary evaporation
	Spray drying
Organic solvents	
Isopropanol, ethanol, acetone	
Chromatography	
Anion exchange chromatography	Gel-filtration chromatography
Cation exchange chromatography	Adsorption chromatography

Table V. Methods for Laboratory-Scale Recovery and Purification of Enzymes of *Aspergillus* Species

Salt fractionation	Affinity chromatography
Solvent fractionation	Immunoaffinity chromatography
Liquid–liquid partitioning	Isoelectric focusing
Ion-exchange chromatography	Chromatofocusing
Gel filtration	Preparative electrophoresis
Hydrophobic interaction	High-performance liquid chromatography
Adsorption chromatography	Fast protein liquid chromatography

tion, concentration can be effected by ultrafiltration (Flaschel *et al.*, 1983; Lee and Hong, 1985; Brummer and Gunzer, 1987; Golker, 1990). After concentration, the broth usually contains suspended solids, and a polishing filtration is normally carried out to clarify the solution. For a dry enzyme preparation, a precipitation step involving salt or solvent is carried out initially, and the remaining drying steps are concerned with obtaining a product with minimum dust formation (Aunstrup, 1977).

When recovering bulk industrial enzymes, a highly purified enzyme is usually not required, and the cost of obtaining the enzyme relative to its end-product value is of major importance. However, recovery of an enzyme on a laboratory scale for research or other purposes is concerned primarily with obtaining a highly purified, homogeneous end product. Low-volume, high-cost enzymes for commercial use may also necessitate the preparation of a homogeneous, highly purified protein. These requirements are achieved by devising a protocol suitable for each particular enzyme, but the techniques in most common usage are chromatographic ones (Table V) (Brummer and Gunzer, 1987; Kopperschlager *et al.*, 1983; Melling and Phillips, 1975; Scawen *et al.*, 1980; Atkinson *et al.*, 1987). These include gel filtration, ion exchange, adsorption chromatography, hydrophobic interaction, affinity chromatography, chromatofocusing, and immunoaffinity.

4. AMYLOLYTIC ENZYMES OF *Aspergillus* SPECIES

Starch-degrading enzymes are ubiquitous in nature (Fogarty and Kelly, 1979, 1980, 1983, 1990) and are widespread in the microbial kingdom. There are essentially six types of enzymes in microorganisms capable of degrading starch or its intermediate hydrolysis products (Table VI). Of these, the principal ones found in *Aspergillus* species are α-amy-

Table VI. Amylolytic Enzymes of Microorganisms

Trivial name	Systematic name	EC number
α-Amylase	α-1,4-D-Glucan 4-glucanohydrolase	3.2.1.1
β-Amylase	α-1,4-D-Glucan maltohydrolase	3.2.1.2
Amyloglucosidase	α-1,4-D-Glucan glucanohydrolase	3.2.1.3
Cyclodextrin-glycosyltransferase	α-1,4-D-Glucan 4-α-D-(1,4-α-D-glucano)-transferase (cyclizing)	2.4.1.19
Debranching enzymes		
1. Pullulanase	α-Dextrin 6-glucanohydrolase	3.2.1.41
2. Isoamylase	Glycogen 6-D-glucanohydrolase	3.2.1.68
3. Iso-pullulanase (α-1,4-pullulanase)	Pullulan 4-glucanohydrolase	3.2.1.57
α-Glucosidase	α-D-Glucoside glucanohydrolase	3.2.1.20

lase (thermolabile α-amylase), amyloglucosidase, and α-glucosidase. The two former enzymes have the ability to degrade the polysaccharide or partially degraded intermediates, whereas the latter generally degrades only low-molecular-weight saccharides. One interesting debranching-type enzyme, called isopullulanase, was reported in *A. niger* (Sakano *et al.,* 1971).

4.1. α-Amylases (Endo-amylases)

α-Amylases are widespread in the genus *Aspergillus* (Table VII). These enzymes hydrolyze α-1,4-glucosidic linkages in amylose, amylopectin, and related structures and release maltooligosaccharides of varying chain lengths having the α-configuration at C_1 of the reducing glucose unit so produced. By their action on the inner regions of the substrate, a rapid reduction in viscosity and iodine staining power of gelatinized starch is effected. These enzymes bypass α-1,6 linkages in amylopectin. Thermostable forms of the enzyme are extremely useful in liquefaction and thinning of starch. Thermostable α-amylases for commercial use are obtained from *Bacillus* species, principally *B. licheniformis, B. amyloliquefaciens,* and *B. stearothermophilus* (Fogarty and Kelly, 1980).

Following liquefaction, thermolabile α-amylase, principally from *A. oryzae,* is used in saccharification processes. This enzyme, while capable of producing syrups having the required composition and properties, is not sufficiently thermostable to permit its use in liquefaction of starch.

Table VII. ***Aspergillus*** **Species That Produce α-Amylase**

Source	References
A. awamori	Hayashida (1975), Watanabe and Fukimbara (1967), Bhella and Altosaar (1985)
A. batatae	Bendetskii *et al.* (1974)
A. foetidus	Hang and Woodams (1977)
A. kawachii	Mikami *et al.* (1987)
A. niger	Aski *et al.* (1971), Minoda *et al.* (1968)
A. oryzae	Norman (1979), Vallier *et al.* (1977), Yabuki *et al.* (1977), Bata *et al.* (1978), Doyle *et al.* (1989)
A. terricola	Aravina and Ponomarera (1977)
A. usamii	Okazaki (1957)

Products of the hydrolysis of starch contain both maltose and maltotriose, and thus it is in the manufacture of syrups high in maltose that these thermolabile enzymes find major application (Allen and Spradlin, 1974; Barfoed, 1976; Fogarty and Kelly, 1983; Norman, 1979). Large amounts of maltotriose are formed initially in the hydrolysis of starch, and as the reaction proceeds, this decreases and the amount of maltose increases to finally reach a maximum of about 60%.

In the commercial production of a high-maltose syrup, liquefied starch (35–40% dissolved solids) is treated with *A. oryzae* α-amylase at about pH 4.8 and 50°C for 36–48 hr. The product normally contains 45–50% maltose. The precise composition of the final product is governed by enzyme dosage rates (Norman, 1979). *Aspergillus oryzae* α-amylase is also used in conjunction with amyloglucosidase in production of high conversion syrups (Underkofler *et al.*, 1965); this subject is dealt with in the following section.

4.2. Amyloglucosidase

Amyloglucosidase (EC 3.2.1.3) is an exo-acting starch-degrading enzyme that produces β-D-glucose (Ono *et al.*, 1965) from the nonreducing chain ends of amylase and amylopectin by consecutive hydrolysis of α-1,4 linkages. It also hydrolyzes α-1,6 and α-1,3 linkages (Pazur and Kleppe, 1962; Fogarty and Kelly, 1983), although at a much slower rate than α-1,4 linkages. Amyloglucosidase occurs extensively in fungi. The enzymes used commercially originate from either *Aspergillus* species or *Rhizopus* species and are used in the conversion of maltodextrins to D-glucose and in the preparation of high conversion syrups. The affinity of the enzyme for different substrates is related to chain length and increases linearly up to maltopentaose. The enzyme from *A. niger* (Fogarty and Benson, 1983) has relative affinities for starch, maltotriose, and maltose of 100, 68, and 31, respectively. It had an unusually high temperature maximum of 70°C for activity. The optima for activity are generally in the range pH 4.0–4.5 and 55–60°C.

Amyloglucosidase is capable of polymerizing glucose in a reaction that is essentially the reverse of hydrolysis. Thus, synthesis of maltose is very rapid, whereas the synthesis of isomaltose and higher oligosaccharides is much slower. The quantity of reversion sugar synthesized is very dependent on substrate concentration. The most significant application of amyloglucosidase is the production of high-glucose syrups containing 96–97% D-glucose, which is used for the production of crystalline D-glucose or as a starting material for high-fructose syrups or directly in fermentation processes.

Another important application of amyloglucosidase is its use in association with *A. oryzae* α-amylase in the production of high conversion syrups (Palmer, 1975). These syrups contain high levels of maltose and glucose—that is, syrups in which both sugars occur in excess of 30–35%. The most important feature of these syrups is that they should have a dextrose equivalent of 60–70 and resist crystallization down to 4°C at 80% dry substance. High conversion syrups are prepared using a mixture of *A. oryzae* α-amylase and amylogluosidase. By adjusting the ratio of the two enzymes, syrups with different compositions can be prepared.

Amyloglucosidase frequently exists in multiple forms in the aspergilli (Table VIII) and other molds; the various forms are frequently

Table VIII. Single and Multiple Forms of Amyloglucosidase in *Aspergillus* Species

Aspergillus species that produce a single form of amyloglucosidase	
A. niger	Abe *et al.* (1982)
A. awamori	Yamasaki *et al.* (1977)
A. oryzae	Ono *et al.* (1988a)
Aspergillus species that produce two or more forms of amyloglucosidase	
A. niger	Svensson *et al.* (1982), Fogarty and Benson (1983), Ono *et al.* (1988b)
A. oryzae	Morita *et al.* (1968), Miah and Ueda (1977a,b)
A. candidus	Manjunath and Raghavendra Rao (1979)
A. hennebergi	Alazard and Baldensperger (1982)
A. saitoi	Takahashi *et al.* (1981)

referred to as AGI, AGII, and so on. Four forms were produced by *A. oryzae* (Morita *et al.*, 1968; Ohga *et al.*, 1966), each with slight differences in sedimentation coefficients, electrophoretic mobilites, and pH stabilities, but having similar enzymatic and other physicochemical properties. Multiple forms have also been reported in *A. niger* (Alazard and Baldensperger, 1982; Ramasesh *et al.*, 1982; Svensson *et al.*, 1982), although some strains may have only one active form (Abe *et al.*, 1982). The enzyme from *A. oryzae* (Ono *et al.*, 1988a) was shown to be homogeneous by gel filtration, polyacrylamide gel electrophoresis (PAGE), sodium dodecyl sulfate–PAGE, ultracentrifugation, and isoelectric focusing. It is considered that medium composition and growth conditions influence production of multiple forms of amyloglucosidase (Hayashida, 1975; Saha *et al.*, 1979; Alazard and Raimbault, 1981). Yoshino and Hayashida (1978) reported that AGII could be formed from AGI in *A. awamori* var. *kawachi* by protease or glycosidase. A protease-negative mutant of this organism produced but one form of amyloglucosidase (Hayashida and Flor, 1981). Four forms of the enzyme showed different susceptibilities to two proteases (Paszczynski *et al.*, 1985) in the culture medium of *A. niger* C. It has been suggested that these two proteases may regulate amyloglucosidase activity and heterogeneity.

Amyloglucosidases are invariably glycoproteins and contain 4–18% carbohydrate. The carbohydrate component generally contains glucose, galactose, glucosamine, and mannose. Relative molecular masses vary between 50,000 and 200,000. The carbohydrate components of AGI of *A. niger* are linked glycosidically to L-threonine and L-serine (Pazur *et al.*, 1980). They play an important part in the stability of the enzyme. Their removal reduces both stability and activity (Pazur *et al.*, 1970).

4.3. Transglucosidase

Transglucosidase (EC 2.4.1.24) is produced by many species of *Aspergillus* and quite frequently is secreted concomitantly with amyloglucosidase. The enzyme may occur in crude preparations of amyloglucosidase (Maher, 1968) unless a transglucosidase-negative mutant is used. It catalyzes the synthesis of nonfermentable glucose saccharides by transfer of a glucosyl residue from an α-1,4 position to an α-1,6 position (Pazur and Ando, 1961); in this way, panose and isomaltose are synthesized. A variety of transfer products formed by transglucosidase in the presence of high concentrations of maltose have been reported (Saroja *et al.*, 1955; Pazur *et al.*, 1978). Fogarty and Benson (1982) demonstrated that glycerol acts as a glucosyl acceptor, but only in the presence of specific donor substrates such as p-nitrophenyl-α-D-glucoside and α-methyl-D-glucoside, though surprisingly not with maltose under the same conditions. A rapid and simple assay procedure has been described (Benson *et al.*, 1982) that overcomes the limitations of earlier methods. It is based on the hydrolysis of α-methyl-D-glucoside, which is not degraded by amyloglucosidase. A highly purified transglucosidase from *A. niger* (W. M. Fogarty and C. P. Benson, unpublished data) had relative activities on maltose, maltotriose, and starch of 100, 38, and 1, respectively. Panose was degraded at about the same rate as maltotriose and isomaltose at less than half this rate.

4.4. Pullulan 4-Glucanohydrolase

Sakano *et al.* (1972) described in *A. niger* a novel and interesting enzyme, pullulan 4-glucanohydrolase, that liberated isopanose from pullulan. The purified enzyme hydrolyzed reducing-end α-1,4 bonds adjacent to α-1,6 bonds in pullulan, 6^3-α-glucosylmaltotriose, 6^2-α-maltosylmaltose, and panose to produce isopanose, isomaltose and maltose, isopanose and glucose, and isomaltose and glucose, respectively. Maximum enzyme activity was at pH 3.0–3.5 and 40°C, and it had an M_r value of 74,000.

4.5. Enzymes of *Aspergilli* and Degradation of Raw Starch

Starch exists in the native state as granules with a crystalline structure. These granules are insoluble in water. Gelatinization, as a result of heating in water, enhances chemical reactivity considerably toward liquefying and saccharifying enzymes. Recently, much interest has developed in enzymes capable of degrading raw starch granules in order to lower costs associated with high temperatures used in gelatinization. *Aspergillus* sp. K-27, a soil isolate selected for its capacity to degrade raw starch (Abe *et al.*, 1988a,b; Bergmann *et al.*, 1988), produces both α-amylase and amyloglucosidase. The latter contained a specific starch binding site, and both enzymes acted synergistically on raw corn and raw potato starch. This phenomenon was also observed with other systems and appears more evident with α-amylases that have low activities for raw starch. It has been suggested that the α-amylase releases new nonreducing end groups on the surface of the granules for the amyloglucosidase. Thus, the latter strips the molecule from the surface and allows the α-amylase access to the next layer of the starch granule (Fujii *et al.*, 1988).

The multiple forms of *A. awamori* var. *kawachi* (Hayashida, 1975; Hayashida *et al.*, 1982) and *A. oryzae* (Miah and Ueda, 1977a,b) varied in their capacity to adsorb onto and digest raw starch. Amyloglucosidase I of an *Aspergillus* species had debranching activity and strong raw-starch adsorption and digestion ability, whereas amyloglucosidase II did not have any such activity, could not be adsorbed onto raw starch, and had weak raw-starch digestion ability (Medda *et al.,* 1982; Saha and Ueda, 1984a).

The raw-starch-degrading amyloglucosidase I of *A. awamori* var. *kawachi* could be degraded by proteases and glycosidases to the non-raw-starch-degrading amyloglucosidases I and II (Hayashida *et al.,* 1976; Hayashida and Yoshino, 1978; Yoshino and Hayashida, 1978). Proteolytic degradation of amyloglucosidase I of *A. awamori* var. *kawachi* with subtilisin gave a glycopeptide that contained the raw starch affinity site essential for raw-starch digestion, but it was enzymatically inactive (Hayashida *et al.,* 1989a,b). A protease-negative mutant of *A. ficum* IFO 4320 produced a raw-starch-adsorbable α-amylase that totally solubilized raw corn starch granules in 6 days (Hayashida and Teramoto, 1986). An inhibiting factor from *A. niger* (Saha and Ueda, 1984b; Towprayoon *et al.,* 1988) inhibited raw-starch digestion by α-amylase and amylogluHcosidase and combinations of both enzymes.

5. PECTINOLYTIC ENZYMES

Pectinolytic enzymes or pectinases occur in many bacteria and fungi and have a variety of commercial applications (Fogarty and Ward, 1974; Fogarty and Kelly, 1983; Pilnik and Rombouts, 1981; Whitaker, 1990). Pectinolytic enzymes include pectin esterase (EC 3.1.1.11), which removes methoxyl residues from pectin, and a range of depolymerizing enzymes that catalyze the cleavage of α-1,4 bonds of pectic polysaccharides. These enzymes are classified under three headings (Fogarty and Kelly, 1983):

1. Whether pectin, pectic acid, or oligo-D-galacturonic acids act as the preferred substrate.
2. Whether the mechanism of action is by transeliminative cleavage or direct hydrolysis.
3. Whether degradation is random (endo) or endwise (exo).

5.1. Production of Pectinolytic Enzymes

Molds are used predominantly for production of pectinolytic enzymes, in particular, enzymes produced by species of the genus *Aspergillus* (Reid, 1961; Nyiri, 1968, 1969; Charley, 1969; Bailey and Pessa,

1990). Suitable organisms include strains of *A. niger, A. wentii, A. awamori, A. foetidus,* and *A. oryzae*. Several different pectinolytic enzymes are secreted by these molds and are involved in pectin degradation in commercial processes. Their collective action results in de-esterification of pectin by pectin methylesterase (EC 3.1.1.11) followed by hydrolysis involving endo- and exo-polygalacturonases. Pectin transeliminases (EC 4.2.2.2), which cleave α-1,4-glycosidic bonds by a transelimination process, are also involved. Economics of commercial production may be enhanced by selection of more productive mutants (Zetalaki-Horvath and Dobra-Seres, 1972) that are not subject to catabolite repression or that synthesize large quantities of enzyme mixture without the necessity of inducer.

Solid-state fermentation for pectinolytic enzymes generally uses low-cost agricultural materials containing pectin as one of the components in the medium. Wheat bran, sugar beet pulp, and rice bran have all been used with considerable success. Additional ingredients used in the wheat bran process have been listed (Beckhorn *et al.,* 1965). Two types of production systems are used in the semisolid fermentation process. In the first type, large closed chambers are fitted with perforated trays containing thin layers of moistened bran medium; in the second type, long horizontal drums—some 3 ft in diameter—are half-filled with a similar medium. After inoculation with a spore suspension of *A. niger,* air is circulated through the systems. Temperature is maintained in the first system using a cold air system and in the rotating drum system by cold water flowing over the cylinder. With both systems, maximal enzyme production is reached in 3–7 days.

In submerged liquid fermentation, the nutrient medium is made up of a number of ingredients (Ueda *et al.,* 1982; Friedrich *et al.,* 1989; Bailey and Pessa, 1990). Most commercial pectinolytic enzymes are inducible, and substrates containing pectin must therefore be used. Studies have shown that very high levels of pectinolytic enzymes may be produced in media containing mixed carbon sources, for example, pectin and glucose, pectin and sucrose, or pectin and lactose (Tuttobello and Mill, 1961; Nyeste and Hollo, 1963; Vasu, 1967; Feniksova and Moldabaeva, 1967). It would appear that pectin, in addition to acting as an inducer substrate, also enhances release of pectinolytic enzymes into the culture liquor (Vasu, 1967; Nyiri, 1968). Material high in pectin, e.g., beet pulp, citrus peel, or apple pomace, is used in media formulation, rather than pectin polysaccharide itself (Zetalaki, 1976; Szajer, 1978).

High-yielding mold strains have been developed for large-scale pectinolytic enzyme production, but few data have been published. Zetalaki-Horvath and Dobra-Seres (1972) reported increased enzyme yields following ultraviolet mutagenesis of *A. niger* and *A. awamori*, while Fiedurek and Ilczuk (1983) reported that strains resulting from crossing high and low polygalacturonase heterokaryons of *A. niger* gave higher activities than the original strains.

5.2. Applications

Pectinolytic enzymes play a major role in nature in the destruction of plants, fruits, and vegetables (Fogarty and Kelly, 1983). In addition, they also have considerable commercial applications in fruit and vegetable processing and manufacturing industries. *Aspergillus niger* is the most widely used microorganism in the commercial production of pectinolytic enzymes. Commercial enzyme preparations contain many pectinolytic enzymes in different ratios (Neubeck, 1975; Ducroo, 1982; Pilnik, 1982). This provides a number of advantages including increased yield of juice and solids from plant materials, considerable reduction in viscosity of concentrates, and solubilization and modification of pectic compounds that effect sedimentation and clarification of juices. Problems in filtration and clarification of fruit juice extracts and concentrates are primarily caused by pectic polysaccharides, which are suspending agents for pulp and other materials and thus cause cloudiness. Following treatment with pectinolytic enzymes, the particles causing turbidity may be separated by settling or filtration. In the processing of crushed fruits, pectinolytic enzymes increase the efficiency of juice extraction, resulting in increased yields. In addition, decrease in viscosity and removal of suspending power increases significantly the rate of clearing and filtration. World production of fruit juices has risen to over 4000 million liters per year, of which some 20% is manufactured in a concentrated form. Details of various uses of pectinolytic enzymes are given by Fogarty and Ward (1974), Fogarty and Kelly (1983), Ducroo (1982), Pilnik (1982), and Peppler and Reed (1987).

6. CELLULASES

Many microorganisms, principally molds (Coughlan, 1990; Stutzenberger, 1990), are capable of secreting a mixture of enzymes that degrade cellulose. Complete hydrolysis of crystalline cellulose to glucose would appear to require synergistic action of a number of enzymes. The hydrolytic and oxidative enzymes associated with cellulose hydrolysis in fungi include cellobiose:quinone dehydrogenase (EC 1.1.5.1), cellobiose oxidase/hydrogenase (EC 1.1.99.18), lactonase (EC 3.1.1.17), endo-β1-4-

glucanase (endoglucanase) (EC 3.2.1.14), β-glucosidase (EC 3.2.1.21), exo-β-1,4-glucanase (EC 3.2.1.74), and cellobiohydrolase (EC 3.2.1.91). The cellulose systems of most soft- and white-rot fungi that degrade crystalline cellulose are composed of endo-glucanase, exo-cellobiohydrolase, and β-glucosidase. In the case of most cellulolytic fungi, their cellulase systems are incomplete in that they lack exo-cellobiohydrolase (Wood and Bhat, 1988). Systems that have been characterized and shown to contain a full complement of cellulolytic enzymes include *Trichoderma reesei* (Berghem and Pettersson, 1973; Berghem *et al.*, 1975; Shikata and Nisizawa, 1975; Gum and Brown, 1976, 1977; Nummi *et al.*, 1983), *Trichoderma koningii* (Halliwell and Griffin, 1973; Wood and McCrae, 1975, 1978; Halliwell and Vincent, 1981), *Talaromyces emersonii* (McHale and Coughlan, 1980, 1981, 1982; Moloney *et al.*, 1985), *Penicillium pinophilum* (Wood and McCrae, 1986; Wood *et al.*, 1989), and *Sporotrichum pulverulentum* (*Phanerochaete chrysosporium*) (Eriksson and Pettersson, 1975a,b; Eriksson, 1978; Eriksson and Wood, 1985; Ljungdahl and Eriksson, 1985). It appears that in most of the systems studied, each of the major cellulolytic enzymes exists in a number of forms (Eriksson and Pettersson, 1975a; Deshpande *et al.*, 1978; McHale and Coughlan, 1981; Moloney *et al.*, 1985). Several possible reasons for such multiplicity have been advanced (Eveleigh, 1987).

Cellulolytic systems of *Aspergillus* species tend to have little activity on native crystalline cellulose (Table IX). They act preferentially on modified cellulose or derivatives of cellulose. In optimization of enzymatic hydrolysis of cellulose, it is important to identify and address the rate-limiting step(s). If the initial treatment is optimal, then the correct balance of enzymes is important to complete the hydrolytic process.

Table IX. ***Aspergillus*** **Species That Produce Cellulolytic Enzymes**

Species	References
A. aculeatus	Murao *et al.* (1979)
A. awamori	Enari *et al.* (1975)
A. fumigatus	Trivedi and Rao (1979), Wase *et al.* (1985a,b)
A. niger	Markkanen *et al.* (1978), Enari *et al.* (1980, 1981)
A. oryzae	Mega and Matsushima (1979)
A. phoenicis	D. Sternberg *et al.* (1977)
A. ustus	Macris and Galiotou-Panayotou (1978)

β-Glucosidase activity is generally the rate-limiting step in the hydrolysis of cellulose. Because it is much less sensitive to inhibition than others, the addition of β-glucosidase from *A. niger* considerably improves overall hydrolysis of cellulose by the *T. reesei* system (Enari *et al.*, 1981).

In cellulase preparations, e.g., from *T. reesei* and its newer mutants (Kubicek, 1981; Saddler *et al.*, 1982), β-glucosidase activity is low and may significantly reduce total hydrolysis of cellulose because the buildup of cellobiose inhibits cellobiohydrolases (Reese and Mandels, 1984). Among molds that produce high levels of β-glucosidase are a number of *Aspergillus* species. Thus, balanced preparations from *Aspergillus* species and *T. reesei* may be used to degrade cellulose more efficiently than a preparation from a single species (Enari, 1983). Comparison of the activities of β-glucosidase and exo-β-glucanase in pure and mixed cultures of *A. phoenicis* and *T. neesei* Rut C30 (Duff, 1985) showed that the mixed culture gave fourfold greater β-glucosidase activity with little loss in exo-β-glucanase activity than that obtained with a pure culture of *T. reesei* Rut C30. Ghose *et al.* (1985) showed that a mixed culture of *A. wentii* and *T. reesei* gave higher activities of three cellulolytic enzymes than pure, single-strain cultures.

Deschamps and Huet (1984a,b) examined β-glucosidase production in *A. phoenicis* in solid-state fermentation using beet pulp and reported a higher temperature for maximum activity than that obtained in liquid submerged fermentation. Comparison of cellulase production by *A. fumigatus* in airlift and stirred tank fermenters (Wase *et al.*, 1985a) gave higher yields with the former system. It was claimed that shear damage to mycelia by the agitator in the stirred tank fermenters caused the lower yields. Trivedi and Rao (1979) claim that cellulosic substrates are essential for optimal production of the cellulolytic enzymes of *A. fumigatus*, although one or more cellulolytic activities are produced on substrates such as starch, pectin, and lactose, but not with glucose, maltose, or sucrose. *Aspergillus fumigatus* produces only β-glucosidase with cellobiose as substrate. Hay and straw were reported to provide higher yields than purified cellulose (Stewart *et al.*, 1983). Wase *et al.* (1985b) used the endo-β-1,4-glucanase of *A. fumigatus* as a model to examine a statistical design for optimization of medium. This resulted in the development of medium containing (g/liter): cellulose, 35; peptone, 0.92; $(NH_4)_2SO_4$, 24; KH_2PO_4, 35; $CaCl_2$, 1.4; $MgSO_4 \cdot 7H_2O$, 0.37.

Vidmar *et al.* (1984) purified endo-β-1,4-glucanases from *A. niger* from citric acid production broth. Crystalline cellulose was not degraded by these enzymes, and cellulolytic enzymes from this source would not appear to have commercial potential. Good β-glucosidase production has been reported for *A. awamori* (Enari *et al.*, 1975), *A. phoenicis* (D. Sternberg *et al.*, 1977), and *A. niger* (Markkanen *et al.*, 1978).

7. XYLANASES

Considerable amounts of xylan—a heterogeneous hemicellulose having a β-1,4-linked xylose structure with branches containing pentoses, hexoses, and uronic acids—occur in natural materials. Enzymes involved in degradation of xylan are of current interest (Reilly, 1981; Visser *et al.*, 1992). Xylan-degrading enzymes include the following types:

ENDO-XYLANSES

Debranching endo-xylanases

Type 1. Enzymes that produce xylobiose and xylose predominantly and hydrolyze branch points in xylan.

Type 2. Enzymes that produce xylooligosaccharides predominantly and hydrolyze branch points in xylan.

Non-debranching endo-xylanases

Type 3. Enzymes that produce xylobiose and xylose predominantly from xylan and are restricted or unable to degrade branch points.

Type 4. Enzymes that produce xylooligosaccharides predominantly and cannot hydrolyze branch points.

EXO-XYLANASES

Enzymes that produce xylose as the predominant end product from xylan and degrade low-molecular-weight xylooligosaccharides slowly.

β-XYLOSIDASES

Enzymes that degrade low-molecular-weight xylooligosaccharides to xylose. Many of these enzymes have considerable transferase activity.

7.1. Endo-xylanases

A number of species of *Aspergillus* produce xylanolytic enzymes (Table X). Debranching endo-xylanases have been reported in strains of *A. niger* (Fukumoto *et al.*, 1970; Rodionova *et al.*, 1977; Tsujisaka *et al.*,

Table X. *Aspergillus* Species That Produce Xylanases

Species	References
A. awamori	Poutanen *et al.* (1986, 1987), Friedrich *et al.* (1987)
A. foetidus	Linko (1981)
A. japonicus	Sharma *et al.* (1985)
A. terreus	Aprasyukhina *et al.* (1985), Abdel-Fattah *et al.* (1987)
A. nidulans	Fernandez-Espinar *et al.* (1992)
A. niger	Frederick *et al.* (1981, 1985), Shei *et al.* (1985), Fournier *et al.* (1985), Fukumoto *et al.* (1970)

1971; Takenishi and Tsujisaka, 1975; John *et al.*, 1979). The latter identified six xylanases with pH optima for activity in the range 4.0–6.5 and having M_r values ranging from 30,000 to 50,000.

Non-debranching endo-xylanases producing xylobiose and xylose are produced by strains of *A. niger* (Frederick *et al.*, 1981; Gorbacheva and Rodionova, 1977a,b; Sinner and Dietrichs, 1975a–c). The *A. niger* enzyme had optima for activity at pH 5.0 and 55°C, respectively, M_r value of 22,000, and a pI of 6.7 (Sinner and Dietrichs, 1975c; Frederick *et al.*, 1981).

Non-debranching endo-xylanases producing xylo-oligosaccharides have also been described in *A. niger*. Frederick *et al.* (1981) described an endo-xylanase that eventually degraded xylan to xylobiose, xylose, and residual branched products. Oligosaccharides, even as short as xylotriose, were degraded to disaccharide and monomer. Two closely related enzymes (endo-xylanases) were also detected (Frederick *et al.*, 1985). It appears that both enzymes required branch points in close proximity to the cleavage point. The end products formed included di- to hexaxylooligosaccharides with tri- and pentasaccharides as the major products. Neither xylose nor arabinose was formed. The preparation had little activity on linear xylooligsaccharides up to xylononaose and was inactive on insoluble xylan that had no branch points. Another endo-xylanase from the same source (Shei *et al.*, 1985) was highly active on both natural soluble xylan and the debranched polymer, but had little activity on insoluble xylan. It was also active on xylopentaose and higher xylooligosaccharides. A fifth endo-xylanase from *A. niger* (Fournier *et al.*, 1985) was active on soluble xylan and on insoluble xylan once the arabinosyl branch points had been removed. It also hydrolyzed xylopentaose and higher xylooligosaccharides. The main end products formed with xylan (soluble and insoluble) were xylooligosaccharides and predominantly tri- and pentasaccharides.

The most outstanding difference in these five endo-xylanases of *A. niger* is in their action patterns. The initial one described (Frederick *et al.*, 1981) was capable of degrading xylooligosaccharides as small as xylotriose. Final products formed with substrates were xylose and xylobiose. The remaining four endo-xylanases were active on xylopentaose and higher saccharides (Frederick *et al.*, 1985; Shei *et al.*, 1985; Fournier *et al.*, 1985). End products formed were intermediate-length xylooligosaccharides with xylobiose as a minor component and no xylose present. The fifth enzyme was unique in that it had considerably enhanced activity on insoluble xylan from which the branch points had been removed (Fournier *et al.*, 1985).

7.2. β-Xylosidases

A number of species of the genus *Aspergillus* produce β-xylosidase (Reese *et al.*, 1973). The enzyme from *A. niger* has been purified and characterized (Fukumoto *et al.*, 1970; Claeyssens *et al.*, 1971; Takenishi *et al.*, 1973; Oguntimein and Reilly, 1980a,b). The enzyme is active on xylooligosaccharides and aryl-β-D-xylanopyranosidases (Fukumoto *et al.*, 1970; Claeyssens *et al.*, 1971).

7.3. Exo-xylanase

The enzyme from *A. niger* has been studied (Rodionova *et al.*, 1977). It hydrolyzes methyl-β-D-xylopyranoside and xylan to give xylose.

8. LIPASES

Lipase (EC 3.1.1.3) has traditionally been obtained from animal pancreas. More recently, interest has been shown in microbial lipases. The development of the concept of interesterification (Macrae, 1983) and the use of lipase as a detergent additive (Nielsen, 1984) has expanded the industrial use of this enzyme (Table XI). In addition to its original use as a digestive and for human consumption, lipase has also been used to modify flavor in foods and in the enzymatic deterioration of serum triglycerides.

Microorganisms that produce lipases are widespread (Johri *et al.*, 1985; Stuer *et al.*, 1986; Godtfredsen, 1990), and include a number of *Aspergillus* species (Table XII). The enzyme from *A. niger* is available

Table XI. Applications of Lipases

Hydrolysis of oils and fats	Enantioselectivity in chemical industry
Interesterification of oils and fats	Leather industry
Esterification of fatty acids	Paper industry
Flavor development in food and dairy products	Health care industry
Detergent additive for cleaning	Analytical reagent

commercially and has been extensively studied (Akita *et al.*, 1986; Andree *et al.*, 1980; Gerlach *et al.*, 1988; Hoq *et al.*, 1985; Kalo, 1988; Linefield *et al.*, 1984; Macrae and Hammond, 1985; Sonnet and Antonia, 1988).

Interesterification, the ability of a lipase enzyme to redistribute fatty acids between the free form and the bound form in triglycerides, is a highly important property. It is used to alter both the composition and the properties of oils and fats and thereby not only upgrade but also increase the value of particular materials.

Lipases possess specificity for both the nature and the structure of the fatty acids present and the position they occupy in the glycerol molecule. For example, some enzymes exhibit specificity for both 1 and 3 positions in glycerol. This type of specificity is found in *A. niger* and pancreatic lipase (Iwai and Tsujisaka, 1984). Other microbial lipases are nonspecific in the sense that they will release fatty acids from all three positions in glycerol (Macrae, 1983). Most lipases are specific for C16 and C18 fatty acid residues, which mirrors their existence in natural fats. The comparative rates of hydrolysis of mono-, di-, and triglycerides are other variants in the specificity of lipases.

The choice of using submerged fermentation or solid-state cultivation as a means of producing lipase from *Aspergillus* species is dependent

Table XII. *Aspergillus* Species That Produce Lipases

A. niger	Kalo (1988), Macrae and Hammond (1985), Hofelmann *et al.* (1985), Miyazawa *et al.* (1988), Sonnet and Antonia (1988), Koritala *et al.* (1987), Godtfredsen (1990), Fukumoto *et al.* (1963)
A. oryzae	Yokozeki *et al.* (1982), Koritala *et al.* (1987)
A. awamori	Yokozeki *et al.* (1982)
A. flavus	Koritala *et al.* (1987), Yeoh *et al.* (1986)
A. japonicus	Werdelmann and Schmid (1982), Vora *et al.* (1988)
A. fumigatus	Satyanarayana and Jori (1981)

on the strain being used (Iwai and Tsujisaka, 1984). A variety of procedures for the production of lipases in submerged fermentation have been reported in which both production conditions and nutrient compositions vary significantly. Pal *et al.* (1978) published extensive details of medium optimisation for production of *A. niger* lipase—the oldest commercial source of the enzyme. However, published details of commercial production methods are not available.

9. GLUCOSE-TRANSFORMING ENZYMES

9.1. Glucose Oxidase

Microorganisms produce a number of enzymes capable of oxidizing glucose. The most important of these from a commercial viewpoint is glucose oxidase (EC 1.1.3.4). It is a flavoprotein that removes two hydrogen atoms from glucose, and the resulting reduced form of the enzyme is reoxidized by molecular oxygen. Hydrogen peroxide is decomposed by catalase. D-Glucono-δ-lactone that is formed is hydrolyzed spontaneously, or with the aid of glucono-δ-lactonase, to gluconic acid. Recent reviews include Miall (1978), Lockwood (1979), Milsom and Meers (1985), and Creuger and Creuger (1990).

Because aerobic oxidases produce hydrogen peroxide, it follows that microorganisms that produce these enzymes require catalase to act as a detoxifying reagent. A distinct advantage in the commercial use of glucose oxidase is the concomitant occurrence of catalase. *Aspergillus niger* is one of two major sources of these enzymes. Large-scale fermentation processes are used to produce concomitantly, and in high yield, glucose oxidase and catalase. In Japan, *Penicillium* species are also used in large-scale production of glucose oxidase, more notably *P. amagaskiense, P. vitale,* and *P. notatum.* Glucose oxidases from *A. niger* and *Penicillium* species, although generally similar, are distinctly different enzymes (Degtyar and Lototskaya, 1968). Detailed analyses and comparisons between the glucose oxidase of *A. niger* and that of *P. amagasakiense* demonstrated considerable similarities in amino acid composition, but distinct differences in a number of properties (Nakamura and Fujiki, 1968). Thus, in the case of the former, the pH for maximal activity lay between 3.5 and 6.0, whereas with the latter it occurred between 4.5 and 5.5. The Michaelis constants (K_m) for glucose oxidase from *A. niger, P. notatum,* and *P. amagasakiense* are 1.1×10^{-2}, 0.96×10^{-2}, and 1.15×10^{-2}, respectively. Glucose oxidase has an extremely high specificity for D-glucose. It oxidizes the β-anomeric form some 160 times faster than the α-form, although this is not important in commercial preparations since they already contain the enzyme mutarotase. Studies on production of glucose oxidase by *A. niger* (Zetalaki and Vas, 1968; Zetalaki, 1970; Lakshminarayanan, 1972) indicate short fermentation times. Maximum yields are obtained after 12–24 hr depending on the strain and growth conditions. Glucose oxidase is an intracellular enzyme in *A. niger* under standard fermentation conditions. The mycelium recovered at the end of the fermentation process for the production of gluconic acid is used to prepare the enzyme (Lockwood, 1975). For highest yields, the ground cell mass is allowed to autolyze in phosphate buffer, pH 5–9. Commercial preparations, depending on the purification efficiency protocol, may contain catalase, mutarotase, gluconolactonase, amyloglucosidase, and invertase.

The primary commercial use of glucose oxidase is in the manufacture of gluconic acid, for which there is a market for approximately 45,000 tons per year (Bigelis, 1985). Other applications include the stabilization of flavor and color in soft drinks, beer, and canned foods by removal of oxygen. It is also used in the removal of glucose in the production of egg powder to inhibit browning caused by the Maillard reaction. Glucose oxidase is also widely used in the quantitative determination of glucose.

9.2. Glucose Dehydrogenases

Glucose dehydrogenases also produce gluconic acid via D-glucono-δ-lactone and a reduced accepter.

9.2.1. D-Glucose:(Acceptor) 1-oxidoreductase

This glucose dehydrogenase (EC 1.1.99.10) occurs as a soluble enzyme in *A. oryzae.* Since it is unable to react with molecular oxygen, oxidation of glucose is catalyzed by quinones and redox dyes. The enzyme contains 1 mole FAD per mole of enzyme and is a glycoprotein with a relative molecular mass of 118,000 (Bak, 1967).

9.2.2. Pyranose:Oxygen 2-oxidoreductase

This enzyme, pyranose oxidase [glucose 2-oxidase-D-carbohydrate oxidase (EC 1.1.3.10)], possesses the capacity to oxidize several carbohydrates at the C_2 position and thus form 2-keto products and H_2O_2. Substrates for this system include D-glucose, D-xylose, L-sorbose, and 1,5-D-gluconolactone. The enzyme was first detected in *A. flavus-oryzae* and *A. parasiticus* (Bond *et al.*, 1937). More recently, the enzyme has been shown to be widely distributed among basidiomycetes (Volc *et al.*, 1985). The enzyme is of considerable interest since D-glucosone is an important intermediate in the synthesis of a variety of substrates, including D-fructose, 2-keto-D-gluconic acid, D-mannitol, and D-sorbitol (Koths and Halenbeck, 1986; Neidleman *et al.*, 1981; Eriksson *et al.*, 1986). The enzyme has also been considered for the determination of blood glucose levels (Taguchi *et al.*, 1985).

10. PROTEINASES

Although primarily considered as essentially degradative enzymes, proteinases also carry out very specific and highly selective modifications of proteins through limited hydrolysis. In addition, they have an important function in clinical and other laboratory analyses as well as in a number of industrial processes. Their introduction as additives to detergents in the 1960s led to widespread commercial applications, and they now represent 60% of industrial enzymes. Proteinases of *A. oryzae* are used in dough modification and soy sauce production, those from *Mucor* species and *Endothia* species in cheese manufacture, and those of *Bacillus* species, principally *B. licheniformis* (alkaline proteinase), as detergent additives.

Extracellular proteinases are produced by many species of fungi (North, 1982). Considerable data are available on the properties of extracellular proteinases, but information is limited on mechanisms of production. Studies by Cohen (1973, 1981) constitute most of the important work on biosynthesis of these enzymes in *Aspergillus* species. Production is governed by derepression in the absence of sulfur, nitrogen, or carbon. Three stable proteinases are secreted. Derepression involves *de novo* synthesis.

10.1. Classification

Proteinases are divided into four groups according to their catalytic mechanisms (Table XIII). This division is based on reactivity toward inhibitors that react with specific moieties in the active site region. *Aspergillus* species produce proteinases in each of the four divisions, i.e. serine, cysteine, aspartic, and metalloproteinases.

10.1.1. Serine Proteinases

Serine proteinases form the biggest group of protein hydrolases in microorganisms (North, 1982) and are present in many *Aspergillus* spe-

Table XIII. Classification of Microbial Proteinases

1. Serine proteinases
2. Cysteine proteinases
3. Aspartic proteinases
4. Metalloproteinases

cies. They possess a reactive serine moiety in the active site and are inhibited by phenylmethylsulfonyl fluoride (PMSF) or diisopropyl fluorophosphate (DFP). Most of these enzymes have maximal activity between pH 7.0 and 11.00, M_r values between 20,000 and 35,000, and broad substrate specificities and extensive esterase activity.

10.1.2. Cysteine Proteinases

These enzymes are not widely distributed in fungi (North, 1982). An extracellular cysteine proteinase is secreted by *A. oryzae*. Most of these enzymes have maximal activity between pH 5.0 and 8.0. They are sensitive to sulphydryl reagents (e.g., *p*-chloromercuribenzoate) and are often activated by reducing agents.

10.1.3. Aspartic Proteinases

The second important industrial proteinase after serine proteinases are the rennet enzymes used in cheese-making. They are essentially acid proteinases with an aspartic acid moiety as a key catalytic element in the active site. The enzymes have maximal activity in the range pH 3.0–4.0, with M_r values in the region 30,000–45,000. They occur widely in fungi and tend to be unstable above pH 7.0. They are specific for aromatic amino acid residues on each side of the cleavage point. Extracellular pepsin-type aspartic proteinases are found in *Aspergillus* species and other molds (North, 1982). They are used in soybean protein hydrolysis in the preparation of soy sauce (Yong and Wood, 1974; Hesseltine, 1983; Yokotsuka, 1985). *Aspergillus candidus* produces an extracellular renin-type proteinase (Morihara, 1974).

10.1.4. Metalloproteinases

Metalloproteinases fall into a number of groups (Table XIV). Maximal activity occurs between pH 5 and 9, and they all respond to chelating agents such as EDTA but are not sensitive to inhibitors that affect serine proteinases or sulphydryl reagents. Zinc, cobalt, calcium, or man-

Table XIV. Classification of Metalloproteinases

Type	Source	pH (maximum activity)	Specificity	References
Acid	*A. sojae* *A. oryzae* *Penicillium* spp.	5–6	Specificity toward synthetic peptides	Gripon *et al.* (1980), North (1982)
Neutral	*Aspergillus* spp. *Bacillus* spp.	7.0	Hydrophobic or large amino acid residues	Morihara (1974)
Alkaline	*Pseudomonas* spp. *Serratia* spp.	7–9	Broad specificity	Morihara (1974)
Myxobacter proteinase I	*Myxobacter* spp.	9.0	Cell walls, Gram-positive bacteria	Morihara (1974)
Myxobacter proteinase II	*Myxobacter* spp.	9.0	Lysine residues	Morihara (1974)

ganese reactivates EDTA-inactivated enzymes (Fogarty and Griffin, 1973; Griffin and Fogarty, 1973). Most metalloproteinases are zinc-containing enzymes, and protein structure is stabilized by the presence of calcium (Gripon *et al.*, 1980). The most outstanding enzyme in this class is probably thermolysin—the extracellular metalloproteinase of *Bacillus thermoproteolyticus* (Morihara, 1974).

10.2. Applications

The major industrial uses of proteinases are as detergent additives, in cheese-making, in the dehairing of hides, and in the food industry. These applications account for some 40% of total enzyme sales. Most industrial enzymes are produced by microorganisms in the genera *Aspergillus* and *Bacillus*. *Aspergillus oryzae* and *A. niger* groups are the most frequently used species of *Aspergillus*. Proteinases of *Aspergillus* species are used as digestive aids and in the treatment of flour by effecting some hydrolysis of gluten and thus improving bread quality and reducing the time required in mixing of dough. Proteinases of *Aspergillus* species, principally *A. sojae* and *A. oryzae*, have been used for many centuries in the preparation of soy products and particularly soy sauce (Hesseltine, 1983; Yong and Wood, 1974; Yokotsuka, 1985). Both alkaline and neutral proteinases of *Aspergillus* species play major roles in the degradation of soybean protein during the preparation of soy sauce. Acid proteinases are also produced (Yokotsuka, 1985). *Aspergillus flavus* var. *columnaris* is used as a fixed inoculum in Thailand in soy sauce production (Impoolsup *et al.*, 1981; Bhumiratana *et al.*, 1980). Both alkaline and neutral proteinases are involved during processing.

Considerable interest has been shown in members of the genus *Aspergillus*, specifically *A. awamori, A. nidulans, A. niger,* and *A. oryzae,* as hosts for heterologous gene expression and production of foreign protein (Cullen *et al.*, 1987; Gwynne *et al.*, 1987; Upshall *et al.*, 1987; Turnbull *et al.*, 1989; Christensen *et al.*, 1988; Ward *et al.*, 1990). Of major interest in this context is the ability to synthesize and secrete bovine chymosin (rennin), which is used in the manufacture of cheese. The increasing demands for cheese and the declining production and increasing costs of calf rennin have stimulated considerable interest in other means of supply. In addition to the approach described here, there is also a huge development in microbial rennin substitutes, notably from *Endothia parasitica* and *Mucor* species.

Aspergillus oryzae preparations show activity in the range pH 4.0–11.0 and contain acid, neutral, and alkaline proteinases. The alkaline enzyme is a serine proteinase, inactivates rapidly at 60°C, and is not thermostable. It is most active between pH 7.0 and 8.5 and is stable between pH 5.0 and 9.0. Calcium has a stabilizing effect on this enzyme (Nasuno, 1972). *Aspergillus oryzae* produces two metalloproteinases. Both enzymes are inhibited by chelating agents. Metalloproteinase I is inactivated rapidly above 50°C, is most active at pH 7.0, and is stable in the pH range 6.0–11.0. Metalloproteinase II is comparatively thermostable and loses only 30% of its activity after 10 min at 90°C. It is most active in the range pH 5.5–6.0 (Nakadai *et al.*, 1973). The acid proteinase of *A. oryzae* had a maximum temperature for activity at 45°C, was most active between pH 4.0 and 4.5, and was stable in the range pH 3.0–6.0. It is similar to pepsin in its active site (Matsubara and Feder, 1971). The black aspergilli, including *A. niger* var. *macrosporus* and *A. saitoi* (*A. phoenicis*), also produce acid proteinases. Two acid proteinases are produced by *A. niger* and have optima at pH 2.0 and 2.5. The acid proteinase of *A. saitoi* is stable between pH 2 and 5 and has maximal activity at 30°C and pH 2.6–3.0. The enzyme has little if any esterase or peptidase activity and has an M_r value of 34,000. It preferentially hydrolyzes peptide bonds having hydrophobic side chains.

11. MISCELLANEOUS ENZYMES

11.1. Catalase

Catalase (EC 1.11.1.6) is produced from *A. niger* as a secondary product in the manufacture of glucose oxidase. Optimization of conditions for production of catalase as a major end product in *A. niger* was described by Chaga *et al.*, (1971). Optimized media contained sucrose (8%) and wheat bran (6%) with nitrate as the preferred source of nitrogen. Catalase is widely distributed in fungi and has been purified to homogeneity by Kamel and Takany (1973) and Jacob and Orme-Johnson (1979) in *Pencilllium notatum* and *Neurospora crassa,* respectively.

11.2. Glycerol Oxidase

Glycerol oxidase is produced by some strains of *Aspergillus, Neurospora,* and *Penicillium* (Uwajima *et al.*, 1980; Uwajima and Terada, 1982). It calalyzes the oxidation of glycerol to glyceraldehyde and hydrogen peroxide and has no requirement for exogenous cofactors. Together with lipase, it has been used in the spectrophotometric assay of triglycerides. The enzyme from *A. japonicus* has been studied in some detail (Uwajima and Terada, 1980, 1982; Uwajima *et al.*, 1980; 1979). Some of its properties are given in Table XV. The enzyme is a hemoprotein and contains protoheme IX and copper ions as prosthetic groups.

11.3. Keratinase

Keratinase degrades the scleroprotein keratin, which is a component of hair, wool, hooves, horns, nails, and mammalian epidermis. Koh and Messing (1963) described the production of keratinases by *A. flavus* and *A. niger* under submerged aerobic conditions. These enzymes have a number of applications including dehairing of hides.

11.4. Acylase

This enzyme catalyzes asymmetrical removal of acyl residues from acyl-D,L-amino acids. Racemic amino acids can be separated into their optically active enantiomorphs by acylating the amino group of a D,L-

Table XV. Properties of Glycerol Oxidase of *Aspergillus japonicus*

Relative molecular mass	400,000
pH for maximum activity	7.0
pI	4.9
Substrate specificity	
Glycerol	100
Dihydroxyacetone	60

amino acid to form the *N*-acyl-D,L-amino acid and then selectively degrading one of the acylated enantiomorphs using acylase. Kirimura and Yoshida (1966) reported that the stability of the acylase of *A. oryzae* could be considerably increased by copolymerizing the enzyme with anhydrides of specific *N*-carboxy-α-amino acids or their derivatives to give polypeptidyl derivatives of acylase.

11.5. β-Galactosidase

β-Galactosidase has applications in the processing of milk products including concentration of milk to a high solids content and addition to ice cream to prevent lactose crystallization. M. Sternberg (1973) reported a method of preparing and purifying β-galactosidase of *A. niger* and *A. foetidus* by mixing the crude enzyme solution with a polyacrylic acid to form an active precipitate that could be readily separated from other impurities in the solution. This process is an extremely useful method of purifying β-galactosidase. Cayle (1973) prepared an acid-stable β-galactosidase from *A. niger.*

11.6. Endo-β-glucanase

Poorly modified malts and barley can cause problems in processing and particularly an increase in time required for filtration of mash and beer. Such increased filtration times are due mainly to increased viscosity attributable to barley β-glucan—a high-molecular-weight polymer of D-glucose containing both β-1,3- and β-1,4-glycosidic linkages. *Aspergillus phoenicis* and *A. saitoi* produce endo-β-glucanase containing little α-amylase or proteinase activity, and the enzyme is suitable for reduction of viscosity of wort or beer (Hjortshoj and Aunstrup, 1978).

11.7. Pterin Deaminase

Aspergillus tamarii, A. oryzae, and *A. gymnosardae* produce pterin deaminase, which effects hydrolytic deamination of pterin, pteroic acid, and folic acid to give the corresponding 2,4-dihydroxy compounds, which possess antitumor activity. The enzyme is also reported to occur in *Pencillium, Mucor,* and *Rhizopus* species (Kusakabe *et al.,* 1976).

11.8. Uricase

Uric acid is a principal product of the catabolism of purine bases including nucleic acids. If degradation or elimination of uric acid does not take place, it can lead to a number of human disorders, e.g., gout, certain forms of rheumatism, calcium in the urinary system, and other problems in the cardiovascular system. Laboureur *et al.* (1974) described the production by a strain of *A. flavus* of uricase, which rapidly eliminates uric acid by conversion to allantoin. The enzyme is also suitable for research and analytical studies.

11.9. Naringinase

Citrus flavonoids may cause considerable bitterness in fruit juices. Flavonoid-hydrolyzing enzymes to reduce bitterness have considerable potential but to date have not been used extensively. The flavonoid found in highest concentration in citrus fruits is naringin—a bitter-tasting substance that is the 7-(2-rhamnosido-β-glucoside) of $4^1,5,7^1$-trihydroxyflavanone. An active commercial preparation of naringinase (Thomas *et al.*, 1958) rapidly debittered grape juice in the region pH 3.5–5.0 at 30–50°C. Okada *et al.*, (1963a–c) described production and properties of the system from *A. niger* grown on bran, soybean meal, and citrus peel.

11.10. Tannase

In the commercial preparation of instant tea soluble in cold water, tannase is used to solubulize the insoluble tea polyphenol–caffeine complex that results when tea extracts are cooled. In this enzyme treatment, tannase from *A. oryzae* is used (Takino, 1976). This tannase may also be used as an additive to augment indigenous green tea enzymes and thus speed up the process of conversion of green tea to black tea (Sanderson and Coggon, 1974). Pectinase (*A. niger*) may also be used in the production of instant tea (Sanderson and Simpson, 1974).

12. CONCLUSIONS

There have been some considerable achievements in the development and use of fungal enzymes since Takamine (1894) prepared the first fungal enzyme from *A. oryzae* for commercial use. Basic and applied research on microbial enzymes has contributed very significantly to new developments in the past two decades. It is difficult to predict where the most significant developments in enzyme technology, and more specifically enzymes of *Aspergillus* species, will take place. Currently, two of the largest applications—use of enzymes in detergents and use of enzymes in preparation of fructose syrups—arose not from planned developments by the enzyme industry but more as a result of circumstance. It is reasonable to predict that research in the development of microbial cellulases, hemicellulases, and lipases, new enzymes in the starch processing industry, and specific enzymes for the biotransformation of steroids, antibiotics, and alkaloids and the resolution of stereoisomers will continue to attract attention. Stimulation of additional research in these and other areas will undoubtedly result in more extensive development in the use of both fungal and bacterial enzymes.

REFERENCES

Abdel-Fattah, A. F., Abdel-Naby, H. A., and Ismail, A. M. S., 1987, Purification and properties of xylan degrading enzyme from *A. terreus* 603, *Biol. Wastes* **20**(2)**:**143–151.

Abe, J.-I., Takeda, Y., and Hizukuri, S., 1982, Action of glucoamylase from *A. niger* on phosphorylated substrate, *Biochim. Biophys. Acta* **703:**26-33.

Abe, J.-I., Bergmann, F. W., Obata, K., and Hizukuri, S., 1988a, Production of the raw starch digestion amylase of *Aspergillus* sp. K-27, *Appl. Microbiol. Biotechnol.* **27:**447–450.

Abe, J.-I., Nakajima, K., Nazano, H., and Hizukuri, S., 1988b, Properties of the raw-starch digesting amylase of *Aspergillus* sp. K-27: A synergistic action of glucoamylase and α-amylase, *Carbohydrate Res.* **175:**85–92.

Akita, H., Matsukura, H., and Oishi, T., 1986, Lipase catalyzed enantioselective hydrolysis of 2-methyl 3-acetoxy esters, *Tetrahedron Lett.* **27:**5241–5244.

Alazard, D., and Baldensperger, J. F., 1982, Amylolytic enzymes from *Aspergillus hennebergi* (*A. niger* group): Purification and characterization of amylases from solid and liquid culture, *Carbohydrate Res.* **107:**231–241.

Alazard, D., and Raimbault, M., 1981, Comparative study of amylolytic enzyme production by *A. niger* in liquid and solid state cultivation, *Eur. J. Appl. Microbiol. Biotechnol.* **12:**113–117.

Allen, W. G., and Spradlin, J. E., 1974, Amylases and their properties, *The Brewers Digest,* July, pp. 48–53, 65.

Andree, H., Muller, W.-R., and Schmid, R. D., 1980, Lipases as detergent compounds, *J. Appl. Biochem.* **2**(3)**:**218–219.

Aprasyukhina, N. I., Tavobilou, I. M., Rodionova, N. A., and Bezborodov, A. M., 1985, Study of the content of extracellular hemicellulases in some mycelial fungi, *Prikl. Biokhim. Mikrobiol.* **21**(6)**:**736–740.

Aravina, L. A., and Ponomarera, V. D., 1977, Influence of the inoculum on the biosynthesis of proteolytic and amylolytic enzymes of *Aspergillus terricola, Microbiologiya* **46:**379–383.

Aski, K., Arai, M., Minoda, Y., and Yamada, K., 1971, Acid stable α-amylase of black *Aspergillus, Agric. Biol. Chem.* **35:**1913–1920.

Atkinson, T., Scawen, M. D., and Hammon, P. M., 1987, Large scale industrial techniques of enzyme recovery, in: *Biotechnology,* Vol. 7A (H.-J. Rehm, G. Reed., and J. F. Kennedy, eds.), VCH Verlagsgesellschaft mbH, Weinheim, Germany, p. 279.

Aunstrup, K., 1977, Production of industrial enzymes, in: *Biotechnology and Fungal Differentiation* (J. Meyrath and J. D. Bu'Lock, eds.), FEMS Symposium No. 4, Academic Press, New York, pp. 157–171.

Axelsson, H. A. C., 1985, Centrifugation, in: *Comprehensive Biotechnology*, Vol. 2 (M. Moo-Young, ed.), Pergamon Press, London, pp. 325–344.
Bailey, M. J., and Pessa, E., 1990, Strain and process for production of polygalacturonase, *Enz. Microbiol. Technol.* **12**(4):266–271.
Bak, T.-G., 1967, Studies on glucose dehydrogenase of *Aspergillus oryzae:* Purification and physical and chemical properties, *Biochim. Biophys. Acta* **139:**277.
Barfoed, H. C., 1976, Enzymes in starch processing, *Cereal Foods World* **21:**588–589, 592–593, 604.
Bata, J., Vallier, P., and Colobert, L., 1978, α-Amylase activity in lysosomes of *Aspergillus oryzae, Experientia* **34:**572–573.
Beckhorn, E. J., Labbee, M. D., and Underkofler, L. A., 1965, Production and use of microbial enzymes for food processing, *J. Agric. Food Chem.* **13:**30–39.
Belter, P. A., Cussler, E. L., and Hu, W.-S., 1988, *Bioseparations,* John Wiley & Sons, New York.
Bendetskii, K. M., Tarovenko, V. L., Korchagina, G. T., Senatovora, T. P., and Khakhanova, T. S., 1974, Amylolytic enzymes from *Aspergillus batatae, Biokimiya* **39:**802–807.
Benson, C. P., Kelly, C. T., and Fogarty, W. M., 1982, Production and quantification of transglucosidase from *Aspergillus niger, J. Chem. Technol. Biotechnol.* **32:**790-798.
Berghem, L. E. R., and Pettersson, L. G., 1973, The mechanism of enzymatic cellulose degradation: Purification of a cellular enzyme from *Trichoderma viride* active on highly ordered cellulose, *Eur. J. Biochem.* **37:**21–30.
Berghem, L. E. R., Pettersson, L. G., and Axio-Frederiksson, U.-B., 1975, The mechanism of enzymatic cellulose degradation: Characterization and enzymatic properties of a β-1,4-glucan cellobiohydrolase from *Trichoderma viride, Eur. J. Biochem.* **53:**55–62.
Bergmann, F. W., Abe, J.-I., and Hizukuri, S., 1988, Selection of microorganisms which produce raw-starch degrading enzymes, *Appl. Microbiol. Biotechnol.* **27:**443–446.
Bhella, R. S., and Altosaar, I., 1985, Purification and some properties of the extracellular α-amylase from *Aspergillus awamori, Can. J. Microbiol.* **31:**149–153.
Bhumiratana, A., Flegel, T. W., Glirisukon, T., and Somporon, W., 1980, Isolation and analysis of molds from soy sauce koji in Thailand, *Appl. Environ. Microbiol.* **39:**430–439.
Bigelis, R., 1985, Primary metabolism and industrial fermentations, in: *Gene Manipulation in Fungi* (J. W. Bennett and L. L. Lasure, eds.), Academic Press, New York, p. 357.
Boing, J. T. P., 1981, Enzyme production, in: *Industrial Microbiology,* Prescott and Dunn, 4th ed. (G. Reed, ed.), AVI Publishing, Westport, Connecticut, pp. 634–708.
Bond, R. C., Knight, E. C., and Walker, T. K., 1937, The production of glucose from carbohydrates by enzymic action, *Biochem. J.* **31:**1033–1040.
Brummer, W., and Gunzer, G., 1987, Laboratory techniques of enzyme recovery, in: *Biotechnology,* Vol. 7A (H.-J. Rehm, G. Reed, and J. F. Kennedy, eds.), VCH Verlagsgesellschaft MbH, Weinheim, Germany, pp. 213–278.
Cannel, E., and Moo-Young, M., 1980, Solid state fermentation systems, *Process Biochem.* **15**(5):2–7.
Cayle, T., 1973, Treating lactase deficiency with an active lactase, United States Patent No. 3,718,739.
Chaga, S., Kamburova, S., and Doncheva, T., 1971, Biological synthesis of acid-resistant catalase, *Nauchni Tr. Nauchnoizsled. Inst. Konservna Promst. Plovoliv* **8:**83–90.
Charley, V. L. S., 1969, Some advances in food processing using pectic and other enzymes, *Chem. Ind.* **20:**635–641.
Christensen, T., Woeldike, H., Boel, E., Mortensen, S. B., Hjortshoej, K., Thim, L., and

Hansen, M. T., 1988, High level expression of recombinant genes in *Aspergillus oryzae, Bio/Technology* **6:**1419–1422.

Claeyssens, M., Loontiens, F. G., Kersters-Hilderson, H., and De Bruyne, 1971, Potential purification and properties of an *Aspergillus niger* β-D-xylosidase, *Enzymologia* **40:**177–198.

Cohen, B. L., 1973, The neutral and alkaline proteases of *Aspergillus nidulans, J. Gen. Microbiol.* **77:**521–528.

Cohen, B. L., 1981, Regulation of protease production in *Aspergillus, Trans. Br. Mycol. Soc.* **76:**447–450.

Coughlan, H., 1990, Cellulose degradation by fungi, in: *Microbial Enzymes and Biotechnology,* 2nd ed. (W. M. Fogarty and C. T. Kelly, eds.), Elsevier, London, pp. 1–37.

Creuger, A., and Creuger, W., 1990, Glucose transforming enzymes, in: *Microbial Enzymes and Biotechnology,* 2nd ed. (W. M. Fogarty and C. T. Kelly, eds.), Elsevier, London, pp. 177–226.

Cullen, D., Gray, G. L., Wilson, L. J., Rey, M. W., Norton, S., and Berka, R. M., 1987, Controlled expression and secretion of bovine chymosin in *Aspergillus nidulans, Bio/Technology* **5:**369–376.

Degtyar, R. G., and Lototskaya, L. S., 1968, Comparative characteristics of glucose oxidase produced by various fungi, in: *Fermenty Med. Pishch. Prom. Setsk Khoz.* (M. F. Gulyi, ed.), Naukova Dumka Naukova, Kiev, pp. 145–148.

Demain, A. L., 1990, Regulation and exploitation of enzyme biosynthesis, in: *Microbial Enzymes and Biotechnology,* 2nd ed. (W. M. Fogarty and C. T. Kelly, eds.), Elsevier, London, pp. 331–368.

Deschamps, F., and Huet, M. C., 1984a, β-Glucosidase production by *Aspergillus phoenicis* in solid state fermentations, *Biotechnol. Lett.* **6:**55–60.

Deschamps, F., and Huet, M. C., 1984b, β-Glucosidase production in agitated solid fermentation, study of its properties, *Biotechnol. Lett.* **6:**451–486.

Deshpande, V., Eriksson, K.-E., and Pettersson, B., 1978, Production, purification and partial characterization of 1,4-β-glucosidase enzymes from *Sporotrichum pulverulentum, Eur. J. Biochem.* **90:**191–198.

Doyle, E. M., Kelly, C. T., and Fogarty, W. M., 1989, The high maltose-producing α-amylase of *Penicillium expansum, Appl. Microbiol. Biotechnol.* **30:**492–496.

Ducroo, F., 1982, Efficacité des préparations pectinolytiques en fonction du type de jus de fruits à traiter, in: *Use of Enzymes in Food Technology,* P. Dupuy, ed.), Technique et Documentation Lavoisier, Paris, pp. 463–469.

Duff, S. J. B., 1985, Cellulose and beta-glucosidase production by mixed culture of *Trichoderma reesei* Rut C30 and *Aspergillus phoenicis, Biotechnol. Lett.* **7**(3)**:**185.

Enari, T.-M., 1983, Microbial cellulases, in: *Microbial Enzymes and Biotechnology* (W. M. Fogarty, ed.), Applied Science Publishers, London, pp. 183–224.

Enari, T.-M., Markkawen, P., and Korhonen, E., 1975, in: *Symposium on Enzymatic Hydrolysis of Cellulose* (M. Bailey, T.-M. Enari, and M. Linko, eds.), Sitra, Helsinki, p. 171.

Enari, T.-M., Niku-Paavola, M.-L., and Nummi, M., 1980, Comparison of cellulolytic enzymes from *Trichoderma reesei* and *Aspergillus niger,* in: *Proceedings of the 2nd International Symposium on Bioconversion and Biochemical Engineering,* Vol. 1 (T. K. Ghose, ed.), IIT, Delhi, pp. 87–95.

Enari, T.-M., Niku-Paavola, M.-L., Harju, L., Lapalainen, A., and Nummi, M., 1981, Purification of *Trichoderma reesei* and *Aspergillus niger* β-glucosidase, *J. Appl. Biochem.* **3:**157–163.

Eriksson, K.-E., 1978, Enzyme mechanisms involved in cellulose hydrolysis by the rot fungus *Sporotrichum pulverulentum, Biotechnol. Bioeng.* **20:**317–332.

Eriksson, K.-E., and Pettersson, B., 1975a, Extracellular enzyme system utilized by the fugus *Sporotrichum pulverulentum* (*Chrysosporium lignorum*) for the breakdown of cellulose, *Eur. J. Biochem.* **51**:193–206.

Eriksson, K.-E., and Pettersson, B., 1975b, Extracellular enzyme system utilized by the fungus *Sporotrichum pulverulentum* (*Chrysosporium lignorum*) for the breakdown of cellulose, *Eur. J. Biochem.* **51**:213–218.

Eriksson, K.-E., and Wood, T. M., 1985, Biodegradation of cellulose in: *Biosynthesis and Biodegradation of Wood Components* (T. Higuchi, ed.), Academic Press, New York, pp. 469–503.

Eriksson, K.-E., Pettersson, B., Volc, J., and Musilek, V., 1986, Formation and partial characterization of glucose-2-oxidase, a H_2O_2 producing enzyme in *Phanaerochaete chrysosporium, Appl. Microbiol. Biotechnol.* **23**:257–262.

Eveleigh, D. E., 1987, Cellulose: A perspective, *Philos. Trans. R. Soc. London* **A321**:435–447.

Feniksova, R. V., and Moldabaeva, R. K., 1967, Effect of carbon sources in the medium on the synthesis of pectolytic enzymes by *Aspergillus niger* deep culture, *Appl. Microbiol. Biochem.* **3**(3):283–290.

Fernandez-Espinar, M. T., Ramon, D., Pinaja, F., and Valles, S., 1992, Xylanase production by *Aspergillus nidulans, FEMS Microbiol. Lett.* **91**:91–96.

Fiedurek, J., and Ilczuk, Z., 1983, Synthesis of pectinolytic enzymes by forced heterokaryons of *Aspergillus niger* in submerged culture, *Acta Alim. Pol.* **9**:101–111.

Flaschel, E., Wandrey, C., and Kula, M. R., 1983, Ultrafiltration for the separation of biocatalysts, *Adv. Biochem. Eng.* **26**:73–139.

Fogarty, W. M., and Benson, C. P., 1982, Measurement of glucose transfer to glycerol by *Aspergillus niger* transglucosidase, *Biotechnol. Lett.* **4**:61-64.

Fogarty, W. M., and Benson, C. P., 1983, Purification and properties of a thermophilic amyloglucosidase from *Aspergillus niger, Eur. J. Appl. Microbiol. Biotechnol.* **18**:271–278.

Fogarty, W. M., and Griffin, P. J., 1973, Production and purification of the metalloprotease of *Bacillus polymyxa, Appl. Microbiol.* **26**(2):185–190.

Fogarty, W. M., and Kelly, C. T., 1979, Starch-degrading enzymes of microbial origin, in: *Progress in Industrial Microbiology,* Vol. 15 (M. J. Bull, ed.), Elsevier, Amsterdam, pp. 87–151.

Fogarty, W. M., and Kelly, C. T., 1980, Amylases, amyloglucosidases and related glucanases, in: *Economic Microbiology,* Vol. 5 (A. H. Rose, ed.), Academic Press, London, pp. 115–158.

Fogarty, W. M., and Kelly, C. T., 1983, Pectic enzymes, in: *Microbial Enzymes and Biotechnology* (W. M. Fogarty, ed.), Applied Science Publishers, London, pp. 131–182.

Fogarty, W. M., and Kelly, C. T., 1990, Recent advances in microbial amylases, in: *Microbial Enzymes and Biotechnology,* 2nd ed. (W. M. Fogarty and C. T. Kelly, eds.), Elsevier, London, pp. 71–133.

Fogarty, W. M., and Ward, O. P., 1974, Pectinases and pectic polysaccharides, in: *Progress in Industrial Microbiology,* Vol. 13 (D. J. D. Hockenhull, ed.), Churchill-Livingstone, Edinburgh, pp. 59–120.

Fournier, A. R., Frederick, M. M., Frederick, J. R., and Reilly, P. J., 1985, Purification and characterization of endo-xylanases from *Aspergillus niger.* III. An enzyme of pI 3.65, *Biotechnol. Bioeng.* **27**(4):539–546.

Frederick, M. M., Frederick, J. R., Fratzke, A. R., and Reilly, P. J., 1981, Purification and characterization of a xylobiose- and xylose-producing endo-xylanase from *Aspergillus niger, Carbohdr. Res.* **97**(1):87–103.

Frederick, M. M., Kiang, C.-H., Frederick, J. R., and Reilly, P. J., 1985, Purification and characterization of endo-xylanases from *Aspergillus niger:* Two isozymes active on xylan backbones near branch points, *Biotechnol. Bioeng.* **27**(4):524–532.

Friedrich, J., Cimerman, A., and Perdih, A., 1987, Mixed culture of *Aspergillus awamori* and *Trichoderma reesei* for bioconversion of apple distillery waste, *Appl. Microbiol. Biotechnol.* **26**(3):299–303.

Friedrich, J., Cimerman, A., and Steiner, W., 1989, Submerged production of pectolytic enzymes by *Aspergillus niger:* Effect of different aeration/agitation regimes, *Appl. Microbiol. Biotechnol.* **31**:490–494.

Frost, G. M., and Moss, D. A., 1987, Production of enzymes by fermentation, in: *Biotechnology,* Vol. 7A (H.-J. Rehm and G. Reed, eds.), VCH Verlagsgesellschaft mbH, Weinheim, Germany, pp. 65–212.

Fujii, M., Homma, T., and Taniguchi, M., 1988, Synergism of α-amylase and glucoamylase on hydrolysis of native starch granules, *Biotechnol. Bioeng.* **32**(7):910–915.

Fukumoto, J., Iwai, M., and Tsugisaka, Y., 1963, Llipase I: Purification and crystallization of a lipase secreted by *Aspergillus niger, J. Gen. Appl. Microbiol.* **9**:353-361.

Fukumoto, J., Tsujisaka, Y., and Takenishi, S., 1970, Hemicellulases. I. Purification and some properties of hemicellulases from *Aspergillus niger, Nippon Nogeikagaku Kaishi* **44**:447–456.

Gerlach, D., Missel, C., and Schreier, P., 1988, Screening of lipases for the enantiomer resolution of R.S.-2-octanol by esterification in organic medium, *Z. Lebens. Unter. Forsch.* **186**(4):315–318.

Ghose, T. K., Panda, T., and Bisaria, V. S., 1985, Effect of culture phasing and mannanase on production of cellulase and hemicellulase by mixed culture of *Trichoderma reesi* D1-6 and *Aspergillus wentii* Pt 2804, *Biotechnol. Bioeng.* **27**(a):1353–1361.

Godtfredsen, S. E., 1990, Microbial lipases, in: *Microbial Enzymes and Biotechnology,* 2nd ed. (W. M. Fogarty and C. T. Kelly, eds.), Elsevier, London, pp. 255–274.

Golker, C., 1990, Isolation and purification, in: *Enzymes in Industry* (W. Gerhartz, ed.), VCH Verlagsgesellschaft mbH, Weinheim, Germany, pp. 43–62.

Gorbacheva, I. V., and Rodionova, N. A., 1977a, Studies on xylan-degrading enzymes. I. Purification and characterization of endo-1,4-β-xylanase from *Aspergillus niger* str. 14, *Biochim. Biophys. Acta* **484**:79–93.

Gorbacheva, I. V., and Rodionova, N. A., 1977b, Studies on xylan-degrading enzymes. II. Action pattern of endo-1,4-β-xylanase from *Aspergillus niger* str. 14 on xylan and xylooligosaccharides, *Biochim. Biophys. Acta* **484**:94–102.

Griffin, P. J., and Fogarty, W. M., 1973, Physicochemical properties of the native, zinc- and manganese-prepared metalloprotease of *Bacillus polymyxa, Appl. Microbiol.* **26**(2):191–195.

Gripon, J. C., Auberger, B., and Lenoir, J., 1980, Metalloproteases from *Penicillium casekolum* and *P. roqueforti:* Comparison of specificity and chemical characterization, *Int. J. Biochem.* **12**:451-455.

Gum, E. K., and Brown, R. D., Jr., 1976, Structural characterization of a glycoprotein cellulase, 1,4-β-D-glucanase cellobiohydrolase C from *Trichoderma viride, Biochim. Biophys. Acta* **446**:371–386.

Gum, E. K., and Brown, R. D., Jr., 1977, Comparison of four purified extracellular 1,4-β-D-glucan cellobiohydrolase enzymes from *Trichoderma viride, Biochim. Biophys. Acta* **492**:225–231.

Gwynne, D. I., Buxton, F. P., Williams, S. A., Garven, S., and Davies, R. W., 1987, Genetically engineered secretions of active human interferon and a bacterial endoglucanase from *Aspergillus nidulans, Bio/Technology* **5**:713–719.

Halliwell, G., and Griffin, M., 1973, The nature and mode of action of the cellulolytic component C_1 of *Trichoderma koningii* on native cellulose, *Biochem. J.* **135:**587–594.

Halliwell, G., and Vincent, R., 1981, The action of cellulose and its derivation of a purified 1,4-β-glucanase from *Trichoderma koningii, Biochem. J.* **199:**409–417.

Hang, T. D., and Woodams, E. E., 1977, Baked-bean waste, a potential substrate for producing fungal amylases, *Appl. Envir. Microbiol.* **33**(6)**:**1293–1294.

Hayashida, S., 1975, Selective submerged productions of three types of glucoamylases by a black-koji mold, *Agric. Biol. Chem.* **39**(11)**:**2093–2099.

Hayashida, S., and Flor, P. Q., 1981, Raw starch-digestive glucoamylase productivity of protease-less mutant from *Aspergillus awamori* var. *kawachi, Agric. Biol. Chem.* **45**(12)**:**2675–2681.

Hayashida, S., and Teramoto, Y., 1986, Production and characteristics of raw-starch-digesting α-amylase from a protease-negative *Aspergillus ficum* mutant, *Appl. Envir. Microbiol.* **52**(5)**:**1068–1073.

Hayashida, S., and Yoshino, E., 1978, Formation of active derivatives of glucoamylase I during the digestion with fungal acid protease and α-mannosidase, *Agric. Biol. Chem.* **42**(5)**:**927–933.

Hayashida, S., Nomura, T., Yoshino, E., and Hongo, M., 1976, The formation and properties of subtilisin-modified glucoamylase, *Agric. Biol. Chem.* **40**(1)**:**141–146.

Hayashida, S., Kunisaki, S.-I., Nakao, M., and Flor, P. Q., 1982, Evidence for raw-starch-affinity site on *Aspergillus awamori* glucoamylase I, *Agric. Biol. Chem.* **46**(1)**:**83–89.

Hayashida, S., Nakahawa, K., Kuroda, K., Miyata, T., and Iwanaga, S., 1989a, Structure of the raw-starch-affinity site on the *Aspergillus awamori* var. *kawachi* glucoamylase I molecule, *Agric. Biol. Chem.* **53**(1)**:**135–141.

Hayashida, S., Nakahawa, K., Kanlayakrit, W., Hava, T., and Teramoto, Y., 1989b, Characteristics and function of raw-starch-affinity site on *Aspergillus awamori* var. *kawachi* glucoamylase I molecule, *Agric. Biol. Chem.* **53**(1)**:**143–149.

Hesseltine, C. W., 1972, Biotechnology report: Solid state fermentations, *Biotechnol. Bioeng.* **14:**517–532.

Hesseltine, C. W., 1983, Microbiology of Oriental fermented foods, *Annu. Rev. Microbiol.* **37:**575–601.

Hjortshoj, K., and Aunstrup, R., 1978, United States Patent No. 4,110,163.

Hofelmann, M., Hartmann, J., Zink, A., and Schreier, P., 1985, Isolation, purification and characterization of lipase isoenzymes from a technical *Aspergillus niger* enzyme, *J. Food Sci.* **50:**1721–1725.

Hoq, M. M., Tagami, H., Yamane, T., and Shimizu, S., 1985, Some characteristics of continuous glyceride synthesis by lipase in a microporous hydrophobic membrane bioreactor, *Agric. Biol. Chem.* **49**(2)**:**335–342.

Impoolsup, A., Bhumiratana, A., and Flegel, T. W., 1981, Isolation of alkaline and neutral proteases from *Aspergillus flavus* var. *columnaris,* a soy sauce koji mold, *Appl. Envir. Microbiol.* **42**(4)**:**619–628.

Iwai, M., and Tsujisaka, Y., 1984, Fungal lipase, in: *Lipases* (B. Borgstrom and H. L. Brockman, eds.), Elsevier, Amsterdam, pp. 443–469.

Jacob, G. S., and Orme-Johnson, W. H., 1979, Catalase of *Neurospora crassa.* 1. Introduction, purification, and physical properties, *Biochemistry* **18**(14)**:**2967–2975.

John, M., Schmidt, B., and Schmidt, J., 1979, Purification and some properties of five endo-1,4-β-D-xylanases and a β-D-xylosidase produced by a strain of *A. niger, Can. J. Biochem.* **57:**125–134.

Johri, B. N., Jain, S., and Chouhan, S., 1985, Enzymes from thermophilic fungi: Proteases and lipases, *Proc. Ind. Acad. Sci. (Plant Sci.)* **94:**175–196.

Kalo, P., 1988, Modification of butter fat by interesterifications catalysed by *Aspergillus niger* and *Mucor miehei* lipases, *Meijeritieteellinen Aikakauskirja,* **46**(1):36–47.
Kamel, M. Y., and Takany, M. M., 1973, *Penicillium notatum* catalase: Purification and properties, *Acta Biol. Med. Ger.* **30**(1):13–23.
Kirimura, J., and Yoshida, T., 1966, United States Patent No. 3,243,356.
Koh, W. Y., and Messing, R. A., 1963, United States Patent No. 3,096,253.
Kopperschlager, G., Lorenz, G., and Usbeck, E., 1983, Applications of affinity partitioning in an aqueous two-phase system to the investigation of triazine dye–enzyme interactions, *J. Chromatogr.* **259:**97–105.
Koritala, S., Hesseltine, C. W., Pryde, E. H., and Mounts, T. L., 1987, Biochemical modification of fats by microorganisms: A preliminary survey, *J. Am. Oil Chem. Soc.* **64**(4):509–513.
Koths, K. E., and Halenbeck, R. F., 1986, *Polyporous obtusus* pyranose-2-oxidase preparation, United States Patent No. 4,569,913.
Kubicek, C. P., 1981, Release of carboxymethyl-cellulase and β-glucosidase from cell walls of *Trichoderma reesei, Eur. J. Appl. Microbiol. Biotechnol.* **13:**226–231.
Kusakabe, H., Kodama, K., Midorikawa, Y., Machida, H., Kuninaka, A., and Yoshino, H., 1976, Neoplasm-inhibiting microbial pterin deaminase, United States Patent No. 3,930,955, German 2,422,580.
Laboureur, P., Brunaud, M. D. P., and Langlois, C., 1974, United States Patent No. 3,810,820.
Lakshminarayanan, K., 1972, Urate oxidase production via microbial fermentation, United States Patent No. 3,701,715.
Lambert, P. W., 1983, Industrial enzyme production and recovery from filamentous fungi, in: *The Filamentous Fungi,* Vol. 4 (J. E. Smith, D. R. Berry, and B. Kristiensen, eds.), Edward Arnold, London, pp. 210–237.
Lee, C. K., and Hong, J., 1985, Concentration, polarization and shear deactivation of protein during ultrafiltration/diafiltration, *Annu. Rep. Ferm. Proc.* **8:**73–91.
Linefield, W. M., Barauskas, R. A., Sivieri, L., Serota, S., and Stevenson, S. R., 1984, Enzymatic fat hydrolysis and synthesis, *J. Am. Oil Chem. Soc.* **61**(2):191–195.
Linko, M., 1981, Biomass conversion program in Finland, *Adv. Biochem. Eng.* **20:**163–172.
Ljungdahl, L. G., and Eriksson, K.-E., 1985, Ecology of microbial cellulose degradation, in: *Advances in Microbial Ecology,* Vol. 5 (K. C. Marshall, ed.), Plenum Press, New York, pp. 237–299.
Lockwood, L. B., 1975, Organic acid production, in: *The Filamentous Fungi* (J. E. Smith and D. R. Berry, eds.), Edward Arnold, London, pp. 140–157.
Lockwood, L. B., 1979, Production of organic acids by fermentation, in: *Microbial Technology* (H. J. Peppler, ed.), Academic Press, New York, pp. 355–387.
Lonsane, B. K., Ghildyal, N. P., Budiatman, S., and Ramakrishna, S. V., 1985, Engineering aspects of solid state fermentation, *Enz. Microb. Technol.* **7**(6):258–265.
Macrae, A. R., 1983, Extracellular microbial lipases, in: *Microbial Enzymes and Biotechnology* (W. M. Fogarty, ed.), Applied Science Publishers, London, pp. 225–250.
Macrae, A. R., and Hammond, R. C., 1985, Present and future applications of lipases, *Biotechnol. Gen. Eng. Rev.* **3:**193–218.
Macris, B. J., and Galiotou-Panayotou, M., 1978, Enhanced cellobiohydrolase production from *Aspergillus ustus* and *Trichoderma harzianum, Enz. Microb. Technol.* **8:**141–144.
Maher, G. G., 1968, Inactivation of transglucosidase in enzyme preparation from *Aspergillus niger, Starke* **20:**228–232.
Manjunath, P., and Raghavendra Rao, M. R., 1979, Comparative studies on glucoamylases from three fungal sources, *J. Bioscience* **1**(4):409–425.

Markkanen, P., Bailey, M., and Enari, T.-M., 1978, Production of cellulolytic enzymes, in: *Proceedings of Symposium: Bioconversion in Food Technology* (P. Linko, ed.), VTT, Espoo, Finland, pp. 111–114.

Matsubara, H., and Feder, J., 1971, Other bacterial, mold, and yeast proteases, in: *The Enzymes,* Vol. 3 (P. D. Boyer, ed.), Academic Press, New York, pp. 721–795.

McHale, A., and Coughlan, M. P., 1980, Synergistic hydrolysis of cellulose by components of the extracellular cellulase system of *Talaromyces emersonii, FEBS Lett.* **117:**319.

McHale, A., and Coughlan, M. P., 1981, The cellulolytic system of *Talaromyces emersonii:* Purification and characterization of the extracellular and intracellular β-glucosidases, *Biochim. Biophys. Acta* **662:**152–159.

McHale, A., and Coughlan, M. P., 1982, Properties of the β-glucosidases of *Talaromyces emersonii, J. Gen. Microbiol.* **128:**2327–2331.

Medda, S., Saha, B. C., and Ueda, S., 1982, Raw starch adsorption and elution behaviour of glucoamylase I of black *Aspergillus, J. Ferm. Technol.* **60:**261–264.

Mega, T., and Matsushima, Y., 1979, Comparative studies of three exo-β-glycosidases of *Aspergillus oryzae, J. Biochem.* **85**(2)**:**335–341.

Melling, J., and Phillips, B. W., 1975, Large-scale extraction and purification of enzymes, in: *Handbook of Enzyme Biotechnology* (A. Wiseman, ed.), Ellis Horwood, Chichester, England, pp. 58–88.

Meyrath, J., 1966, Fungal amylase production, *Proc. Biochem.,* July, pp. 234–238.

Miah, M. N. N., and Ueda, S., 1977a, Multiplicity of glucoamylase of *A. oryzae.* Part 1. Separation and purification of three forms of glucoamylase, *Starch* **29:**191–196.

Miah, M. N. N., and Ueda, S., 1977b, Multiplicity of *A. oryzae.* Part 2. Enzyme and physiochemical properties of three forms of glucoamylase, *Starch* **29:**235–239.

Miall, L. M., 1978, Gluconic acid, in: *Primary Products of Metabolism* (A. H. Rose, ed.), Academic Press, London, pp. 99–105.

Mikami, S., Iwano, K., Shinoki, S., and Shimada, T., 1987, Purification and some properties of acid-stable α-amylases from shochu koji (*Aspergillus kawachii*), *Agric. Biol. Chem.* **51:**2495–2501.

Milsom, P. E., and Meers, J. L., 1985, Gluconic and l-taconic acids, in: *Comprehensive Biotechnology,* Vol. 3 (M. Moo-Young, ed.), Pergamon Press, Oxford, pp. 681–697.

Minoda, Y., Arai, M., Torigoe, Y., and Yamada, K., 1968, Acid-stable α-amylase of black *Aspergillii.* Part III. Separation of acid-stable α-amylase and acid-unstable α-amylase from the same mold amylase preparation, *Agric. Biol. Chem.* **32:**110–113.

Miyazawa, T., Takitani, T., Uegi, S., Yamada, T., and Kuwata, T., 1988, Optical resolution of unusual amino-acids by lipase-catalysed hydrolysis, *J. Chem. Soc. Chem. Commun.* **17:**1214–1215.

Moloney, A. P., McCrae, S. I., Wood, T. M., and Coughlan, M. P., 1985, Isolation and characterization of the 1,4-β-D-glacan glucanohydrolases of *Talaromyces emersonii, Biochem. J.* **225:**365–374.

Moo-Young, M., Moreira, A. R., and Tengeroly, R. P., 1983, Principles of solid-substrate fermentation, in: *The Filamentous Fungi,* Vol. 4 (J. E. Smith, D. R. Berry, and B. Kristiansen, eds.), Edward Arnold, London, pp. 117–144.

Morihara, K., 1974, Comparative specificity of microbial proteinases, *Adv. Enzymol.* **41:**179–243.

Morita, Y., Ogha, M., and Shimizu, K., 1968, *Memoirs of the Research Institute of Food Science,* Kyoto University, **29:**191.

Murao, S., Kanamoto, J., and Arai, M., 1979, Isolation and identification of a cellulolytic enzyme-producing microorganism, *J. Ferm. Technol.* **57:**151–156.

Nakadai, T., Nasuno, S., and Iguchi, N., 1973, Purification and properties of neutral proteinase II from *Aspergillus oryzae, Agric. Biol. Chem.* **37**(12):2703–2708.
Nakamura, S., and Fujiki, S., 1968, Comparative studies on the glucose oxidase of *Aspergillus niger* and *Penicillium amagasakiense, J. Biochem.* **63**:51–58.
Nasuno, S., 1972, Differentiation of *Aspergillus sojae* from *Aspergillus oryzae* by polyacrylamide gel disc electrophoresis, *J. Gen. Microbiol.* **71**:29–33.
Neidleman, S. L., Amon, W. F., and Geigert, J., 1981, Fructose from glucose by enzymic conversion, European Patent No. 42,221.
Neubeck, C. E., 1975, Fruits, fruit products, and wines, in: *Enzymes in Food Processing,* 2nd ed. (G. Reed, ed.), Academic Press, New York, pp. 397–442.
Nielsen, R. T., 1984, Australian Patent No. 29,764/84.
Norman, B. E., 1979, The application of polysaccharide degrading enzymes in the starch industry, in: *Microbial Polysaccharides and Polysaccharidases* (R. C. W. Berkeley, G. W. Gooday, and D. C. Ellwood, eds.), Academic Press, London, pp. 339–376.
North, M. J., 1982, Comparative biochemistry of the proteinases of eucaryotic microorganisms, *Microbiol. Rev.* **46**(3):308–340.
Nummi, M., Niku-Paavola, M.-L., Lappalainen, A., Enari, T.-M., and Raunio, V., 1983, Cellobiohydrolase from *Trichoderma reesei, Biochem. J.* **215**(3):677–683.
Nyeste, L., and Hollo, J., 1963, Polygalacturonase in molds. III. The influence of the culture conditions on the synthesis of polygalacturonase in *Aspergillus niger, Z. Allg. Mikrobiol.* **3**:37–41.
Nyiri, L., 1968, Manufacture of pectinases. Part 1, *Process Biochem.* **3**(8):27–30.
Nyiri, L., 1969, Manufacture of pectinases, *Process Biochem.* **4**(8):27–30.
Oguntimein, G. B., and Reilly, P. J., 1980a, Purification and immobilization of *Aspergillus niger* β-xylosidase, *Biotechnol. Bioeng.* **22**(6):1127–1142.
Oguntimein, G. B., and Reilly, P. J., 1980b, Properties of soluble and immobilized *Aspergillus niger* β-xylosidase, *Biotechnol. Bioeng.* **22**(6):1143–1154.
Ohga, M., Shimizu, K., and Morita, Y., 1966, Studies on amylases of *Aspergillus oryzae* cultured on rice. Part II. Some properties of glucoamylases, *Agric. Biol. Chem.* **30**(10):967–972.
Okada, S., Kishi, K., Higashihara, M., and Fukumoto, J., 1963a, Flavonoid-hydrolyzing enzymes. I. Crystallisation of naringinase I and hesperidinase I and their actions, *Nippon Nogei Kagaku Kaishi* **37**(2):84–89.
Okada, S., Kishi, K., Higashihara, M., and Fukumoto, J., 1963b, Flavonoid-hydrolyzing enzymes. II. Substrate specification of naringinase I and hesperidinase I, *Nippon Nogei Kagaku Kaishi* **37**:142–145.
Okada, S., Kishi, K., Itaya, K., and Fukumoto, J., 1963c, Flavonoid-hydrolysing enzymes. III. Purification of purinin and hesperetin 7-glucose-hydrolyzing enzyme, *Nippon Nogei Kagaku Kaishi* **37**:146–150.
Okazaki, H., 1957, Mold amylase systems, *Proc. Int. Symp. Enz. Chem.,* Tokyo and Kyoto, **2**:494–499.
Ono, K., Hiromi, K., and Hamauzu, Z.-I., 1965, Quantitative determination of anomeric forms of sugar produced by amylases. I. Gluc-amylase-phenyl-α-maltoside or maltose system, *J. Biochem.* **57**(1):34–38.
Ono, K., Shigeta, S., and Oka, S., 1988a, Effective purification of glucoamylase in koji, a solid culture of *Aspergillus oryzae* on steamed rice, by affinity chromatography using an immobilised acarbose (BAY-g5421), *Agric. Biol. Chem.* **52**:1707–1714.
Ono, K., Shintani, K., Shigeta, S., and Oka, S., 1988b, Comparative studies of various molecular species in *Aspergillus niger* glucoamylase, *Agric. Biol. Chem.* **52**:1699–1706.

Pal, N., Das, S., and Kundu, A. K., 1978, Influence of culture and nutritional conditions on the production of lipase by a submerged culture of *Aspergillus niger, J. Ferm. Technol.* **56:**593–598.

Palmer, T. J., 1975, Glucose syrups in food and drink, *Process Biochem.* **10:**19–20.

Paszczynski, A., Fiedurek, J., Ilczuk, Z., and Ginalska, G., 1985, The influence of proteases on the activity of glucoamylase for *Aspergillus niger* C, *Appl. Microbiol. Biotechnol.* **22:**434–437.

Pazur, J. H., and Ando, T., 1961, The isolation and the mode of action of a fungal tranglucosylase, *Arch. Biochem.* **93:**43–49.

Pazur, J. H., and Kleppe, K., 1962, The hydrolysis of α-D-glucosides by amyloglucosidase from *Aspergillus niger, J. Biol. Chem.* **237**(4)**:**1002–1006.

Pazur, J. H., Knull, H. R., and Simpson, D. L., 1970, Glycoenzymes: A note on the role for the carbohydrate moieties, *Biochem. Biophys. Res. Commun.* **40:**110–116.

Pazur, J., Tominaga, Y., De Brosse, C., and Jackman, L., 1978, The synthesis of 1,6-anhydro-α-D-glucopyranose and D-glucosyl oligosaccharides from maltose by a fungal glycosyltransferase, *Carbohydr. Res.* **61:**279–291.

Pazur, J. H., Tominaga, Y., Forsberg, L. S., and Simpson, D. L., 1980, Glycoenzymes: An unusual type of glycoprotein structure for a glucoamylase, *Carbohydr. Res.* **84**(1)**:**103.

Peppler, H. J., and Reed, G., 1987, Enzyme technology: Enzymes in food and food processing, in: *Biotechnology,* Vol. 7a (J. F. Kennedy, ed.), VCH Verlagsgesellschaft, Weinheim, Germany, pp. 547–603.

Pilnik, W., 1982, Enzymes in the beverage industry, in: *Use of Enzymes in Food Technology,* (P. Dupuy, ed.), Technique et Documentation Lavoisier, Paris, pp. 425–450.

Pilnik, W., and Rombouts, F. M., 1981, Pectic enzymes, in: *Enzymes and Food Processing* (G. Birch, N. Blakebrough, and K. J. Parker, eds.), Applied Science Publishers, London, pp. 105–128.

Poutanen, K., Puls, J., and Linko, M., 1986, The hydrolysis of birchwood hemicellulose by enzymes produced by *Trichoderma reesei* and *Aspergillus awamori, Appl. Microbiol. Biotechnol.* **23**(6)**:**487–490.

Poutanen, K., Ratto, M., Puls, J., and Viikari, L., 1987, Evaluation of different microbial xylanolytic systems, *J. Biotechnol.* **6:**49–60.

Ramasesh, N., Sreekantish, K. R., and Murthy, V. S., 1982, Studies on the two forms of amyloglucosidase of *Aspergillus niger* van Tieghem, *Starch* **34:**346–351.

Ramesh, M. V., and Lonsane, B. K., 1990, Critical importance of moisture content of the medium in α-amylase production by *Bacillus licheniformis* M27 in a solid state fermentation system, *Appl. Microbiol. Biotechnol.* **33:**501–505.

Reese, E. T., and Mandels, M., 1984, Rolling with the times: Production and applications of *Trichoderma reesei* cellulase, in: *Annual Reports of Fermentation Processes,* Vol. 7 (G. T. Tsao, ed.), Academic Press, New York, pp. 1–20.

Reese, E. T., Maguire, A., and Parrish, F. W., 1973, Production of β-D-xylopyranosidases by fungi, *Can. J. Microbiol.* **19:**1065–1074.

Reid, W. W., 1961, Preparation, properties, and assay of pectin degrading enzymes, *Soc. Chem. Industry Monogr.* **11:**35–47.

Reilly, P. J., 1981, Xylanases: Structure and function, in: *Basic Life Sciences,* Vol. 18 (A. Hollander, ed.), Plenum Press, New York, pp. 111–130.

Rodionova, N. A., Gorbacheva, I. V., and Buivid, V. A., 1977, Fractionation and purification of endo-1,4-β-xylanases and exo-1,4-β-xylosidases of *Aspergillus niger, Biokhimiya* **42**(4)**:**659–671.

Saddler, J. N., Brownell, H. H., Clermont, L. P., and Levitin, N., 1982, Enzymatic hydrolysis of cellulose and various pretreated wood fractions, *Biotechnol. Bioeng.* **24:**1389–1402.

Saha, B. C., and Ueda, S., 1984a, Raw starch degradation by glucoamylase II by black *Aspergillus, J. Jpn. Soc. Starch Sci.* **31:**8–13.

Saha, B. C., and Ueda, S., 1984b, Production and characteristics of inhibitory factor of raw starch digestion from *Aspergillus niger, Appl. Microbiol. Biotechnol.* **19:**341–346.

Saha, B. C., Mitsue, T., and Ueda, S., 1979, Glucoamylase produced by submerged culture of *Aspergillus oryzae, Starch* **31:**307–314.

Sakano, Y., Masuda, N., and Kobayashi, T., 1971, Hydrolysis of pullulan by a novel enzyme from *Aspergillus niger, Agric. Biol. Chem.* **35**(6)**:**971–973.

Sakano, Y., Higuchi, M., and Kobayashi, T., 1972, Pullulan 4-glucohydrolase from *Aspergillus niger, Biochem. Biophys.* **153:**180–187.

Sanderson, G. W., and Coggon, P., 1974, Manufacture of instant tea, United States Patent No. 3,812,266; German Patent No. 2304073.

Sanderson, G. W., and Simpson, W. S., 1974, Pectinase enzyme treating process for preparing high bulk density tea powders, United States Patent No. 3,787,582.

Saroja, K., Venkataraman, R., and Giri, K. V., 1955, Transglucosidation in *Penicillium chrysogenum* W-176: Isolation and identification of the oligosaccharides, *Biochem. J.* **60:**399.

Satyanarayana, T., and Jori, B. N., 1981, Lipolytic activity of thermophilic fungi of paddy straw compost, *Curr. Sci.* **50:**680–682.

Scawen, M. D., Atkinson, A., and Darbyshire, J., 1980, Large-scale enzyme purification, in: *Applied Protein Chemistry* (R. A. Grant, ed.), Applied Science Publishers, London, pp. 281–324.

Sharma, A., Milstein, O., Vered, Y., Gressel, J., and Flowers, H. M., 1985, Effects of aromatic compounds on hemicellulose degrading enzymes in *A. japonicus, Biotechnol. Bioeng.* **27**(8)**:**1095–1101.

Shei, J. C., Fratzke, A. R., Frederick, M. M., Frederick, J. R., and Reilly, P. J., 1985, Purification and characterisation of endo-xylanases from *Aspergillus niger.* II. An enzyme of pI 4–5, *Biotechnol. Bioeng.* **27:**533–538.

Shikata, S., and Nisizawa, K., 1975, Purification and properties of an exo-cellulase component of novel type from *Trichoderma viride, J. Biochem.* **78:**499–512.

Silman, R. W., 1980, Enzyme formation during solid-substrate fermentation in rotating vessels, *Biotechnol. Bioeng.* **22:**411–420.

Sinner, M., and Dietrichs, H. H., 1975a, Enzymic hydrolysis of hardwood xylans. I. Study of xylanases and other polysaccharide-decomposing enzymes from commercial fungal enzyme preparations, *Holzforschung* **29:**123–130.

Sinner, M., and Dietrichs, H. H., 1975b, Enzymic hyrolysis of hardwoods. II. Isolation of five β-1-4 xylanases from three commercial enzyme products, *Holzforschung* **29**(5)**:**168–177.

Sinner, M., and Dietrichs, H. H., 1975c, Enzymic hydrolysis of hardwood xylans. III. Characterization of five isolated β-1-4 xylanases, *Holzforschung* **29:**207–214.

Sonnet, P., and Antonia, E., 1988, Synthesis and evaluation of pseudolipids to characterize lipase selectives, *J. Agric. Food Chem.* **36:**856–862.

Sternberg, D., Vijayakumar, P., and Reese, E. T., 1977, β-Glucosidase: Microbial production and effect on enzymatic hydrolysis of cellulose, *Can. J. Microbiol.* **23:**139–147.

Sternberg, M., 1973, Purification of lactase, United States Patent No. 3,737,377.

Stewart, J. C., Lester, A., Milburn, B., and Parry, J. B., 1983, Xylanase and cellulase production by *Aspergillus fumigatus* Fresenius, *Biotechnol. Lett.* **5:**543–548.

Stuer, W., Jaeger, K. E., and Winkler, V. K., 1986, Purification of extracellular lipase from *Pseudomonas aeruginosa, J. Bacteriol.* **168:**1070–1074.

Stutzenberger, F., 1990, Bacterial cellulases, in: *Microbial Enzymes and Biotechnology,* 2nd ed. (W. M. Fogarty and C. T. Kelly, eds.), Elsevier, London, pp. 37–70.

Svensson, B. T., Pedersen, T. G., Svendsen, I., Sakai, T., and Ottesen, M., 1982, Characterisation of two forms of glucoamylase from *Aspergillus niger, Carls. Res. Commun.* **47:**55–69.

Szajer, I., 1978, Effect of various nitrogen salts in beet pulp medium on polygalacturonase activity of *Penicillium* sp. 7/4B/EI 1 mutant, *Acta Microbiol. Pol.* **27:**237–242.

Taguchi, T., Ohwaki, K., and Okuda, J., 1985, Glucose-2-oxidase (*Coriolus versicolor*) and its application to D-glucose colorimetry, *J. Appl. Biochem.* **7:**289–295.

Takahashi, T., Inokuchi, N., and Irie, M., 1981, Purification and characterisation of a glucoamylase from *Aspergillus saitoi, J. Biochem.* **89:**125–134.

Takamine, J., 1984, United States Patent No. 525,823.

Takenishi, S., and Tsujisaka, Y., 1975, On the modes of action of three xylanases produced by a strain of *Aspergillus niger* van Tieghem, *Agric. Biol. Chem.* **39**(12)**:**2315–2323.

Takenishi, S., Tsujisaka, Y., and Fukumoto, J., 1973, Purification and properties of the β-xylosidase produced by *Aspergillus niger* van Tieghem, *J. Biochem.* **73:**335–343.

Takino, Y., 1976, United States Patent No. 3,959,497.

Thomas, D. W., Smythe, C. V., and Labbee, M. D., 1958, Enzymic hydrolysis of naringin, the bitter principle of grapefruit, *Food Res.* **23:**591–598.

Towprayoon, S., Saha, B. C., Fujio, Y., and Ueda, S., 1988, Some characteristics of a raw starch digestion inhibitory factor from *Aspergillus niger, Appl. Microbiol. Biotechnol.* **29:**289–291.

Trivedi, L., and Rao, K. K., 1979, Production of cellulolytic enzymes by *Aspergillus fumigatus, Ind. J. Exp. Biol.* **17:**671–674.

Tsujisaka, Y., Takenishi, S., and Fukumoto, J., 1971, Hemicellulases. II. Mode of action of three hemicellulases produced from *Aspergillus niger.* 1. *Nippon Nogeikagaku Kaishi* **45**(6)**:**253–259.

Turnbull, I. F., Rand, K., Willets, N. S., and Hynes, M. J., 1989, Expression of the *Escherichia coli* enterotoxin subunit B gene in *Aspergillus nidulans* directed by the AMDS promoter, *Bio/Technology* **7**(2)**:**169–174.

Tuttobello, R., and Mill, P. J., 1961, The pectic enzymes of *Aspergillus niger.* I. The production of active mixtures of pectic enzymes, *Biochem. J.* **79:**51–57.

Ueda, S., Fujio, Y., and Lim, J. Y., 1982, Production and some properties of pectic enzymes from *Aspergillus oryzae* A-3, *J. Appl. Biochem.* **4:**524–532.

Underkofler, L. A., Denault, L. J., and Hov, E. F., 1965, Enzymes in the starch industry, *Starke* **17**(6)**:**179–84.

Upshall, A., Kumer, A. A., Bailey, M. C., Parker, M. D., Favreau, M. A., Lewison, K. P., Joseph, M. L., Maraganore, J. M., and McKnight, G. L., 1987, Secretion of active human tissue plasminogen activator from the filamentous fungus *Aspergillus nidulans, Bio/Technology* **5**(12)**:**1301–1304.

Uwajima, T., and Terada, O., 1980, Properties of a new enzyme glycerol oxidase from *Aspergillus japonicus* AT 008, *Agric. Biol. Chem.* **44:**2039–2045.

Uwajima, T., and Terada, O., 1982, Glycerol oxidase from *Aspergillus japonicus,* in: *Methods in Enzymology,* Vol. 89 (W. A. Wood, ed.), Academic Press, New York, pp. 243–248.

Uwajima, T., Akita, H., Ito, K., Mihara, A., Aisaka, K., and Terada, O., 1979, Some characteristics of a new enzyme "glycerol oxidase," *Agric. Biol. Chem.* **43:**2633–2634.

Uwajima, T., Akita, H., Ito, K., Mihara, A., Aisaka, K., and Terada, O., 1980, Formation and purification of a new enzyme, glycerol oxidase, and stoichiometry of the enzyme reaction, *Agric. Biol. Chem.* **44**(2)**:**399–406.

Vallier, P., Bata, J., and Colobert, L., 1977, Optimal conditions of α-amylase production by *Aspergillus oryzae* in liquid medium, *Ann. Microbiol.* **128B**(3)**:**359–371.

Vasu, S., 1967, Pectolytic enzymes of *Aspergillus niger.* II. Intracellular distribution of polygalacturonases, pectinethylesterases and pectintranseliminases, *Rev. Roum. Biochem.* **4**(1)**:**67–74.

Vidmar, S., Turk, V., and Kregar, I., 1984, Cellulolytic complex of *Aspergillus niger* under conditions for citric acid production: Isolation and characterisation of two β-(1,4)-glucan hydrolases, *Appl. Microbiol. Biotechnol.* **20:**326–330.

Visser, J., Beldman, G., Kusters-van-Someren, M. A. and Voragen, A. G. J., 1992, *Xylans and Xylanases,* Elsevier, Amsterdam.

Volc, J., Denisova, N. P., Nerud, F., and Musilek, V., 1985, Glucose 2-oxidase activity in mycelial cultures of basidiomycetes, *Fol. Microbiol.* **30**(2)**:**141–147.

Vora, K. A., Bhandara, S. S., Pradhan, R. S., Amin, A. R., and Modi, V. V., 1988, Characterisation of extracellular lipase produced by *Aspergillus japonicus* in response to *Calotropis gigentea* latex, *Biotechnol. Appl. Biochem.* **10:**465-472.

Ward, M., Wilson, L. J., Kodama, K. H., Rey, M. W., and Berka, R. M., 1990, Improved production of chymosin in *Aspergillus* by expression as a glucoamylase–chymosin fusion, *Bio/Technology* **8**(5)**:**435–440.

Wase, D. A. J., McManamey, W. J., Raymahasay, S., and Vaid, A. K., 1985a, Comparison between cellulase production by *Aspergillus fumigatus* in agitated vessels and in an air-lift fermentor, *Biotechnol. Bioeng.* **27:**1166–1172.

Wase, D. A. J., Vaid, A. K., and McDermott, C., 1985b, Increases in endo-1,4-β-D-glucanase titres produced by *Aspergillus fumigatus* through application of sample statistical methods, *Enz. Microb. Technol.* **7:**134–138.

Watanabe, K., and Fukimbara, T., 1967, Saccharogenic amylase produced by *Aspergillus awamori.* VIII. Inhibition of acid-stable and less acid-stable saccharogenic amylase, *J. Ferm. Technol.* **45:**226–232.

Werdelmann, B. W., and Schmid, R. D., 1982, The biotechnology of fats—a challenge and an opportunity, *Fette Seifen Anstrichmittel* **84:**436–443.

Whitaker, J. R., 1990, Microbial pectolytic enzymes, in: *Microbial Enzymes and Biotechnology,* 2nd ed. (W. M. Fogarty and C. T. Kelly, eds.), Elsevier, London, pp. 133–176.

Wimpenny, J. W. T., 1967, Breakage of microorganisms, *Process Biochem.* **2**(7)**:**41–44.

Wood, T. M., and Bhat, K. M., 1988, Methods for measuring cellulase activity, *Methods Enzymol.* **160:**87–112.

Wood, T. M., and McCrae, S. I., 1975, Cellulase complex of *Trichoderma koningii,* in: *Symposium on Enzymatic Hydrolysis of Cellulose* (M. Baileym, T.-M. Enari, and M. Linko, eds.), SITRA, Helsinki, pp. 231–254.

Wood, T. M., and McCrae, S. I., 1978, The cellulase of *Trichoderma koningii, Biochem. J.* **171:**61–72.

Wood, T. M., and McCrae, S. I., 1986, Studies of two low-molecular weight endo-(1,4)β-D-xylanases constitutively synthesized by the cellulolytic fungus *Trichoderma koningii, Carbohdr. Res.* **148:**321–331.

Wood, T. M., McCrae, S. I., and Bhat, K. M., 1989, The mechanism of fungal cellulase action, *Biochem. J.* **260:**37–43.

Yabuki, M., Ono, N., Hoshino, K., and Fukui, S., 1977, Rapid induction of α-amylase by non-growing mycelia of *Aspergillus oryzae, Appl. Envir. Microbiol.* **34:**1–6.

Yamasaki, Y., Suzuki, Y., and Ozawa, J., 1977, Three forms of α-glucosidase and a glucoamylase from *Aspergillus awamori, Agric. Biol. Chem.* **41:**2149–2161.

Yeoh, H. H., Wong, F. M., and Lim, G., 1986, Screening for fungal lipase using chromogenic lipid substrates, *Mycologia* **78:**298–300.

Yokotsuka, T., 1985, Fermented protein foods in the Orient, with emphasis on shoyu and miso in Japan, in: *Microbiology of Fermented Foods,* Voil. I (B. J. B. Wood, ed.), Elsevier, London, pp. 197–247.

Yokozeki, K., Tanaka, T., Yamanaka, S., Takinami, T., Hirose, Y., Sonomoto, K., Tanaka, A., and Fuki, S., 1982, Ester exchange of triglyceride by entrapped lipase in organic solvent, *Enz. Eng.* **6:**151–152.

Yong, F. M., and Wood, B. J. B., 1974, Microbiology and biochemistry of soy sauce fermentation, *Adv. Appl. Microbiol.* **17:**157–194.

Yoshino, E., and Hayashida, S., 1978, Enzymic modification of glucoamylae of *Aspergillus awamori* var. *kawachi, J. Ferm. Technol.* **56**(4)**:**289–295.

Zetalaki, K., 1970, The role of aeration and agitation in the production of glucose oxidase in submerged culture II, *Biotechnol. Bioeng.* **12:**379–397.

Zetalaki, K., 1976, Optimal carbon source concentration for the pectolytic enzyme formation by *Aspergilli, Process Biochem.* **11**(7)**:**11–18.

Zetalaki, K., and Vas, K., 1968, The role of aeration and agitation in the production of glucose oxidase in submerged culture, *Biotechnol. Bioeng.* **10**(1)**:**45–59.

Zetalaki-Horvath, K., and Dobra-Seres, I., 1972, Improvement of polygalacturonase production of *Aspergillus* strains by UV-irradiation, *Acta Aliment. Acad. Sci. Hung.* **1**(2)**:**139–151.

Health-Related Aspects of the Genus *Aspergillus*

8

C. W. LEWIS, J. G. ANDERSON, and J. E. SMITH

1. INTRODUCTION

While the genus *Aspergillus* is now recognized to contain approximately 200 species and varieties (Samson and van Reenen-Hoekstra, 1988), fewer than half a dozen are considered to be important implicating factors in human disease, notably *A. fumigatus, A. flavus, A. niger, A. nidulans,* and *A. terreus* (Evans and Gentles, 1985). *Aspergillus wentii, A. amstelodami, A. repens, A. ruber, A. niveus, A. candidus, A. conicus, A. carneus, A. ustus, A. oryzae, A. glaucus, A. clavatus, A. restrictus, A. parasiticus, A. sydowii, A. tamarii, A. ochraceus, A. flavipipes, A. fischeri,* and *A. versicolor* (Rogers and Kennedy, 1991) are among the second, less important group of health-threatening species, largely through their ability either to produce toxic secondary metabolites or to cause infections in humans and other animals or to do both.

A "pneumomycosis" described in 1847 is probably the first recorded instance of a human disease attributable to the genus; three years later, the term "aspergillosis" was coined, the causative agent being named *Aspergillus fumigatus.* Aspergillosis is primarily a pulmonary disease, caused most frequently by *A. fumigatus,* which, in its most serious manifestation, involves invasion of the lung tissues from which dissemination to other organs may result. More commonly, inhalation of *Aspergillus* conidia results in either colonization of existing lung cavities (aspergilloma) or a hypersensitivity reaction (allergic aspergillosis). In the United States, aspergillosis is the second most common fungal infection requiring hospitalization.

C. W. LEWIS, J. G. ANDERSON, and J. E. SMITH • Department of Bioscience and Biotechnology, University of Strathclyde, Glasgow G1 1XW, Scotland.

Aspergillus, edited by J. E. Smith. Plenum Press, New York, 1994.

However, *Aspergillus* strains can cause disease in a number of different ways, sometimes in combination with each other, with the pathological responses variable in severity and clinical course. The clinical manifestations are determined mostly by the local or general immunological and physiological state of the host and can be divided into hypersensitivity syndromes, usually induced by the presence of conidia, such as extrinsic asthma, extrinsic allergic alveolitis, and allergic bronchopulmonary aspergillosis; local infections, as in aspergilloma, otomycosis (external ear canal infection), keratitis (corneal infection), onychomycosis (nail infection), subcutaneous mycetoma (skin infection), and so forth; invasive inflammatory, granulomatous, necrotizing infections of the lungs, and systemic and fatal disseminated disease; and a response to toxicity normally due to ingestion of foods contaminated with mycotoxins or other metabolites (Evans and Gentles, 1985; Cohen, 1991; Rogers and Kennedy, 1991; Smith and Henderson, 1991). There are at least 14 food-borne toxigenic species of *Aspergillus* (Frisvad, 1988), some of which can produce ten or more different toxic compounds. The aflatoxins, produced by *A. flavus* and *A. parasiticus,* are the best known, with aflatoxin B_1 being hepatotoxic and one of the most potent animal carcinogens yet discovered.

Increasingly, the detrimental effects of molds, including the aspergilli, on human health in the domestic environment are being realized (Martin *et al.,* 1987; Hunter *et al.,* 1988; Burr *et al.,* 1988; Lewis *et al.,* 1989; Platt *et al.,* 1989), while the potential occupational hazards posed by large numbers of fungal conidia or mycotoxins or both, particularly in the agricultural, food, fermentation, and timber industries, are now well recognized (Sorenson, 1990; Land *et al.,* 1987). The relationship between the presence of high levels of molds, including aspergilli, and the volatile compounds produced by them within the built environment, and the so-called "sick building syndrome" or "building-related illnesses" experienced by their occupants is a subject of some controversy, but evidence suggests a positive correlation and has been observed in many countries (Nolard *et al.,* 1988; Samson, 1985; Kemper *et al.,* 1990). The reader is referred to Section 4.5.1 for a detailed appraisal of the health hazards posed by aspergillosis in various industrial situations, while Smith (1985) provides a comprehensive review of mycotoxin hazards as a whole to animal and human health.

2. ASPERGILLOSIS

2.1. Clinical Conditions

2.1.1. Acute Invasive Aspergillosis

The respiratory tract is the most common route of infection of *Aspergillus* conidia, and where the conidia are less than 4 μm in diameter, penetration to the alveolar air space can occur (Cohen, 1991; Institute of Environmental Health Officers, 1985). *Aspergillus fumigatus*, the most common cause of this type of aspergillosis, falls into this category, as does *A. flavus*, and these two species are predominant in cases of severely immunocompromised individuals in which this condition is most, if not exclusively, prevalent. Widespread growth of the fungus ensues in the lung tissue and may spread via the invasion of blood vessels to other organs, notably the kidneys and brain, in a systemic mycosis (Evans and Gentles, 1985). The prognosis for invasive aspergillosis is poor, and there have been few medical advances made over the last two decades, since diagnosis presents great difficulties and is usually achieved only postmortem. The increasing utilization of empirical antifungal therapy may be reducing the severity of infection, but any such improvement cannot yet be quantified (Cohen, 1991). Clinical manifestations are varied and include fever, cough, dyspnea, and other pleural chest pains. Occasionally, the organism can be isolated or necrotic skin lesions will appear.

Systemic mycoses have been reported in individuals exposed to air from contaminated ventilation systems. Autopsies on patients from one particular hospital demonstrated immunosuppressed individuals with invasive pulmonary aspergillosis, with 84% of the patients yielding isolates of *A. flavus*, the causative organism (Canadian Public Health Association, 1987).

Phagocytic cells are believed to have an important role in the eradication of fungal spores from the lung. However, *A. fumigatus* spores are known to inhibit phagocytosis and killing by macrophages and polymorphonuclear leukocytes. Results of experiments aimed at elucidating the mechanisms involved suggest that when spores of *A. fumigatus* become bound to the surface of phagocytes, they are able to release a substance (mol. wt. < 14,000) that inhibits their ingestion, while having little or no effect on surface binding (Robertson *et al.*, 1987a). Spore diffusates of the fungus in fact inhibit movement of the phagocytic cell membrane (Robertson *et al.*, 1987c) by failing to trigger and also inhibiting the production of reactive oxygen intermediates by the cells (Robertson *et al.*, 1987b). Thus, there are indications that *A. fumigatus* has specific mechanisms that protect against the lung's defenses. These mechanisms appear to include the rapid diffusion from the spore surface of a low-molecular-weight substance that is able to inhibit chemotaxis, as well as phatocytosis by macrophages and other phagocytic cells of man and other animals (Seaton and Robertson, 1989).

2.1.2. Chronic Necrotizing Aspergillosis

This is a second, slightly less serious, pulmonary aspergillosis, also referred to as "chronic granulomatous aspergillosis" or even "semiinvasive aspergillosis." Again, *A. fumigatus* is usually implicated and is generally isolated easily from the sputum. Biopsies reveal branching septate hyphae with local invasion and destruction of lung parenchyma, which happens progressively. Most at risk are middle-aged or elderly patients, often with a history of lung disease or immunosuppression or both, with the most frequent symptoms being cough, fever, and sputum production, which may last years before diagnosis, though dissemination does not occur. Cavities are usually formed in the lung, often in the upper lobe, in which mycetomas may later appear, and death can occur 6–12 months after onset (Cohen, 1991).

2.1.3. Aspergilloma

Aspergillomas or "fungus balls" occur more or less as sporadic chance infections among people with preexisting pulmonary lesions such as old tubercular cavities and cysts (Seeliger and Tintelnot, 1988). A compact mass of mycelium is formed, often surrounded by a dense fibrous wall, of variable size up to 8 cm in diameter. An aspergilloma is a local pulmonary infection and usually solitary in nature. Patients may have chronic cough, exhibit weight loss and fever, and produce small amounts of sputum; they may be symptom-free, or (in 50–70% of cases) hemoptysis may occur, possibly leading to massive hemorrhaging [10% according to Hay (1988)], should a major blood vessel be invaded by the fungus. Surgical removal is often practiced, but an aspergilloma can often be safely left untreated and the fungus may itself die (Evans and Gentles, 1985). In extreme cases, however, the onset of hemoptysis may render surgery impossible, in which case amphotericin or other efficacious drugs must be administered directly into the cavity as a matter of extreme urgency (De Coster *et al.*, 1988).

2.1.4. Miscellaneous Local Infections

Much of the information presented in this section is derived from Young *et al.* (1970), updated and put in overall context by Cohen (1991).

2.1.4a. Endocarditis. *Aspergillus* has two main methods of involvement in heart disease, either as a complication of open-heart surgery, characterized by abscess formation, septal perforation, and septal emboli with large mycelial masses forming on the heart valves, or as myocardial aspergillosis in association with a disseminated infection.

2.1.4b. Head and Neck Infections. Various regions of the head and neck are susceptible to infection. Cerebral aspergillosis usually occurs as a result of disseminated infection from the lung, and in fact the underlying pathology is identical to the pattern observed in pulmonary disease.

2.1.4c. Skin Infections. Primary cutaneous aspergillosis usually arises at the site of a wound or intravenous infusion site. Lesions may develop as macules, nodules, or an area of cellulitis and are slowly progressive. *Aspergillus fumigatus, A. flavus,* and *A. niger* have all been implicated in such infections. Skin lesions can also develop as part of discriminate disease that has usually originated in the thoracic region.

2.1.4d. Other Infections. Bone infections are very infrequent, and fungal osteomyelitis is usually an opportunistic infection of the immunosuppressed host. *Aspergillus fumigatus, A. flavus, A. niger, A. nidulans,* and *A. terreus* have, however, all caused osteomyelitis. Involvement of the gastrointestinal tract is usually regarded as a minor and infrequent manifestation of disseminated aspergillosis. Nevertheless, in some groups of patients, the gut can be second only to the lung as the most frequent site of infection, with lesions occurring at sites from tongue to colon and the most common clinical features being esophageal ulceration or abscess formation and large bowel ulcers. Involvement of the liver or spleen or both complicates 10–15% of disseminated aspergillosis cases. Finally, the urinary tract, kidneys, and bladder are all possible sites of aspergillosis, but occurrence in the reproductive organs is very rare.

2.1.5. Allergic Aspergillosis

2.1.5a. Overview. As well as being agents of pulmonary infection, notably the pathogenic *A. fumigatus,* aspergilli are potent respiratory allergens via the inhalation of conidia. Hypersensitivity may be either extrinsic or intrinsic. In the former, inhalation of spores results in disease; in the latter, symptoms are produced by hypersensitivity to hyphae growing in the air passages (Hay, 1988). Most common is the classic immunoglobulin E (IgE) section with symptoms of bronchospasm ultimately associated with rhinitis and conjunctivitis. *Aspergillus* species may also give rise to a group of diseases classified as hypersensitivity pneumonitis or allergic alveolitis, due to inhalation of organic dust, and are often occupational diseases, while the more severe form of allergy induced by aspergilli is allergic bronchopulmonary aspergillosis (ABPA), characterized by moderate to severe asthma (Stevens, 1988).

Essentially, the allergic symptoms are related to the sites within the respiratory tract where the spores are deposited: rhinitis, where the nasal passages are affected; asthma, where the trachea, bronchi, or bronchioles are affected; and alveolitis, where the alveoli of the lung are affected. There is an approximate relationship between spore size, probable site of deposition, and the resulting allergic symptoms, although this relationship is not clear-cut. *Aspergillus flavus* spores may be between 2 and 10 μm and able to give rise to rhinitis and asthma; *A. terreus* spores can be from less than 2 μm to 4 μm and cause rhinitis and asthma; *A. fumigatus* and *A. clavatus* produce spores of 4–10 μm and cause rhinitis, asthma, and alveolitis (Institute of Environmental Health Officers, 1985).

Samson (1985) has cited examples of dangers to health posed by exposure to mold spores in the industrial environment, including allergic symptoms caused by industrially prepared feed protein, though mostly by genera other than *Aspergillus*. Again, Section 4.5.1 provides a lengthier account of aspergillosis health hazards in industry, including those during citric acid production via *A. niger*. Al-Doory and Domson (1984) and Kurup and co-workers (Kurup 1984) mention hypersensitivity reactions to mold spores in the workplace.

Aspergillus species are used extensively in chemical synthesis within industrial biotechnology. A danger of such processes is the sensitization of workers to airborne fungi and culture fluid material, which may result in either extrinsic allergic alveolitis or asthma. Cases of sensitization to *A. niger* have been reported among workers in the fermentation industry producing citric acid via the fermentation of molasses. Particularly interesting is the fact that specific antigens involved in one of these occurrences was not present in commercially available abstracts (Nolard *et al.*, 1988). In some cases, it is possible, however, to employ nonsporulating strains.

While Oriental fermentations involving *A. oryzae* and *A. sojae*, close relatives of the aflatoxigenic *A. flavus* and utilized particularly for soy sauce manufacture, have repeatedly tested negative for mycotoxin production, as reviewed by Smith (1985), certain anomalies do seem to exist. For example, a 19-year-old Japanese girl was diagnosed as having contracted ABPA from her father's *A. oryzae* starter culture for making soybean paste (Akiyama *et al.*, 1987), while mutant strains of a soy sauce koji mold, *A. flavus* var. *columnaris*, have been found acutely toxic to weanling rats though testing negative for aflatoxin production (Kalayanamitr *et al.*, 1987). Thus, vigilance must be exercised within the fermentation industry, since even a single mutation can give rise to a potentially toxic organism.

2.1.5b. Extrinsic Asthma. Asthmatic hypersensitivity to aspergilli may manifest itself as an uncomplicated allergy—an intermittent disease of acute onset characterized by wheezing and dyspnea usually observed in atopic subjects (10–20% of the population). Uncomplicated asthma (reversible airway obstruction) is often associated with allergy to other inhalant allergens such as grass pollen allergy and house dust mites (*Dermatophagoides pteranyssinus*) and sufferers who have raised IgE levels. Late summer is the time of highest spore concentrations in most parts of Europe and North America, so the symptomatology can mimic grass pollen or seed pollen allergy. In such patients, precipitins to *Aspergillus* are mostly absent or weakly positive. Approximately 10–20% of asthmatics react to *A. fumigatus*, though while asthma is fairly common in the United Kingdom and the United States, it is less so in some other countries. In a Belgian survey, for instance, specific anti-*A. fumigatus* IgE antibodies were found in only 8% of allergic children and adults (Stevens, 1988). In some studies, 56% of asthmatic patients are sensitive to common airborne molds such as *Aspergillus* and *Alternaria*, as well as to other allergens (Hay, 1988).

In clinical terms, the symptoms are mediated by type I (immediate type) hypersensitivity, and in later stages, permanent lung damage may result and chronic breathlessness occurs. Pulmonary emphysema may develop in the chronically affected in whom the main symptoms may occur at any time of the year (Hay, 1988).

2.1.5c. Eosinophilic Pneumonitis. Asthma with eosinophilia of sputum or blood is a more chronic form, resulting in progressive lung damage and loss of lung function. It is associated with transient pulmonary shadows and blood eosinophilia. Symptoms are often striking, with high fever, cough, and night sweats. Approximately 75% of patients have fungus growing in the larger airways, enlarging the bronchi, for example (bronchiectasis), to produce mycelial plugs and mucus that block off segments of lung tissue and even entire lobes. These plugs, which are often coughed up, are a diagnostic feature, as is the presence of antibodies to aspergilli (IgE as well as other classes) (Evans and Gentles, 1985; Hay, 1988). Mucal impaction is the syndrome whereby large plugs of mucus, some containing *Aspergillus* hyphae, are the dominant clinical feature.

2.1.5d. Bronchocentric Granulomatosis. Hay (1988) considers this a rare disease, perhaps showing minimal asthmatic symptoms, but with fever, weight loss, cough, and malaise predominating.

2.1.5e. Allergic Bronchopulmonary Aspergillosis. ABPA is a characteristic intrinsic asthmatic condition caused mainly by *A. fumigatus*. The organisms survive and grow within the bronchi and bronchioles, giving rise to a more complex immunopathological process that may produce chronic pulmonary damage with secondary bronchiectasis, fibrosis, and progressive dyspnea (difficult breathing). It may also originate in patients with extrinsic *Aspergillus*-related asthma. Types I and III hypersensitivity reactions are presumed to be involved in the pathogenesis of ABPA (Evans and Gentles, 1985; Hay, 1988).

2.1.5f. Extrinsic Alveolitis. Hypersensitivity pneumonitis or allergic alveolitis is a chronic relapsing form of immunologically mediated pulmonary disease. Immune damage is believed to be mediated by type III as well as delayed type (IV) hypersensitivity, and the inflammatory reaction around inhaled spores is the cause of lung damage. Particularly heavy and repeated exposure to aspergilli conidia results in breathlessness, fever, and malaise some hours (4–6) after exposure. Serum antibodies to the appropriate organism are usually present. Lesions resemble those in ABPA and are thought to be associated with immune complex and cellular immunity reactions to the inhaled antigens. The disease is associated with the development of precipitating antibodies to the responsible fungi, whereas IgE-mediated responses are mostly absent (Stevens, 1988). A well-known example is maltster's lung, which occurs in workers who handle barley on which *A. clavatus* has sporulated during the malting process. A survey in Scotland showed that 5% of malt workers in distilleries had such symptoms (Evans and Gentles, 1985). More rarely, *A. fumigatus* has been implicated in alveolitis in patients with farmer's lung (Hay, 1988) and in mushroom worker's lung (van Griensven *et al.*, 1991), where high IgG-type antibody types were found.

3. MYCOTOXICOSIS

3.1. Overview

In addition to being physical agents of disease, certain aspergilli have the ability to produce toxic secondary metabolites that have been strongly implicated in animal and human ill health. Human mycotoxicoses have been mainly associated with ingestion, and certainly the aflatoxins have been especially associated with food contamination, though the respiratory route through the inhalation of spores, mycelial fragments, and other airborne toxigenic particles (particularly associated with agricultural materials) cannot be discounted (Lacey, 1989). High levels of aflatoxins (200,000 ppb) have been determined in both conidia and sclerotia of *A. flavus* and *A. parasiticus* (Wicklow and Shotwell, 1983), while conidia of *A. niger* and *A. fumigatus* have been shown to contain significant quantities of aurasperone C and fumigaclavine C, respectively (Palmgren and Lee, 1986). Since many toxigenic species have spores capable of reaching the alveoli of the lung and because of the solubility and low molecular weight of most mycotoxins, such toxins are likely to be absorbed via the respiratory epithelium and translocated to other sites (Flannigan *et al.*, 1991). Nonetheless, it is the gastrointestinal route of toxin absorption associated with contaminated foods and feeds that continues to attract the most attention (Dvorackova, 1990; Moss *et al.*, 1989; Yourtee *et al.*, 1989; Smith and Henderson, 1991). Table I gives an indication of the range of toxins potentially producible by food-borne aspergilli, though many substrates are poor for toxin production or may give rise to only small amounts or those that are chemically unstable. The chemical structures of 12 of the most important *Aspergillus* toxins are presented in Fig. 1.

The possibility of industrial strains of *Aspergillus* producing toxins constitutes a real hazard to biotechnological processes from the workers' health point of view as well as the purity of the product (and hence consumers' health). Soya products produced by *A. oryzae, A. sojae,* and *A. tamarii* must clearly be toxin-free, but other industrial products such as glucose (using *A. niger* glucoamylase saccharification), acidulants like citric acid (fermentation utilizing *A. niger*), malic acid (from *A. flavus* var. *columnaris*), itaconic acid (*A. terreus*), and gluconic acid (*A. niger*), as well as proteinase (*A. oryzae*) and lipase (various *Aspergillus* species), respectively, for Gouda and blue cheese manufacture are all food-oriented. *Aspergillus* is therefore a crucial genus for industrial biotechnology, particularly with respect to food biotechnology, and the fact that several spe-

Table I. Production of Mycotoxins by Food-Borne Aspergilli[a]

Species	Toxins
A. alutaceus (*ochraceus*)	Emodin, kojic acid, neoaspergillic acids, ochratoxins,[b] penicillic acid,[b] secalonic acid A,[c] viomellein,[b] xanthomegnin[b]
A. candidus	Candidulin, terphenyllin, xanthoascin[c]
A. clavatus	Ascladiol, clavatol, cytochalasin E,[c] kojic acid, kotanine, patulin,[c] trytiquivalins,[c] xanthocillins[d]
A. flavus	Aflatoxins,[b] aflatrem,[c] aflavinin, aspergillic acids, cyclopiazonic acid,[b] 3-nitropropionic acid, paspalinin
A. fumigatus	Fumigaclavines,[c] fumigallin, fumigatin, fumitoxins,[c] fumitremorgins,[c] gliotoxin,[c] kojic acid, spinulosin, tryptiquivalins, verruculogen[c]
A. niger	Malformins, naphthoquinones, nigragillin
A. oryzae	Aspergillomarasmin, 3-nitro-propionic acid, oryzacidin, maltoryzin
A. parasiticus	Aflatoxins,[b] aspergillic acids, kojic acid
A. sydowii	Nidulatoxin, sterigmatocystin,[c] griseofulvin
A. tamarii	Cyclopiazonic acid,[c] fumigaclavine A
A. terreus	Citreoviridin,[c] citrinin,[c] gliotoxin,[d] patulin,[c] terrein, terreic acid, terretonin, territrems,[c] cytochalasin E,[d] terredionol
A. ustus	Austamid,[c] austdiol,[c] austins,[c] austocystins,[c] kojic acid, sterigmatocystin,[d] xanthocillin X[d]
A. versicolor	Nidulotoxin, sterigmatocystin[b]
A. wentii	Emodin, kojic acid, 3-nitropropionic acid, wentilacton,[c] physicon

[a]Adapted from Frisvad (1988).
[b]Toxins of great significance in foods.
[c]Toxins of possible significance in foods, since they are produced in great quantities in pure culture.
[d]Toxins of questionable significance in foods.

cies are known toxin producers demands extreme vigilance from those in the various industries involved in their utilization. Raw material feedstocks such as soybeans and molasses should therefore be vigorously screened for traces of mycotoxin contamination. In particular, the yellow-green *Aspergillus* strains used in Asian soybean fermentations have long been subjects of controversy. *Aspergillus oryzae,* previously the most commonly utilized mold for this purpose, for example, does not produce aflatoxins, but is certainly capable of producing other toxic metabolites including aspergillic acid and so should not blithely be referred to as a "non-toxin-producer" (Flegel, 1988).

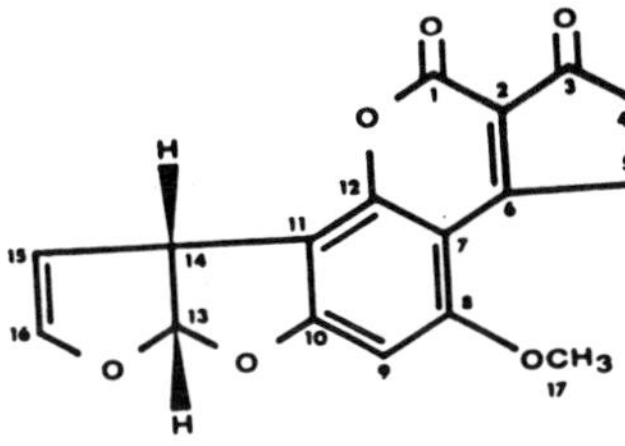

Aflatoxin B_1

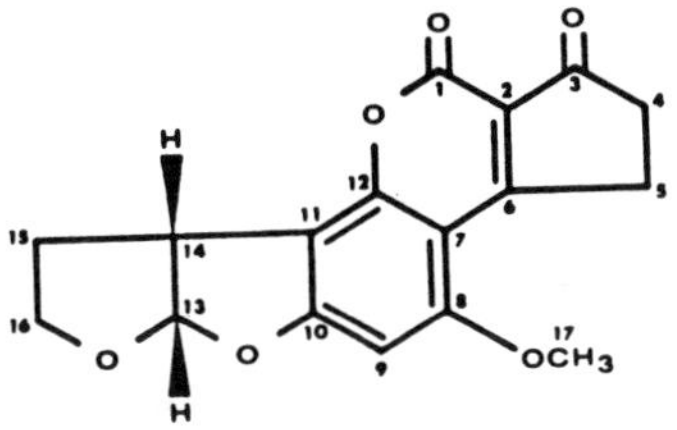

Aflatoxin B_2

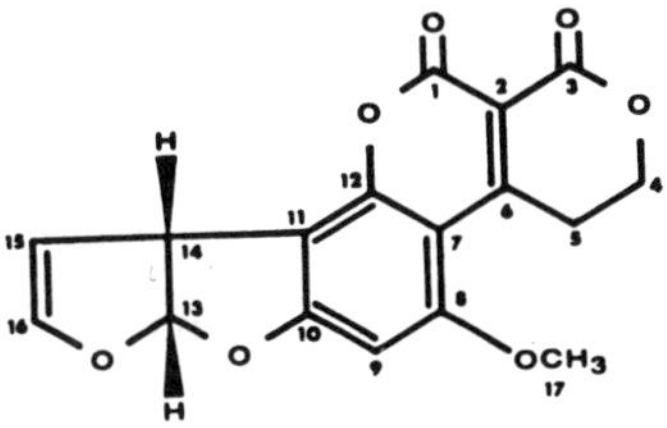

Aflatoxin G_1

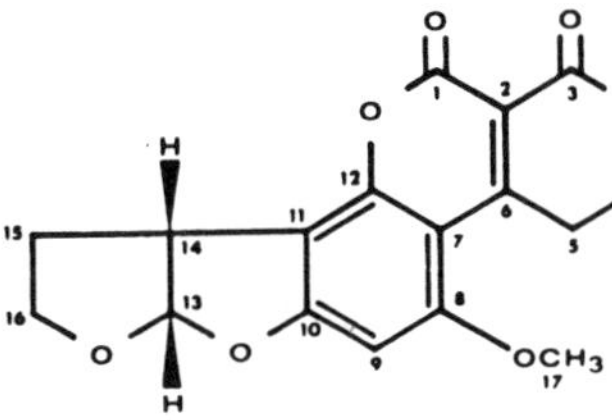

Aflatoxin G_2

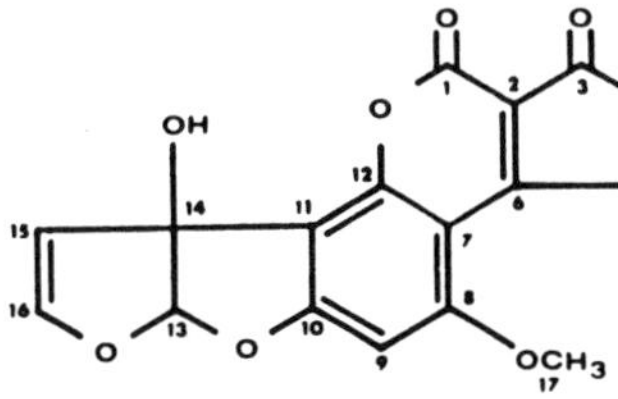

Aflatoxin M_1

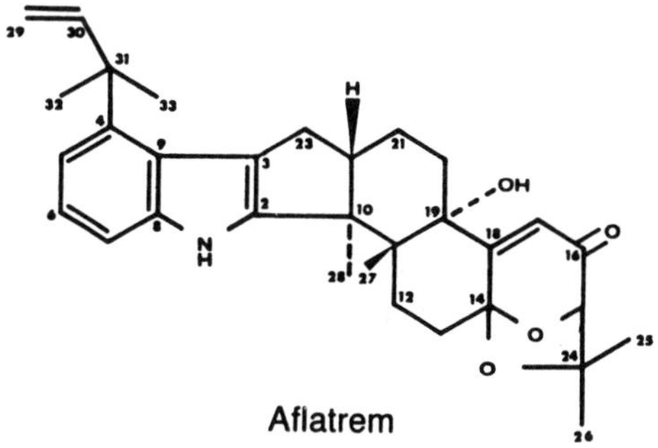

Aflatrem

Figure 1. Twelve important *Aspergillus* toxins. From CAST (1989).

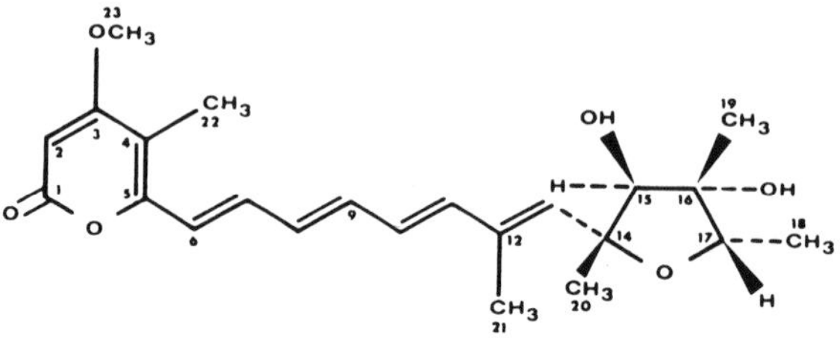

Citreoviridin

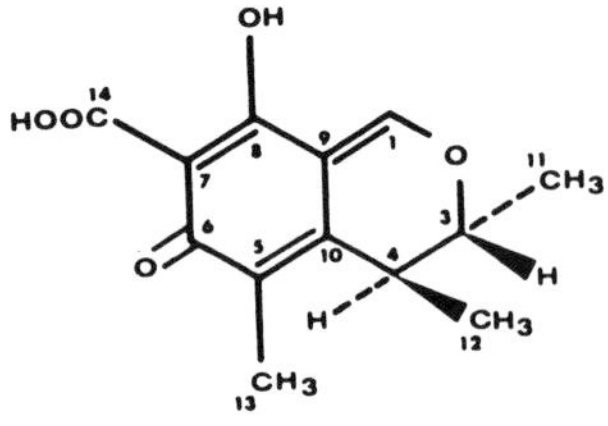

Citrinin

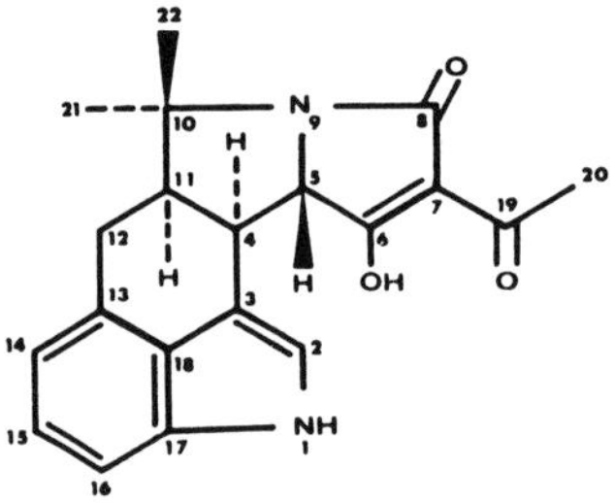

Cyclopiazonic Acid

Figure 1. (*Continued*)

3.2. Aflatoxins

3.2.1. Discovery

The aflatoxins are by far the most widely studied of the mycotoxins and have been since their discovery in turkey poults and ducklings in the early 1960s. Sargeant and colleagues (Sargeant *et al.*, 1961) identified *A. flavus* Link ex Fries as the toxin-producing agent, the toxin itself becoming known as aflatoxin.

During the subsequent few years, several papers appeared, particularly in the prestigious journal *Nature*, outlining improved detection methods, differentiating the aflatoxins into various related chemical

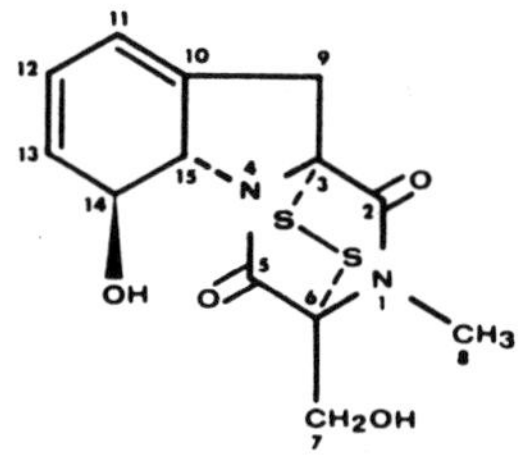

Gliotoxin

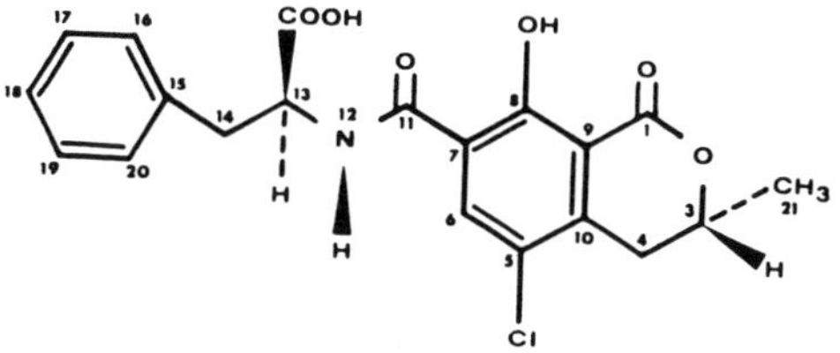

Ochratoxin A

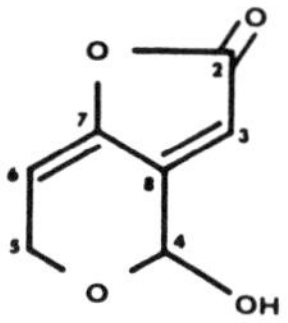

Patulin

Figure 1. (*Continued*)

structures, proposing metabolic pathways for their formation, and confirming their great toxicity. The possibility that aflatoxins are responsible for the high incidence of liver cancer in parts of Africa, particularly among the Bantu and other communities consuming groundnut products, was suggested (Anon, 1962), and the mechanism of toxicity by the compounds' interaction with DNA was proposed (Sporn *et al.*, 1966; Clifford *et al.*, 1967). Yet today in the 1990s, more than a generation after such studies, the evidence linking aflatoxicosis and human disease, particularly cancers, remains largely circumstantial, though considerable (Smith and Moss, 1985; Mackenzie, 1988; Bosch and Peers, 1991). The role of aflatoxins, indeed of mycotoxins as a whole, in human disease has yet to be fully defined.

The aflatoxins are highly substituted coumarins containing a double furan ring and are produced by *A. flavus* and *A. parasiticus*, with aflatoxin B_1 being hepatotoxic and one of the most potent animal carcinogens known. All animals tested have been susceptible, with the LD_{50} being 0.4 mg/kg in the duckling, the most sensitive, and 17 mg/kg in the female rat, in which the disease is characterized by periportal necrosis and bile duct proliferation. In human liver cell tissue culture, doses of 10 μg/ml are lethal (Conning, 1983). Such findings have stood the test of time until the present day.

3.2.2. Human Aflatoxicosis—Possible Mechanisms and Probable Evidence

The aflatoxins now number 15, and of those found naturally, while aflatoxin B_1 has the most potent biological effects, its metabolite, aflatoxin M_1, found in meat and milk, is also biologically very active. Aflatoxin has the ability to bind with both DNA and RNA and thereby inhibit macromolecular synthesis by interfering with transcription and other aspects of protein formation. Clinically, aflatoxicosis is manifested as acute hepatic damage and failure of liver function, while mutagenic, carcinogenic, and teratogenic effects are well documented in some animal species (Smith and Henderson, 1991). Aflatoxins also cause a number of changes in the immune system of most recipients, namely, impairment of protein formation and ability to cause involution or hypoplasia (underdevelopment) of the thymus (Pier, 1991).

It must be remembered, however, that the conclusions presented above were derived largely from nonhuman animal studies. Nevertheless, it is generally accepted that metabolism of aflatoxin B_1 consists of two stages. In the first, the toxin is converted by cytochrome P_{450}-mediated mixed-function oxidase (or monooxygenase) in the hepatic endoplasmic reticulum into active electrophilic intermediates that are then conjugated covalently with nucleophiles in the hepatocyte. On intoxication, the active intermediates attack cellular nucleophiles such as DNA, RNA, and protein. By contrast, detoxification occurs when nucleophiles in the cytosol, such as reduced glutathione and other SH-containing compounds, conjugate enzymatically with the activated intermediates, the complex being excreted as water-soluble mercapturic acids. Morphological evidence that the smooth-surfaced endoplasmic reticulum (SER) proliferates after the administration of aflatoxin B_1 correlates well with the molecular pharmacological evidence. It is postulated that the toxicity of aflatoxin B_1 is accounted for by the formation of a 2,3-epoxide intermediate, which seems to have been confirmed by experimental evidence. This has led to a further postulation that aflatoxin B_1 is metabolized by the hepatic microsomal mixed-function monooxygenase system (MFO) to electrophilic 2,3-oxide, the causative agent for the toxicity or carcinogenicity or both (Terao and Ohtsubo, 1991).

Since studies to identify human aflatoxin metabolites would provide essential information for understanding the clinical manifestations of these toxicants, greater efforts are now being made in this direction. Yourtee and Kirk-Yourtee (1989) mention that ten aflatoxins have been isolated from human tissues: four mold products and six human metabolites including an aflatoxin B_1-guanine adduct. Such metabolites have appeared in urine, blood, and tissues in the picogram to nanogram per milliliter range and are compiled in Table II. In addition to urine and blood, aflatoxins have been observed in liver, umbilical cord blood, and milk. Isolation of aflatoxin G_1 is important, since it occurs commonly in foods and feeds and verifies exposure of humans to the aflatoxins. Even though association between aflatoxins and human disease has been made in several studies, some controversy in the role of aflatoxins and human health remains. Latency is one problem, since the amount of aflatoxin ingested may not be known at the onset of disease, though aflatoxins are found in specimens from diseased individuals. Limitations of analytical sensitivity must also be considered, as must the possible causative or promotional factors of liver disease of synthetic and natural carcinogens and the hepatitis B virus.

More specifically, in a study conducted with pregnant African women (Maxwell *et al.*, 1989), aflatoxins were detected in 37 (37%) of Sudanese, 53 (28%) of Kenyan, and 163 (32%) of Ghanaian breast milk samples; detection was positive in 86 (31%) of Ghanian, 9 (12%) of Nigerian, and 37 (37%) of Kenyan cord blood samples. The latter provides firm, not just circumstantial, evidence that aflatoxins cross the human placental membrane and, with the concentrations found, that aflatoxins may well accumulate in the fetus when exposed to the toxins *in utero.* The consequences of prenatal exposure to aflatoxins are unkown in humans, though a decrease in the mean birth weight of females was recorded in Kenya and three unexplained stillbirths (two in Kenya and one in Nigeria) were recorded where aflatoxins were detected in maternal and cord blood at delivery. Furthermore, in 161 human urine samples collected in Lagos, Nigeria (a society with high liver cancer incidence), high-pressure liquid chromatography (HPLC) detected aflatoxins B_{2a} in 32.7% of the samples, B_1 in 3.1%, M_1 in 8.7%, G_1 in 9.9%, and L in 9.3% (Bean *et al.*, 1989). Aflatoxin G_1 was present in the highest mean concentration at 12 ng/100 ml urine. In a further study (Wilkinson *et al.*, 1989), an enzyme-linked immunosorbent assay (ELISA) was

Table II. Examples of Aflatoxin Metabolites Reported in Humans[a]

	Year reported	Method of analysis[b]	Body fluid	Metabolite	Concentration
1.	1970	TLC	Urine	M_1	>0.2 ng/ml
2.	1980	HPLC	Serum	B_1	2.4–4.7 ng/ml
			Urine	B_1	5–61 pg/ml
				M_1	50–170 pg/ml
3.	1981	HPLC	Serum	B_1	2–12 ng/ml
4.	1982	TLC	Urine	M_1	>0.03 ng/ml
5.	1983	TLC	Urine	M_1	avg. 0.62 ng/ml
6.	1984	HPLC	Milk	B_1	<8.218 ng/ml
7.	1984	HPLC, RIA	Serum	B_1	20–1168 pg/ml
8.	1985	HPLC	Urine	Total AFL	0.258–0.490 ng/ml
			Serum	Total AFL	0.077–0.154 ng/ml
9.	1985	HPLC	Liver	B_1	0.062–4.409 ng/g
				L	0.012–0.99 ng/g
10.	1985	RIA	Urine	B_1	0–0.67 ng/equiv./ml
11.	1985	RIA	Urine	M_1	0.1–10 ng/ml
				B_1-guanine	7–10 ng/ml
12.	1986	ELISA	Urine	B_1	<4.25 ng/ml
13.	1987	RIA	Urine	B_1	0.1–10 ng/ml
14.	1987	ELISA, TLC	Urine	M_1	0–3.2 ng/ml
15.	1987	RIA	Liver tissue	B_1	13.9–148.4 ng/ml
16.	1987	TLC	Urine	B_1	1 ng/ml
				B_2	1 ng/ml
				M_1	4.2 ng/ml
				G_1	9 ng/ml
				G_2	24 ng/ml
17.	1988	ELISA	Serum	B_1 equiv.	<20 pg/ml
18.	1988	HPLC	Cord blood	B_1	0.185–43.822 pg/ml
				B_2	0.011–0.925 pg/ml
				M_1	0.034–7.320 pg/ml
				G_1	0.354–1.354 pg/ml
				G_2	0.037 pg/ml
				L	0.0117 pg/ml
			Milk	M_1	0.002–1.816 pg/ml
				M_2	0.016–2.075 pg/ml
				B_2	0.049–0.050 pg/ml
				L	0.064–0.270 pg/ml
19.	1988	RIA	Serum albumin	B_1	50 mg/g albumin
20.	1988	RIA	Urine	B_1	0.2 ng–eqv./ml
				M_1	67 ng/ml
21.	1988	ELISA	Serum	B_1	20 pg–3.1 ng/ml
22.	1988	HPLC	Urine	B_2a	7 pg/ml
				M_1	89 pg/ml
				G_1	120 pg/ml
				B_1	70 pg/ml
				L	6 pg/ml

[a]Adapted from Yourtee and Kirk-Yourtee (1989).
[b](TLC) Thin-layer chromatography; (HPLC) high-pressure liquid chromatography; (RIA) radioimmunoassay; (ELISA) enzyme-linked immunosorbent assay.

adapted to rapidly analyze human serum for aflatoxin. While none was found in sera from United Kingdom subjects, 76% and 100%, respectively, of Nigerian and Nepalese serum samples were deemed positive. In the case of maternal and cord sera from Thai subjects, only 6% of maternal blood had detectable aflatoxin levels compared to 49% of cord sera, thus reinforcing the findings of Maxwell and colleagues (Maxwell *et al.*, 1989) above. Finally, utilizing a competitive ELISA technique, one or more substances recognized by a monoclonal antibody against aflatoxin B_1 were detected in 92% of 78 human urine samples collected in Denmark from individuals eating a normal Danish diet (Autrup, 1989). The concentration of the urinary aflatoxinlike substance was equivalent to 0–8 μg aflatoxin B_1/adult per day, with the amount varying with the type of food ingested 24–48 hr prior to urine collection. Increases were observed when the diet contained beer, dairy products, and meat. HPLC analysis of the antigenic material concentrated on an affinity column showed the presence of several antigenic compounds different from human aflatoxin metabolites. It was therefore suggested that the antigenic material was produced by enzymatic degradation of animal proteins from livestock fed aflatoxin-B_1-contaminated feed.

3.3. Other Mycotoxins

3.3.1. Ochratoxin A

Of the toxins produced by the aspergilli other than the aflatoxins, ochratoxin A (OTA) is probably the most important. Originally isolated from *A. ochraceus* (and *Penicillium viridicatum*) as a hepatotoxin, it is now considered to be more of a nephrotoxin, with *A. alutaceus* (*A. ochraceus*) as the major source from *Aspergillus* species. It occurs widely on wheat, apparently causes porcine nephrosis, and has been shown to induce tubular necrosis. There is no evidence to definitely associate this toxin with human disease, but it is probably a contributing factor in Balkan nephropathy (Conning, 1983). Animal studies have shown OTA to be a potent nephrotoxin, immune suppressant, teratogen, and carcinogen, while there have been several recent reports of a high incidence (up to 54%) of OTA in human blood specimens in some northern European countries. Contamination of raw agricultural products is region-dependent and shows far higher occurrence rates and levels than in human foods (Pohland *et al.*, 1992).

OTA is the most toxic of isocoumarin compounds in whole animals, embryos, and cultured cells. In animals, notably in the case of porcine nephropathy, OTA causes degeneration of the proximal tubules followed by interstitial fibrosis and hybridization of some glomeruli. It inhibits intramitochondrial phosphate transport (Berry, 1988) and protein synthesis (translation) through the blocking of phenylalanine + RNA synthetase. The relationship between this established biochemical evidence and other toxicological effects such as depression of renal function, tumorigenicity, and immunological response remains to be classified (Ueno, 1991). The major effects of ochratoxins on the immune response comprise diminutions of the two major circulating immunoglobulins, IgG and IgM, as well as significant reductions in antibody titer. A reduction in macrophage motility and in phagocytosis of foreign particles by granulocytic phagocytes is also commonly found (Pier, 1991).

Nephropathy occurs endemically in the Balkan states of Bulgaria, Romania, and Yugoslavia in 3–8% of the population, is more prevalent in females between 30 and 50 years old where it progresses to death, and occurs mainly in the rural areas where the food consumed is home grown. There is a great reduction in kidney size, tubular degeneration, interstitial fibrosis, and hyalinization of glomeruli, particularly those in the superficial part of the cortex. A significant association is evident between the number of deaths over a two-year period and any excess rainfall during the previous two harvests, when the levels of ochratoxin would have risen. Of blood samples taken in an endemic area of Yugoslavia, 6.5% contained OTA at 3–5 ng/g serum; thus, while not conclusive, the evidence points strongly to OTA being the cause of Balkan nephropathy. Furthermore, in the Bulgarian and Serbian endemic regions, the incidence of urinary tract tumors in patients with nephropathy can reach 40% (Berry, 1988).

3.3.2. Miscellaneous Toxins

As stated in Section 3.1 and listed in Table I (Frisvad, 1988), the aspergilli are capable of producing a wide range of toxins, though none as medically important as the aflatoxins and ochratoxin A. The sterigmatocystins and versicolorins, produced by *A. versicolor,* for example, have been found only rarely in foods and are therefore considered not to be significant health hazards, though the carcinogenicity of the former, at a potency two orders of magnitude less than that of aflatoxin B_1, requires vigilance (Pohland and Wood, 1987). Sterigmatocystin contamination from *Chaetomium* species was, in fact, responsible for the reassessment of the Canadian Waterloo single cell protein (SCP) production process. It is also associated with hepatic necrosis. Of the remainder, gliotoxin, a secondary metabolite of *A. fumigatus,* has been linked with lung disease and acquired immune deficiency syndrome (AIDS); austdiol, with gastro-intestinal disturbances; patulin, with hemorrhages in the lung and brain; citrinin, with renal damage, vasodilation, bronchial constriction, and increased muscular tone; and emodin and secalonic acid, with reduced cellular oxygen uptake. Citreoviridin is a neurotoxic mycotoxin that causes paralysis and respiratory arrest, and the neurotoxic tremorgens, such as verruculogen and fumitremorgins A and C, produce trembling in animals as a result of their neurotoxicity (Hsieh, 1987).

Gliotoxin is interesting since it is a potent inhibitor of phagocytosis by macrophages and has strong immunosuppressive effects. In fact, it has been hypothesized that gliotoxin production occurs *in situ* in lung disease caused by *A. fumigatus* and that similar fungal metabolites may be implicated in the etiology of AIDS (Canadian Public Health Association, 1987). For further information, see Section 4.1.

It should be remembered that with regard to human health, the associations enumerated above are almost certainly circumstantial and should be viewed as such. Also, Table I is by no means complete for mycotoxins produced by all the aspergilli. For example, *A. giganteus* produces patulin, *A. niveus* citrinin, *A. nidulans* sterigmatocystin, *A. niger* aurasperone C, *A. parasiticus* norsolorinic acid, and so on.

3.4. Mycotoxicosis and Animal Health

While this chapter concentrates on human health problems posed by the genus *Aspergillus,* much of the knowledge gained concerning modes of action, pathological conditions, degrees of toxicity, detection, analysis, and control of the various mycotoxins produced by the aspergilli has inevitably been derived from animal studies.

The aflatoxins are the most often reported group of mycotoxins to occur in raw agricultural commodities, much of which goes to provide animal feed, but other toxins that can emanate from various *Aspergillus* species can also be implicated in diseases incurred by a wide range of animals. Some examples are presented in Table III, though it must be remembered that mycotoxins such as ochratoxin A, citrinin, patulin, and penicillic acid are also produced by other genera, notably the penicillia.

However, Table III is by no means exhaustive. For example, tremorgenic syndrome in cattle can occur when corn silage from a trench-type site is improperly unloaded using a front-end loader. This aerates the silage, causing a fermentation involving toxigenic strains of *A. fumigatus* that are capable of producing alkaloids and tremorgens belonging to the verriculogen–fumitremorgin group. The disease is characterized

Table III. Effects of Some *Aspergillus* Mycotoxins in Animals, and Commodities in Which Contamination Has Occurred[a]

Mycotoxins	Effects in animals: Animals affected	Effects in animals: Pathological effects	Commodity found contaminated
Aflatoxins B_1, B_2, G_1, G_2, M_1, M_2	Birds Duckling, turkey poult, pheasant chick, mature chickens, quail Mammals Young pigs, pregnant sows, dogs, calves, mature cattle, sheep, cat, monkey, humans Fish Laboratory animals	Hepatotoxin Liver damage Hemorrhage Intestinal tract Kidneys Bile duct Hyperplasia Carcinogen Liver tumors	Peanuts, corn, wheat, rice, cottonseed, copra, nuts, various foods, milk, eggs, cheese
Sterigmatocystin	Mouse Rat	Carcinogen	Green coffee, wheat, Dutch cheeses
Ochratoxin A	Swine, dogs, ducklings, chickens, rats Humans(?)	Nephrotoxin Tubular necrosis of kidney Mild liver damage Enteritis Porcine nephropathy Teratogenic	Cereal grains Wheat, barley, oats, corn Dry beans
Citrinin	Swine, dogs, laboratory animals	Nephrotoxin Porcine nephropathy, acute kidney damage, swelling of kidney, tubular necrosis of kidney	Cereal grains Wheat, barley, corn, rice
Patulin	Birds Chickens, chicken embryo, quail Mammals Cat, cattle, mouse, rabbit, rat	Edema Brain Lungs Hemorrhage Lungs Capillary damage Liver Spleen Kidney	Moldy feed, rotted apples, apple juice, wheat straw residue

(continued)

Table III. (*Continued*)

Mycotoxins	Effects in animals: Animals affected	Effects in animals: Pathological effects	Commodity found contaminated
	Others Brine shrimp, guppies, zebra fish larvae	Paralysis of motor nerves Convulsions Carcinogen Antibiotic	
Penicillic acid	Mouse, rat, chicken embryo, quail, brine shrimp	Liver damage Fatty liver cell necrosis Kidney damage Digitalis-like action on heart Dilates blood vessels Antidiuretic Edema in rabbit skin Carcinogenic Antibiotic	Stored corn, grains, dried beans, moldy tobacco

[a]Adapted from Bullerman (1986).

by a general deterioration typical of protein deficiency and malnutrition, even though ample pasture and supplemental feed may be available. Other prominent clinical signs are diarrhea, irritability, and abnormal behavior.

Other examples include aflatoxin B_1's association with bovine abortions, while ochratoxin A and sterigmatocystin are reported to effect embryonic removal in various animal species. The biological effects of cyclopiazonic acid (CPA) from *A. flavus* in rats, dogs, pigs, and chickens are well known. Clinical signs of intoxication include anorexia, diarrhea, pyrexia, dehydration, weight loss, ataxia, immobility, and extensive spasm at the time of death. Finally, immune suppression by some aflatoxins and ochratoxin A is known to occur in domesticated and laboratory test animals. Mechanisms are analogous to those in humans as described elsewhere in this chapter, with the major effects on cellular responses being reduced phagocytosis by macrophages, reduced delayed cutaneous hypersensitivity, reduced lymphoblastogenesis (response to mitogens), and reduced graft vs. host response. The effects on humoral factors are principally reductions in immunoglobulin (IgG and IgA) concentrations in the serum, reduced complement activity, and reduced bactericidal activity of the serum (CAST, 1989).

The impact of mycotoxins on animals not only is a physical or health-related one, but also has economic consequences. Lower productivity, reduced weight gain, reduced feed efficiency, less meat and egg production, greater disease incidence owing to immune system depression, subtle damage to vital body organs, and interferences with reproduction have a far greater economic impact than that of immediate morbidity and lethality (CAST, 1989). Furthermore, it is in the less developed countries of Africa and Asia where the proportion of contaminated grain or feeds consumed is greatest that the economic effects are most seriously felt.

4. EPIDEMIOLOGY, HOST SUSCEPTIBILITY, AND HIGH-RISK GROUPS

4.1. Compromised Host

4.1.1. Aspergillosis

Secondary mycoses, normally effected by *A. fumigatus* in the case of invasive pulmonary aspergillosis, are often direct results of a previous disease condition, often in individuals who have severe defects in their own defense system or who are neutropenic (decreased neutrophil count in the blood). Neutropenia may occur in a wide variety of diseases including aplastic anemias, agranulocytosis, and acute leukemias. The lowering of the host defense mechanism puts at risk host groups such as people with diabetes mellitus, alcoholics, drug addicts, the aged, and chronically ill individuals. Clinical observations also suggest that chronic granulomatous disease as a primary defect of polymorphonuclear leukocytes is a main risk factor for disseminated aspergillosis—in other words, an inherited host factor leading to increased susceptibility (Seeliger and Tintelnot, 1988).

Aspergillus can sometimes cause primary cutaneous infection in immunosuppressed patients, with the infection often recurring at the site of the adhesive tape securing intravenous infusion sets in hospital patients. Invasive aspergillosis is seen most commonly, however, as a complication of organ transplantation or neoplasia or in the setting of chemotherapy-induced (see Section 4.2) neutropenia in patients with leukemia or bone marrow transplants (Cohen, 1991).

The link with AIDS—essentially a basic alteration of the T-cell-linked defense mechanism—has seemingly been resolved, though it remains something of a controversial topic (Seeliger and Tintelnot, 1988). Numerous viruses, the most notable of which is the human immunodeficiency virus (HIV), can subvert or impair one or more components of the immunological response, rendering the host less resistant to other forms of infection. The mechanisms remain ill-defined, but appear to result from alterations in macrophage or T-cell regulator or effector function. What can be said is that fungal infection is a major problem in HIV-positive individuals, in whom there is a gradual depletion of an important class of circulating lymphocytic cell, the CD_4-antigen-bearing helper T-cell, which secretes factors that regulate macrophage function and the proliferation and differentiation of effector T-cells in normal individuals. Loss of T-cell help eventually results in impaired immunoglobulin production. Other adverse effects on the immune system are also observed and predispose the affected individuals to a number of fungal infections including candidiasis, histoplasmosis, and cryptococcosis. However, neutrophil function is generally well preserved, probably accounting for the now recognized low incidence of aspergillosis in patients with AIDS. Among 130 AIDS patients autopsied in Frankfurt, however, 11 cases of aspergillosis were detected, so all is not cut and dried in this instance (Warnock, 1991). In alcoholics, poor nutrition, cirrhosis, and repeated aspiration of stomach contents are infection risks, and alcoholism is also associated with impairment of T-cell numbers and function.

4.1.2. Mycotoxicosis

Aflatoxins have oncogenic and immunosuppressive properties and therefore may induce infections by their admittance to human tissues. For example, users of heroin (Hendrickse *et al.*, 1989) and marijuana (Cohen, 1991), particularly users of intravenous heroin, are prone to infection, because of the linkage with hepatitis B (HPB) and HIV viruses, which appear to be peculiarly aggressive in these drug abusers. Both marijuana and heroin are plant products susceptible to aflatoxin contamination. In effect, therefore, aflatoxins are responsible for producing compromised hosts in these individuals.

Of 13 samples of "street" heroin analyzed for aflatoxins in Merseyside, England, 4 were positive. Subsequently, 20% of 133 heroin addicts in England and Holland tested positive for aflatoxins B_1, M_1, B_2, M_2, and aflatoxicol in concentrations ranging from 25 to 9541 pg/ml. The results reveal a previously unsuspected group who are regularly or intermittently exposed to aflatoxins injected into the systemic circulation. Thus, after absorption from the gut, some or all of the aflatoxin is detoxified by the liver. Intravenous-heroin users thus risk direct systemic exposure to aflatoxin B_1, one consequence of which could be suppression of cell-mediated immunity. It is also postulated that systemic toxicity is enhanced by aflatoxin injection that bypasses the liver and that aflatoxins may contribute to pathology in heroin addicts and babies born to addicted mothers (Hendrickse *et al.,* 1989).

4.2. Drug Treatment

Treatment of lymphopathies, leukemia, Hodgkin's disease, and many other malignancies necessitates the prolonged administration of corticosteroids, antimetabolic drugs, and antibiotics, thus increasing the risk of development of visceral mycoses, particularly by *Aspergillus* species (Seeliger and Tintelnot, 1988). In fact, it is thought that *Aspergillus* infection may become more common as more AIDS patients are rendered neutropenic after treatment with drugs such as zidovudine (Warnock, 1991). Corticosteroid therapy is often a cause of impaired macrophage function, and since the pulmonary alveolar macrophages (which ingest inhaled spores and inhibit germination) along with the hyphae-destroying neutrophils are the two main components of the host defenses responsible for containing *Aspergillus* infection (Cohen, 1991), their suppression is often the cause of pulmonary and systemic aspergillosis as common complications in immunosuppressed organ transplant recipients.

4.3. Hospital Infection Complications

Paradoxically, such is the complex epidemiology of aspergillosis that, for example, agglomerations of conidia originating from potted plants at the bedside of susceptible hospital patients exposed to their inhalation may constitute a danger, as might the release of large numbers of conidia through hospital reconstruction work. There are many accounts of exogenous infection as a consequence of environmental infection with *Aspergillus* spores, often in large numbers (Cohen, 1991). Spore contamination of an operating theater led to seven cases of prosthetic valve endocarditis, and four cases of infection in patients undergoing open heart surgery have been attributed to aspergilli found in pigeon excretion on windowsills and moss on the hospital roof, but construction work seems to be the most frequent cause of such incidents of aspergillosis (Cohen, 1991).

Several reports document outbreaks of invasive aspergillosis in high-risk groups, such as patients with hematological malignancy or transplant recipients, exposed to high levels of spores generated in the environment by construction work. One account monitored air spore counts before and after the demolition of one of the hospital buildings; a fourfold order of magnitude increase was observed in *Aspergillus* conidia alone (Cohen, 1991).

The ubiquitous saprophytic existence of molds, including the aspergilli, makes their existence on potted plants in close proximity to severely ill patients a danger, though to establish a definite causal link between masses of *Aspergillus* conidia in the soil around such plants and infection in the victims is difficult. Strong evidence relates to a patient with chronic pulmonary tuberculosis developing aspergillosis (at home) due to *A. niger* spores produced over many years by growth of the fungus on violets, while from 1978 to 1982, 33 patients with aspergillosis of the lungs or respiratory tract provided an epidemiological link to potted plants and moldy soil. There were three instances of fatal invasive aspergillosis among five liver transplant patients treated in the same hospital room, probably linked to potted plants, while a female with kidney failure and pneumonia was found to produce both *A. flavus* and *A. fumigatus* in biological specimens provided by her, these species being identical to the species isolated from a potted plant at her bedside (Seeliger and Tintelnot, 1988).

The failure of ventilation or defective mechanical ventilation and air filtration systems has been the cause of outbreaks of aspergillosis in hospital patients in some United States cities. Invasive pulmonary aspergillosis following renal transplantation occurred in four cases because of a contaminated air intake system. Much of the evidence is circumstantial, and sometimes there may be more than one contributory factor. For instance, in a bone marrow transplant unit between April 1982 and March 1983, there were 10 of 26 (38%) cases of invasive aspergillosis. This compared with only 3 of 46 (6%) compromised patients developing aspergillosis during a period roughly five years previously. The 1982 outbreak was explained by the low efficiency of the airflow filters and by excessive duct movement, which, in turn, was connected with an inspection at the hospital site and new construction on the hospital grounds west of the hospital, where the prevailing winds were generally from the southwest.

It is impossible to cite all incidents of aspergillosis in hospitals, but another common cause is exposure of patients to infected pigeon droppings. *Aspergillus fumigatus* and *A. nidulans* have both been implicated in this case. In fact, bird droppings, contaminated foodstuffs, and potted plants seem to be the main focus of *Aspergillus* epidemiology in European hospitals, while construction work is the main focus in the United States (Seeliger and Tintelnot, 1988).

4.4. Dietary and Geographic Factors

Diet and the geographic region inhabited tend to be strongly correlated, and so here also mycotoxicosis is a main area of concern. Nevertheless, the frequency with which the various *Aspergillus* species cause aspergillosis in its different forms may vary with the fungal distribution pattern. Paranasal granuloma caused by *A. flavus* is seen frequently, for example, in the Sudan, while opportunistic iatrogenic infections produced by *A. fumigatus* occur mostly in developed countries where advanced medical procedures involving immunosuppressive drugs and complex surgery are more commonly practiced (Evans and Gentles, 1985). *Aspergillus* conidia may well play a role in outdoor winter allergy, particularly in alpine environments (Ebner *et al.*, 1992).

Sections 3.2.2 and 3.3.1 dealt in part with the probable occurrence of, respectively, aflatoxin and ochratoxin poisoning in various parts of the world. Suspected incidence of aflatoxicosis in particular has been extensively documented. For instance, a survey in Thailand revealed that in an area with a high incidence of liver cancer (6 new cases/100,000 population per year), the average consumption of aflatoxin was 9–10 times greater than in an area with a low cancer incidence. A correlation was also found between a high incidence of hepatoma (15 cases/100,000 per year) and aflatoxin consumption in a district of Uganda where almost half the food samples contained over 1 ppm. Table IV shows the

Table IV. Summary of Available Data on Liver Cancer Frequency and Aflatoxin Ingestion Levels[a]

Country	Area	Estimated average aflatoxin consumption (ng/kg body wt. per day)	Incidence of registered liver cancer (cases/100,000 population per yr)
Kenya	Murang (high altitude)	3.5	1.2
Thailand	Songkhla	5.0	2.0
Swaziland	High veldt	5.1	2.2
Kenya	Murang (middle altitude)	5.9	2.5
Swaziland	Middle veldt	8.9	3.8
Kenya	Murang (low altitude)	10.0	4.0
Swaziland	Lebombo	15.4	4.3
Thailand	Rarburi	45.0	6.0
Swaziland	Low veldt	43.1	9.2
Mozambique	Imhambone	222.1	13.0
Uganda	Karamajo	400.0	15.0

[a]Adapted from Grasso (1983) and Dvorackova (1990).

results of various individual studies that demonstrate a dose–response relationship. The Kenyan survey also demonstrated that in humans as well as laboratory animals, the male appears to be more susceptible to the carcinogenic effects of the toxin than does the female.

Significant geographic differences in the incidence of primary liver cancer have been demonstrated in China, with such areas where exposure to aflatoxin is highest having the greatest incidence of the cancer (Dvorackova, 1990). However, there are major difficulties in actually proving conclusively a link between hepatocarcinoma and aflatoxin ingestion. First, as Bosch and Peers (1991) point out, exposure to aflatoxins in Africa and parts of Asia and Latin America might begin very early in life (perhaps in the uterus) and occur episodically thereafter. The number of episodes and the degree of exposure vary greatly by country and region, by agricultural and crop storage practices, by season, and by other factors difficult to control in any questionnaire-based study. Second, long-lasting biological markers for aflatoxins are still not available. Third, there is a high geographic correlation between exposure to aflatoxins and to hepatitis B virus.

Certainly, however, correlation studies at the population level have demonstrated that average aflatoxin contamination of diets and foodstuffs is high in areas where primary liver cancer is frequent. Table V compiles the proportion of aflatoxin-contaminated foodstuffs in various surveys and demonstrates the considerable sample variation. In some studies, attempts have been made to adjust the estimates of the aflatoxin–liver cancer association for the prevalence rates of hepatitis B virus infection in the same population in regions of both Africa and China. Conclusions tend to indicate that the incidence or mortality of primary liver cancer correlates more closely to estimates of aflatoxin

Table V. Percentage of Aflatoxin-Contaminated Foodstuffs in Various Surveys[a]

		Food					
Country	Year(s)	Cassava	Peanuts	Corn	Rice	Alcohol	Beans
Philippines	1982	100	99	84	38	46	56
Thailand	1967, 1975, 1980, 1982, 1987	—	28–82	33–80	2–6	—	20–40
Swaziland	1987	—	21	3.5	—	6.5	—
Gambia	1988	—	100	—	—	—	—
Hong Kong	1972	—	0	25	0	—	13
Guatemala	1980	—	—	21.7	14.3	—	77.8

[a]Adapted from Bosch and Peers (1991).

exposure than to the prevalence of markers of exposure to hepatitis B virus, notably the hepatitis B surface antigen (HBs Ag). On a global level, however, prevalence of HBs Ag is strongly correlated with primary liver cancer (Bosch and Peers, 1991).

Furthermore, although there is a clear epidemiological association between the level of staple-food contamination and the incidence of primary liver cancer, this association does not invariably hold, and where ascertainment is better—for example, in the United States—the links are less clear. The Food and Drug Administration has estimated a daily food intake of 2.73 ng/kg body weight in the United States, while in some developing countries, a range of 3.5 to several hundred ng/kg body weight has been estimated, as seen above. However, such "global" data are not particularly useful, especially when it is considered that 500 μg/kg or more has been found in the liver and other tissues of individuals in Europe and North America (Berry, 1988).

Although linkage between aflatoxins and hepatomas has captured most of the headlines, these toxins also give rise to acute poisoning when high intakes occur. Hepatoxicity and perhaps cardiotoxicity are the results, and as much as 2–6 mg/day intake for a month has been recorded. The first major epidemic of aflatoxicosis in man occurred simultaneously in 1974 in approximately 150 villages in the Indian states of Gujarat and Rajasthan through the ingestion of infected locally grown maize. Levels up to 15 mg/kg maize were consumed, and in the whole outbreak, some 1000 cases and 300 deaths occurred (Krishnamachuri *et al.*, 1975; Berry, 1988).

In 1982, an outbreak of acute hepatitis in 20 patients thought to be caused by aflatoxin occurred in Kenya. There were 12 deaths in patients, and aflatoxin was found in 2 liver samples at autopsy, at levels of 39 and 88 ppb. Two families, from which 8 people died, were eating maize containing as much as 12,000 ppb of aflatoxin B_1 probably caused by storage from a poor harvest into a wet period that undoubtedly contributed to its deterioration.

Aflatoxin has also been associated with Reye's syndrome, a virus-associated biphasic disease causing acute encephalopathy with fatty degeneration of the viscera, almost exclusively in children. Studies in Czechoslovakia (Dvorackova, 1990), for example, showed the presence of aflatoxin in the livers of 60% of Reye's syndrome cases and in the food consumed by the patients shortly before the onset of the disease. However, in some observers' view, the association between aflatoxin and Reye's syndrome is not tenable, and neither is that suggested between aflatoxin and juvenile cirrhosis found in India (Berry, 1988).

Toxigenic aspergilli have been isolated from outdoor and indoor air in several studies. During 1971 and 1972, 72% of 397 isolates of *A. flavus* from outdoor air and the indoor air of a poultry shed in Mysore, India, were toxigenic, with the incidence of toxigenic strains being greater in winter (Rati and Ramalingam, 1979). In Egypt, 12 isolates of *A. flavus,* of which 2 produced detectable aflatoxins, were obtained from 20 air-dust samples (Abdel-Hafez *et al.,* 1986); in Saudi Arabia, 20 dust samples yielded 31 fungal genera and 70 species, with *A. niger, A. flavus,* and *A. flavus* var. *columnaris* being among the most common. Both aflatoxins B_1 and B_2 were identified as being produced from the aspergilli (Abdel-Hafez and Shoreit, 1985).

4.5. Occupational Disease

4.5.1. Aspergillosis

According to Samson (1985), health hazards involving molds are becoming more frequent both in the working environment and in the environment in general. Most cases involve exposure to allergens and *Aspergillus* species. Exposure to large amounts of conidia may either be due to the nature of the occupation itself or be incidental, as when Samson (1985) found *A. niger, A. flavus,* and *A. ochraceus* in ornamental-plant containers in a bank office. However, agriculture seems to be the main industrial source of potentially harmful fungi, and for obvious reasons, though this is by no means the only one. Occupations or workplaces, or both, that may involve exposure to fungi include grain harvesting, storage, and processing; sawmills and wood pulp mills; mushroom cultivation; waste treatment; and even office workers and library and museum employees. In fact, almost any occupation that brings workers into close association with biodeterioration processes presents the possibility of exposure to fungi and their spores. Over 5 million United States grain handlers and lumber and wood workers, plus farm workers, are thought to be at risk of developing occupational asthma and rhinitis. Similarly, workers at risk of various hypersensitivity pneumonitides involving fungi include approximately 3 million United States farm workers, mushroom workers, and malt workers combined (Sorenson, 1990).

Although there is no exposure to true or opportunistically pathogenic fungi, classic examples of occupational or avocational mycoses include aspergillosis among wig makers and pigeon fanciers (Seeliger and Tintelnot, 1988). As mentioned previously, maltster's lung occurs in workers who handle barley on which *A. clavatus* has sporulated during the malting process. A Scottish survey showed that around 5% of distillery malt workers had symptoms of allergic alveolitis (Evans and Gentles, 1985). More than 200,000 spores of *A. clavatus*/m^3 in a malthouse have been recorded after a grain handler had developed an alveolitis, with the strain isolated showing strong antigenic properties (Nolard *et al.*, 1988). Citric acid production via the mass cultivation of *A. niger* has led to cases of sensitization to the organism leading to either extrinsic allergic alveolitis or asthma. Such fermentation workers can suffer from tracheobronchitis, chronic forms of bronchopneumonia, and external otitis. Subsequent air sampling and cultivation revealed up to 1000 colony-forming units of *A. niger* in 100 liters of air (Seeliger and Tintelnot, 1988), and skin tests proved sensitivity against *A. niger* among the exposed.

Aspergillus fumigatus has been implicated in alveolitis in patients with farmer's lung and mushroom worker's lung. The two diseases are related and believed to be caused by the repeated inhalation of large concentrations of thermophilic actinomycetes. In farmer's lung, allergy occurs to fungal spores grown in inadequately dried stored hay. An acute reversible form can develop in a few hours after exposure; a chronic form, with the gradual development of irreversible breathlessness, occurs with or without preceding acute attacks. Mushroom worker's lung (*Agaricus bisporus*) is a type III IgG-mediated pneumonitis or extrinsic allergic alveolitis in which the actinomycetes grow on phase I compost and are subsequently inhaled. Symptoms are fever, dyspnea, cough, malaise, increase in leukocyte numbers, and restrictive changes of lung function, with possible permanent damage caused by fibrosis in the lung. Spores of the oyster mushroom, *Pleurotus* species, and the shitake mushroom, *Lentinus edodes,* are also reported to be causal agents (van Griensven *et al.*, 1991). There is also now recognized a new syndrome, organic dust toxic syndrome (ODTS), possibly associated with the presence of fungi, frequently misdiagnosed as farmer's lung, and a common and substantial respiratory hazard to young farm workers. ODTS has been reported to be common on dairy farms in New York State when the dry uppermost silage in concrete-stone silos is removed after several weeks to months of storage. Total dust levels of more than 100 mg/m^3 and respirable dust levels of more than 20 mg/m^3 were reported in some silos, with numbers of microorganisms in excess of 109/g dust. Fever, myalgia, chest tightness, cough, and headache were the major symptoms 5.3 hr on average after inhalation of organic dust and mold by 24 men and 1 woman with a mean age of 29 years (May *et al.*, 1986).

Aspergillus fumigatus is often found in decaying organic material and has an unusual ability to adapt to increased temperatures, thus rendering itself a potential human pathogen in composting processes. Hypersensitivity pneumonitis due to *A. fumigatus* in compost has been reported by Vincken and Roels (1984), while Anderson and Smith (1987) cite an epidemiological survey of sewage sludge composting workers in the United States in which tests indicated an increased incidence of skin, nose, and ear conditions relative to unexposed control groups, with *A. fumigatus* frequently detected in nose and throat swabs of the exposed individuals.

4.5.2. Mycotoxicosis

Among workers in the Swedish timber industry, acute alveolitis, known as "woodtrimmer's disease," has been recognized for some time. One of the signs in the acute phase of the disease is transitory tremor assumed to be associated with wood dust containing fungal spores, dominant of which at sawmills are those of *A. fumigatus*. Also, the importance of *A. fumigatus* on the surface of kiln-dried wood is accentuated by its ability to produce tremorgenic mycotoxins. Extracts given orally to rats produced varying degrees of tremorgenic reactions, with HPLC analysis revealing the tremorgenic mycotoxins verruculogen and fumitremorgin C, thus implying that "woodtrimmer's disease" is at least in part a mycotoxicosis (Land *et al.*, 1987).

Most occupational diseases linked with aflatoxins occur in the agricultural and food processing industries. One good example was provided in a Dutch peanut oil crushing plant where the average estimated respiratory exposure of the 55 exposed workers was associated with a relative risk of 3.2 times the chance of developing cancer over the normal population rate, and indeed two fatal cases of liver disease were subsequently reported, though the exposure was much less than, for example, in cases of oral intake in certain parts of the Third World. The mortality occurring between 1963 and 1980 in the group that had been exposed to aflatoxin, primarily via the respiratory route, from 1961 to 1969 (including 16 cancer deaths) was higher than for a group of nonexposed workers (7 cancer deaths) for both total cancer and respiratory cancer. The greatest difference between observed and expected mortality was in the period between 1963 and 1968, that is, during the period of exposure (Hayes *et al.*, 1984).

The significance of the possible toxigenic air spores plus associated dust in cereal-crop and peanut production and other farming operations is now recognized, particularly, but not exclusively, in the southeastern states of the United States, where climatic conditions are favorable for fungal growth. Grain elevators are especially frequent sources of quite high levels of toxigenic fungi, including aspergilli, and there have been many studies carried out on both the qualitative and the quantitative nature of toxins present in fungal extracts from agricultural sources (Sorenson *et al.*, 1984; Zennie, 1984; Burg and Shotwell, 1984). Burg and co-workers (Burg *et al.*, 1982) demonstrated that corn dust collected in the field during the Georgian corn harvest was contaminated with 12.7–142.8 ppb aflatoxin B_1; the same authors (Burg *et al.*, 1981) had previously measured an average of 138 ppb in corn dust and also 191.4 ppb at a grain elevator. Their results indicated that the dust generated when handling contaminated commodities also may be contaminated and represent a potential inhalation hazard. This fact, together with the extreme toxicity and carcinogenicity of aflatoxins already demonstrated in animal studies, suggests that appropriate measures should be taken to prevent worker exposure during the handling of contaminated materials. Of 31 *A. flavus* isolates collected by air samples from corn production sites in Missouri in 1977, 20 (65%) produced aflatoxin (Holtmeyer and Wallin, 1980), and Zennie (1984) found up to 3.5 ppb of alfatoxin B_1 at a grain elevator in Illinois, showing that the problem is most certainly not confined to the southeastern states.

Rather, it is potentially a worldwide problem, as exemplified by the discovery of aflatoxins B_1, B_2, G_1, and G_2 in the airborne dust during rice and wheat harvesting operations in the Punjab (Kiran *et al.*, 1985) and a toxigenic *A. flavus* strain in a rice mill in Ahmedabad district (Desai and Ghosh, 1989), both in India. Sorenson and colleagues (Sorenson *et al.*, 1984) found contaminated peanuts giving airborne dust aflatoxin B_1 levels up to 612 ppb; if one assumes a breathing rate of 3 m^3/hr and an airborne aflatoxin concentration of 1 ng/m^3 (100 ppb at a dust concentration of 50 mg/m^3), this level would mean that a worker inhales 15 ng/hr and therefore 120 ng during an 8-hr work shift (Sorenson, 1990). A Danish study (Olsen *et al.*, 1988) involving livestock feed processing workers in a plant where the average concentration of organic dust in the environment was approximately 100 mg/m^3, imported crops for feed had an average aflatoxin level of 140 ppb, and the estimated daily pulmonary exposure was about 170 ng showed elevated risks for liver cancer and cancer of the biliary tract that increased by 2- to 3-fold following a 10-year latency period.

It is possible that aflatoxins may be implicated in tumors other than those of the liver. For instance, two British scientists developed adenocarcinomas of the colon in the early 1970s after exposure to purified aflatoxins over two- and one-year periods, respectively, while two chemical engineers who had worked with *A. flavus*-contaminated peanut meal were reported to have died from alveolar carcinoma (Dvorackova, 1990).

Ochratoxin production by *A. ochraceus* is occasionally, but rarely, found in koji cultivation as a result of contamination in the southern part of Japan (Yokotsuka, 1985), and soybeans have also been demonstrated to be contaminated with aflatoxins on occasion, such that the aflatoxins would not be destroyed during the fermentation process for soy sauce manufacture (Jarvis, 1976).

4.6. Indoor Environment

While *Aspergillus* is considered to be one of the main four fungal genera causing type I allergy (immediate reaction), these genera being present in large numbers in both outdoor and indoor air (Gravesen, 1985), the genus is generally regarded as being primarily an "indoor" mold. In various surveys carried out in Europe and North America, a combined percentage of *Aspergillus* and *Penicillium* in the outdoor air fungal spora on an annual basis is 1–6% (Nolard *et al.*, 1988). Gravesen (1985), on the other hand, found aspergilli in 15% of indoor air samples and 48% of house dust. Hunter and co-workers (Hunter *et al.*, 1988) found *Aspergillus* in 74.5% of 47 Scottish homes, it being the fourth (at 6.7%) most abundant fungus in the air spora of these houses. An earlier Danish study (Gravesen, 1972) also placed *Aspergillus* fourth in the airborne spora of 44 Danish houses at 11%, making *Aspergillus* an excellent candidate for causing allergic responses among the occupants.

Air contamination of private houses by spores of aspergilli and other fungi plays an important role in both acute and chronic respiratory disease morbidity. Sensitivity to *Aspergillus* is probably the mold allergy that produces the most serious effects, particularly bronchopulmonary aspergillosis (Burr *et al.*, 1988). Cellulose-containing paper and paintings have been known to be contaminated by the *A. glaucus* group and *A. versicolor* in very humid houses, while *A. fumigatus* often contaminates wicker or straw material like chairs or baskets. Dust—for example, in mattresses—is often replete with spores, particularly of the *A. glaucus* and *A. restrictus* groups and also xerophilic aspergilli like *A. penicilloides*. Section 4.3 emphasized the importance of potted plants as sources of aspergilli, while kitchens and cellars are also very fertile areas for various species, some of which have been known to induce allergic asthma (Nolard *et al.*, 1988). Hunter and colleagues (Hunter *et al.*, 1988) indicated that the air spora may be affected by type of carpeting, the keeping of pets, and the raising of dust during vacuum cleaning, and that the number of viable fungal colony-forming units, including yeasts, could be as high as 449,800/m^3. Xerophilic fungi are always seen in association with house-dust mite communities and likely form a symbiotic relationship, the mites themselves being allergenic organisms.

Evidence is now gradually accumulating that the presence of mold growth on the internal surfaces of damp dwellings often results in a high airborne fungal spore count, which, in turn, presents a distinct health risk to the occupants (Institute of Environmental Health Officers, 1985; Canadian Public Health Association, 1987; Malrtin *et al.*, 1987). Although these adverse health conditions tend to be principally of a respiratory or allergenic nature or both in atopic individuals, mainly, but not exclusively, children (Burr *et al.*, 1988; Strachan, 1988), a prevalence of other symptoms including nausea and vomiting, blocked nose, breathlessness, backache, fainting, and nervous complaints was reported by more adults, and respiratory symptoms (wheeze, sore throat, running nose), headaches, and fever by more children living in damp, moldy dwellings compared with those inhabiting dry dwellings (Platt *et al.*, 1989; Morris *et al.*, 1989). Table VI shows the relationship between surface mold growth and children's symptoms found in a study of over 500 council houses, mostly in Edinburgh and Glasgow. It must be remembered, however, that the contribution of aspergilli to these symptoms may be far less than, for example, that of the penicillia. *Aspergillus*

Table VI. Surface Mold Severity and Children's Symptoms in Damp Scottish Dwellings[a]

	Visible mold				
Symptoms	None $n = 317$ (%)	Slight $n = 100$ (%)	Moderate $n = 96$ (%)	Severe $n = 76$ (%)	p
Aches and pains	16.1	17.0	13.5	15.8	
Diarrhea	19.9	18.0	16.7	21.1	
Wheeze	17.4	24.0	34.4	21.2	<0.01
Vomiting	14.8	17.0	21.9	17.1	
Sore throat	28.1	42.0	43.8	40.8	<0.001
Irritability	15.8	18.0	20.8	23.7	<0.05
Tiredness	16.4	16.0	21.9	14.5	
Headaches	12.9	19.0	17.7	27.6	<0.02
Fever	14.2	22.0	28.1	21.1	<0.02
Unhappiness	13.9	12.0	16.7	17.1	
Temper tantrums	23.0	22.0	31.3	28.9	
Bedwetting	22.1	29.0	17.7	22.4	
Poor appetite	21.1	24.0	31.3	18.4	
Persistent cough	33.0	44.0	50.0	30.3	
Runny nose	40.1	53.0	53.1	43.4	<0.05
Any symptoms	82.6	93.0	89.6	88.2	<0.02

[a] Adapted from Morris *et al.* (1989).

spores were found in the air of 21% of the 24 dwellings concentrated on in Glasgow and Edinburgh compared to 100% frequency in the case of the penicillia and 96% in that of the cladosporia (Lewis *et al.*, 1989).

It is now known that both atopic and nonatopic individuals may develop extrinsic allergic alveolitis as a result of response to high concentrations of spores of allergenic fungi within the home and that—as well as the risk of this disease, asthma, and bronchitis—respiratory mycotoxicosis is also a potential health hazard (Flannigan *et al.*, 1991). As has been noted, however, although there have been many instances of associations between mycotoxins and human disease, both inside and outside the home environment, the evidence is circumstantial rather than direct, and while it is tempting to assume that mycotoxins are a causative factor, this temptation must, for the moment, be resisted (Samson, 1985). Yet it is clear that the risks cannot be ignored, particularly in association with aflatoxin food contamination (Berry, 1988). It is also conceivable that toxins within the spore, in the presence of specific allergens, could play a role in sensitization to the allergen (Sorenson, 1990). The aspergilli would almost certainly have a part to play here, since one of the rooms of a Scottish dwelling sampled (Lewis *et al.*, 1989) gave a total airborne spore count of over 21,000/m^3, while a count, including yeasts, of 449,800/m^3 (Hunter *et al.*, 1988) has already been noted.

Sections 2 and 4 provided instances of the indoor environment of hospitals giving rise to respiratory or invasive *Aspergillus* infections acquired from exogenic sources, particularly in immunocompromised patients. In such patients, the risk of contracting pulmonary aspergillosis is related to the degree and duration of neutropenia. High-efficiency particulate air filters have been advocated as a preventive measure, but if these or other suitable safeguards cannot be implemented, then air sampling should be undertaken. Even so, there are no accepted standards, and *A. fumigatus* spores are found in normal air at concentrations of around 1 colony-forming unit/m^3 (Humphreys, 1992).

5. OTHER MEDICAL ASPECTS

5.1. Sick Building Syndrome

Over a decade or so, increasing attention has been paid to the role of microorganisms, including molds such as the aspergilli, in what has become known as "sick building syndrome" (SBS) or "building-related illness" (BRI). In general, the condition is defined as the excessive reporting by a building's occupants of one or more of the symptoms of headaches, burning eyes, fatigue, dizziness, flulike symptoms, and upper respiratory complaints (White, 1990). It is often associated with air-conditioned buildings, and increases in the number of complaints have coincided with the energy-efficient procedures of reduced ventilation and the use of new building materials.

In a clinical study of 14 town halls employing 4369 people in Greater Copenhagen, Skov and colleagues (Skov *et al.*, 1987) showed that this population of office employees had a high prevalence of work-related mucosal irritation (28%) and of general symptoms such as headache and abnormal fatigue (36%). Women had a higher prevalence of these symptoms than men and complained more often about the quality of the indoor environment. Airborne microfungi as well as microfungi in dust were thought to make a contribution to these health effects. In another study (Austwick *et al.*, 1989) of the aerobiology of 15 buildings affected with SBS, while there was little quantitative or qualitative difference between the airborne bacterial levels in affected and nonaffected buildings, a marked difference did appear in the recovered mycoflora. Naturally ventilated (unaffected) buildings had up to an order of magnitude higher levels of spores per cubic meter of air. However, the species range was not unusual, and the symptoms reported did not seem to be associated with any particular aspect of the airborne microflora, though it was interesting to note that *A. fumigatus* appeared in one air-conditioned building at 45 colony-forming units/m^3 air.

5.2. Fungal Volatiles

The study cited above (Skov *et al.*, 1987) listed volatile organic compounds as possible factors contributing to "sick building syndrome," and one source of these compounds is fungi. Health problems associated with the fungal production of volatile compounds within the built environment are now becoming increasingly evident (Samson, 1985). In countries such as Canada, the Netherlands, and Sweden, symptoms including headaches, eye, nose, and throat irritations, and fatigue have been reported, apparently connected with the presence of fungal odors; furthermore, while some people do not react to moldy smells, others become nauseated and may exhibit even more severe symptoms. Alcohols, aldehydes, esters, hydrocarbons, and aromatic compounds all contribute to fungal volatiles, and a wide range of fungi have been implicated in their production. Examples of fungal volatiles include 3-methyl-1-butanol, 3-octanone and 1-octen-3-ol (Abramson *et al.*, 1983; Tuma *et al.*, 1989), styrene, 2-pentylfuran, 3-methylanisole, 2-(2-fungal) pentanal, 2-ethyl-5-methylphenol and 2-methylacetophenone, 1-octanol (Abramson *et al.*, 1980), 1,5-octadien-3-ol, 1,5-octadien-3-one, 3-octanol,

3-methylisoborneol and 2-methoxy-3-isopropylpyrazine, 2-methylfuran, 2-methyl-2-propanol plus smaller quantities of alkanes and terpenes, 3-methylfuran, 1-propanol, 2-methyl-2-propanol; 3-methyl-1-butanol, 3-octen-2-ol, 1-octen-3-ol, acetic acid, and sesquiterpenes (Borjesson *et al.,* 1989). Genera noted in the aforecited studies for their ability to produce volatiles, mostly on grain or other foods or feeds, include *Aspergillus, Fusarium, Penicillium,* and *Alternaria*—all of which are found in the domestic environment. 3-Methyl-1-butanol and 1-octen-3-ol seem to be the most common fungal volatiles, though others such as 2-octen-1-ol and geosmin (1,10-dimethyl-9-decalol) are cited by Samson (1985) as being frequent sources of unpleasant earthy odors. The main message he conveys is that "research on moldy odors is worthwhile continuing" since there is much yet to be discovered, particularly in the indoor environments, in relation to ill health.

Of the aspergilli, *A. versicolor, A. repens, A. flavus, A. glaucus,* and *A. amstelodami* are well-known producers of volatiles, but inhalation toxicity data are either lacking or very poor. Certainly, aldehydes and terpenes can cause health problems, with the former often being irritants producing tissue damage at the site of application. Terpenes, too, can have similar effects, and both sets of compounds are potentially carcinogenic to mice and rats (Feron *et al.,* 1992).

6. CONCLUSIONS

Species of *Aspergillus,* as stated in Section 1, can be a danger to health through various routes. Tissue invasion, normally of the pulmonary region by the pathogenic *A. fumigatus,* is particularly hazardous, especially in the compromised host. Hypersensitivity reactions involving the immune system, again affecting the respiratory tract to different degrees of severity, are common and are generally caused by the presence of large amounts of various species. Aspergillosis can often be considered an occupational disease, particularly in the agricultural, food, and timber industries, and occasionally becomes systemic, with usually fatal results.

The production of mycotoxins has strong linkages, in the case of the aflatoxins, with liver disease and possible carcinomas, though firm evidence for the latter in humans is still lacking. Various studies have yielded positive linkages between the incidence of aflatoxicoses and liver disease, especially in individuals who have ingested appreciable amounts of aflatoxin-contaminated grain in tropical countries, but the evidence is generally circumstantial.

Attention is now being increasingly paid to the harmful effects of *Aspergillus* and its spores (together with other fungi) within the built environment, particularly in damp houses and workplaces, where many kinds of symptoms have been reported by the occupants, including wheeze, backache, headache, vomiting, fatigue, nervousness, and irritability. The problem of volatile organic compounds produced by fungi such as *A. versicolor* is worthy of further investigation, since these compounds too are potential health hazards and may well contribute to what has become known as "sick building syndrome."

Although much has been discovered about the medical relationship of *Aspergillus* over the past few decades, much evidence remains not proven. Clearly, more research is required concerning the epidemiology, biochemistry, physiology, toxigenicity, and immunology of these organisms in relation to human, rather than experimental animal, disease before firm conclusions can be made with a greater degree of confidence than at present.

REFERENCES

Abdel-Hafez, S. I. I., and Shoreit, A. A. M., 1985, Mycotoxin-producing fungi and mycoflora of air-dust from Taif, Saudi Arabia, *Mycopathologia* **92:**65–71.

Abdel-Hafez, S. I. I., Shoreit, A. A. M., Abdel-Hafez, A. I. I., and Maghraby, O. M. O., 1986, Mycoflora and mycotoxin-producing fungi of air-dust particles from Egypt, *Mycopathologia* **93:**25–32.

Abramson, D., Sinha, R. N., and Mills, J. T., 1980, Mycotoxins and odor formation in moist cereal grain during granary storage, *Cereal Chem.* **57:**346–351.

Abramson, D., Sinha, R. N., and Mills, J. T., 1983, Mycotoxin and odor formation in barley stored at 16 and 20% moisture in Manitoba, *Cereal Chem.* **60:**350–355.

Akiyama, K., Takizawa, H., Suzuki, M., Miyachi, S., Ichinohe, M., and Yanagihara, Y., 1987, Allergic bronchopulmonary aspergillosis due to *Aspergillus oryzae, Chest* **91:**285–286.

Al-Doory, Y., and Domson, J. F. (eds)., 1984, *Mould Allergy,* Lea and Febiger, Philadelphia.

Anderson, J. G., and Smith, J. E., 1987, Composting, in: *Biotechnology of Waste Treatment and Exploitation* (J. M. Sidwick and R. S. Holdom, eds.), Ellis Horwood, Chichester, England, pp. 301–325.

Anon, 1962, Progress in cancer research: potent liver toxin and carcinogen produced in groundnuts as a result of infection with the fungus *Aspergillus flavus, Br. Med. J.* **2**(5298)**:**172–174.

Austwick, P. K. C., Little, S. A., Lawton, L., Pockering, C. A. C., and Harrison, J., 1989, Microbiology of sick buildings, in: *Airborne Deteriogens and Pathogens* (B. Flannigan, ed.), Biodeterioration Society, Kew, Surrey, England, pp. 122–128.

Autrup, H., 1989, Exposure to aflatoxins in Denmark, *Toxin Rev.* **8:**53–67.

Bean, T. A., Yourtee, D. M., Akanda, B., and Ogunlewe, J., 1989, Aflatoxin metabolites in the urine of Nigerians: comparison of chromatographic methods, *Toxin Rev.* **8:**43–52.

Berry, C. L., 1988, The pathology of mycotoxins, *J. Pathol.* **154:**301–311.

Borjesson, T., Stollman, U., Adamek, P., and Kasperson, A., 1989, Analysis of volatile compounds for detection of molds in stored cereals, *Cereal Chem.* **66:**300–304.

Bosch, F. X., and Peers, F., 1991, Aflatoxins: data on human carcinogenic risk, in: *Relevance to Human Cancer of N-Nitroso Compounds, Tobacco and Mycotoxins* (I. K. O'Neill, J. Chen, and H. Bartsch, eds.), International Agency for Research on Cancer, Lyons, France, pp. 48–53.

Bullerman, L. B., 1986, *Mycotoxins and Food Safety,* Institute of Food Technologists, Chicago, pp. 4–5.

Burg, W. R., and Shotwell, O. L., 1984, Aflatoxin levels in airborne dust generated from contaminated corn during harvest and at an elevator in 1980: hazards to agricultural workers, *J. Assoc. Off. Anal. Chem.* **67:**309–312.

Burg, W. R., Shotwell, O. L., and Saltzman, B. E., 1981, Measurement of airborne aflatoxins during the handling of contaminated corn, *J. Am. Ind. Hyg. Assoc.* **42:**1–11.

Burg, W. R., Shotwell, O. L., and Saltzman, B. E., 1982, Measurements of airborne aflatoxins during the handling of 1979 contaminated corn, *J. Am. Ind. Hyg. Assoc.* **43:**580–586.

Burr, M. L., Mullins, J., Merrett, T. G., and Stott, N. C. H., 1988, Indoor moulds and asthma, *J. R. Soc. Health* **3:**99–101.

Canadian Public Health Association, 1987, *Significance of Fungi in Indoor Air: Report of a Working Group, Can. J. Public Health* **78**(2)**:**S1–S14.

CAST, 1989, *Mycotoxins: Economic and Health Risks,* Council for Agricultural Science and Technology, Ames, Iowa.

Clifford, J. I., Rees, K. R., and Stevens, M. E. M., 1967, The effect of the aflatoxins B_1, G_1 and G_2 on protein and nucleic acid synthesis in rat liver, *Biochem. J.* **103:**258–261.

Cohen, J., 1991, Clinical manifestations and management of aspergillosis in the compromised patient, in: *Fungal Infection in the Compromised Patient,* 2nd ed. (D. W. Warnock and M. D. Richardson, eds.), John Wiley & Sons, Chichester, England, pp. 118–152.

Conning, D. M., 1983, Systemic toxicity due to foodstuffs, in: *Toxic Hazards in Food* (D. M. Conning and A. B. G. Lansdown, eds.), Croom Helm, London, pp. 5–22.

De Coster, A., Dierck, P., and Grivegnee, A., 1988, Aspergilloma, in: *Aspergillus and Aspergillosis* (H. Vanden Bosche, D. W. R. Mackenzie, and G. Cauwenbergh, eds.), Plenum Press, New York, pp. 107–119.

Desai, M. S., and Ghosh, S. K., 1989, Aflatoxin related occupational hazards among rice mill workers, *Toxin Rev.* **8:**81–87.

Dvorackova, I., 1990, *Aflatoxins and Human Health,* CRC Press, Boca Raton, Florida.

Ebner, M. R., Haselwandter, K., and Frank, A., 1992, Indoor and outdoor incidence of airborne fungal allergens at low- and high-altitude alpine environments, *Mycol. Res.* **96:**117–124.

Evans, E. G. V., and Gentles, J. C., 1985, *Essentials of Medical Mycology,* Churchill Livingstone, Edinburgh.

Feron, V. J., Til, H. P., de Vrijer, F., and Van B Inderon, P. J., 1992, Toxicology of volatile organic compounds in indoor air and strategy for further research, *Indoor Environ.* **1:**69–81.

Flannigan, B., McCabe, E. M., and McGarry, F., 1991, Allergenic and toxigenic microorganisms in houses, in: *Pathogens in the Environment* (B. Austin, ed.), *J. Appl. Bacteriol. Symp. Suppl.,* No. 20, pp. 61S–73S.

Flegel, T. W., 1988, Yellow-green *Aspergillus* strains used in Asian soybean fermentations, *ASEAN Food J.* **4:**14–30.

Frisvad, J. C., 1988, Fungal species and their specific production of mycotoxins, in: *Intro-*

duction to Food-Borne Fungi, 3rd ed. (R. A. Samson and E. S. van Reenen-Hoekstra, eds.), Centraalbureau voor Schimmelcultures, Baarn, The Netherlands, pp. 239–249.

Grasso, P., 1983, Carcinogens in food, in: *Toxic Hazards in Food* (D. M. Conning and A. B. G. Lansdown, eds.), Croom Helm, London, pp. 122–144.

Gravesen, S., 1972, Identification and quantitation of indoor airborne micro-fungi during 12 months from 44 Danish homes, *Acta Allergol.* **27:**337–354.

Gravesen, S., 1985, Indoor airborne mould spores, *Allergy* **40**(Suppl. 3)**:**21–23.

Hay, R. J., 1988, Pulmonary aspergillosis—the clinical spectrum, in: *Aspergillus and Aspergillosis* (H. Vanden Bosche, P. W. R. Mackenzie, and E. Cauwenbergh, eds.), Plenum Press, New York, pp. 97–105.

Hayes, R. B., van Nieuwenheize, J. P., Raatgever, J. W., and ten Kate, R. J. W., 1984, Aflatoxin exposures in the industrial setting: an epidemiological study of mortality, *Food Chem. Toxic.* **22:**39–43.

Hendrickse, R. G., Maxwell, S. M., and Young, R., 1989, Aflatoxins and heroin, *Toxin Rev.* **8:**89–94.

Holtmeyer, M. G., and Wallin, J. R., 1980, Identification of aflatoxin-producing atmospheric isolates of *Aspergillus flavus, Phytopathology* **70:**325–327.

Hsieh, D. P. H., 1987, Modes of action of mycotoxins, in: *Mycotoxins in Food* (P. Krogh, ed.), Academic Press, London, pp. 149–176.

Humphreys, H., 1992, Microbes in the air—when to count! (The role of air sampling in hospitals), *J. Med. Microbiol.* **37:**81–82.

Hunter, C. A., Grant, C., Flannigan, B., and Bravery, A. F., 1988, Mould in buildings: the air spora of domestic dwellings, *Intern. Biodeterioration* **24:**81–101.

Institute of Environmental Health Officers, 1985, *Mould Fungal Spores—Their Effects on Health and the Control, Prevention and Treatment of Mould Growth in Dwellings,* IEHO, London, p. 13.

Jarvis, B., 1976, Mycotoxins in food, in: *Microbiology in Agriculture, Fisheries and Food* (F. A. Skinner and J. G. Carr, eds), Academic Press, London, pp. 251–267.

Kalayanamitr, A., Bhamiratna, A., Flegel, T. W., Glinsukon, T., and Shinmyo, A., 1987, Occurrence of toxicity among protease, amylase, and color mutants of a nontoxic soy sauce koji mold, *Appl. Environ. Microbiol.* **53:**1980–1982.

Kemper, R. A., White, W. C., and Gettings, R. L., 1990, Sustained aeromicrobiological reductions utilizing silane-modified quaternary amines applied to carpeting: preliminary data from an observational study of commercial buildings, *Dev. Indust. Microbiol.* **31:**237–244.

Kiran, U., Garcha, H. S., and Khanna, P., 1985, Fungal and actinomycetes spores in farm air during cereal crop operation, *Ann. Biol.* **1:**64–70.

Krishnamachuri, K. A. V. R., Bhat, R. V., Nagarajan, V., and Tilak, T. B. G., 1975, Hepatitis due to aflatoxicosis, *Lancet* **1:**1061–1063.

Kurup, V. P., Barboriak, J. J., and Fink, J. N., 1984, Hypersensitivity pneumonitis, in: *Mould Allergy* (Y. Al-Doory and J. F. Domson, eds.), Lea and Febiger, Philadelphia, pp. 216–243.

Lacey, J., 1989, Airborne health hazards from agricultural materials, in: *Airborne Deteriogens and Pathogens* (B. Flannigan, ed.), Biodeterioration Society, Kew, Surrey, England, pp. 13–28.

Land, C. J., Hult, K., Fuchs, R., Hagelburg, S., and Lundstroem, H., 1987, Tremorgenic mycotoxicosis from *Aspergillus fumigatus* as a possible occupational health problem in sawmills, *Appl. Environ. Microbiol.* **53:**787–790.

Lewis, C. W., Anderson, J. G., Smith, J. E., Morris, G. P., and Hunt, S. M., 1989, The incidence of moulds within 525 dwellings in the United Kingdom, *Int. J. Environ. Studies* **35:**105–112.

Mackenzie, D. W. R., 1988, *Aspergillus* in man, in: *Aspergillus and Aspergillosis* (H. Vanden Bosche, D. W. R. Mackenzie, and G. Cauwenbergh, eds.), Plenum Press, New York, pp. 1–8.

Martin, C. J., Platt, S. D., and Hunt, S. M., 1987, Housing conditions and ill health, *Br. Med. J.* **294:**1125–1127.

Maxwell, S. M., Apeagyei, F., de Vries, H. R., Mwanmut, D. D., and Hendrickse, R. G., 1989, Aflatoxins in breast milk, neonatal cord blood and sera of pregnant women, *Toxin Rev.* **8:**19–29.

May, J. J., Stallones, L., Darrow, D., and Pratt, D. S., 1986, Organic dust toxicity (pulmonary mycotoxicosis) associated with silo unloading, *Thorax* **41:**919–923.

Morris, G. P., Murray, D., Gilzean, I. M., Anderson, J. E., Lewis, C. W., Smith, J. E., and Hunt, S. M., 1989, A study of dampness and mould in Scottish public sector housing, in: *Airborne Deteriogens and Pathogens* (B. Flannigan, ed.), Biodeterioration Society, Kew, Surrey, England, pp. 163–173.

Moss, M. O., Jarvis, B., and Skinner, F. A. (eds.), 1989, *Filamentous Fungi in Foods and Feeds, J. Appl. Bacteriol. Symp. Suppl.*, No. 18.

Nolard, N., Detandt, M., and Beguin, H., 1988, Ecology of *Aspergillus* species in the human environment, in: *Aspergillus and Aspergillosis* (H. Vanden Bosche, D. W. R. Mackenzie, and G. Cauwenberg, eds.), Plenum Press, New York, pp. 35–41.

Olsen, J. H., Dragsted, L., and Autrup, H., 1988, Cancer risk and occupational exposure to aflatoxins in Denmark, *Br. J. Cancer* **58:**392–396.

Palmgren, M. S., and Lee, L. S., 1986, Separation of mycotoxin-containing sources in grain dust and determination of their mycotoxin potential, *Environ. Health Perspect.* **66:**105–108.

Pier, A. C., 1991, The influence of mycotoxins on the immune system, in: *Mycotoxins and Animal Foods* (J. E. Smith and R. S. Henderson, eds.), CRC Press, Boca Raton, Florida, pp. 489–497.

Platt, S. D., Martin, C. J., Hunt, S. M., and Lewis, C. W., 1989, Damp housing, mould growth, and symptomatic health state, *Br. Med. J.* **298:**1673–1678.

Pohland, A. E., and Wood, G. E., 1987, Occurrence of mycotoxins in food, in: *Mycotoxins in Food* (P. Krogh, ed.), Academic Press, London, pp. 35–64.

Pohland, A. E., Nesheim, S., and Friedman, L., 1992, Ochratoxin A: A review, *Pure Appl. Chem.* **64:**1029–1045.

Rati, E., and Ramalingam, A., 1979, Toxic strains among air-borne isolates of *Aspergillus flavus* Link, *Indian J. Exp. Bot.* **17:**97–98.

Robertson, M. D., Seaton, A., Milne, L. J. R. and Raeburn, J. A., 1987a, Resistance of spores of *Aspergillus fumigatus* to ingestion by phagocytic cells, *Thorax* **42:**466–472.

Robertson, M. D., Seaton, A., Milne, L. J. R., and Raeburn, J. A., 1987b, Suppression of host defences by *Aspergillus fumigatus, Thorax* **42:**19–25.

Robertson, M. D., Seaton, A., Raeburn, J. A., and Milne, L. J. R., 1987c, Inhibition of phagocyte migration and spreading by spore diffusates of *Aspergillus fumigatus, J. Med. Vet. Mycol.* **25:**389–396.

Rogers, A. L., and Kennedy, M. J., 1991, Opportunistic hyaline lyphomycetes, in: *Manual of Clinical Microbiology,* 5th ed. (A. Balows, W. J. Hausler, Jr., K. L. Herrmann, H. D. Isenberg, and H. J. Shadomy, eds.), American Society for Microbiology, Washington, D.C., pp. 659–673.

Samson, R. A., 1985, Occurrence of moulds in modern living and working environments, *Eur. J. Epidemiol.* **1:**54–61.

Samson, R. A., and van Reenen-Hoekstra, E. S. (eds.), 1988, *Introduction to Food-Borne Fungi,* 3rd ed., Centraalbureau voor Schimmelcultures, Baarn, The Netherlands.

Sargeant, K., Sheridan, A., O'Kelly, J., and Carnaghan, R. B. A., 1961, Toxicity associated with certain samples of groundnuts, *Nature (London)* **192:**1098.

Seaton, A., and Robertson, M. D., 1989, *Aspergillus,* asthma, and amoebae, *Lancet* **4573:**893–894.

Seeliger, H. P. R., and Tintelnot, K., 1988, Epidemiology of aspergillosis, in: *Aspergillus and Aspergillosis* (H. Vanden Bosche, D. W. R. Mackenzie, and G. Cauwenbergh, eds.), Plenum Press, New York, pp. 23–34.

Skov, P., Valbjorn, O., and the Danish Indoor Climate Study Group, 1987, The "Sick" Building Syndrome in the office environment: the Danish town hall study, *Environ. Int.* **13:**339–349.

Smith, J. E., 1985, Mycotoxin hazards in the production of fungal products and by-products, in: *Comprehensive Biotechnology,* Vol. 4 (M. Moo-Young, ed.), Pergamon, Oxford, pp. 543–560.

Smith, J. E., and Henderson, R. S. (eds.), 1991, *Mycotoxins and Animal Foods,* CRC Press, Boca Raton, Florida.

Smith, J. E., and Moss, M. O., 1985, *Mycotoxins: Formation, Analysis and Significance,* John Wiley & Sons, Chichester, England.

Sorenson, W. G., 1990, Mycotoxins as potential occupational hazards, *Dev. Indust. Microbiol.* **31:**205–211.

Sorenson, W. G., Jones, W., Simpson, J., and Davidson, J. I., 1984, Aflatoxins in respirable airborne peanut dust, *J. Toxicol. Environ. Health* **14:**525–533.

Sporn, M. B., Dingman, C. W., Phelps, H. L., and Wogan, G. N., 1966, Aflatoxin B_1: Binding to DNA *in vitro* and alteration of RNA metabolism *in vivo, Science* **151:**1539–1541.

Stevens, W. J., 1988, Allergic aspergillosis, in: *Aspergillus and Aspergillosis* (H. Vanden Bosche, D. W. R. Mackenzie, and G. Cauwenbergh, eds.), Plenum Press, New York, pp. 87–95.

Strachan, D. P., 1988, Damp housing and childhood asthma: Validation of reporting of symptoms, *Br. Med. J.* **297:**1223–1226.

Terao, K., and Ohtsubo, K., 1991, Biological activities of mycotoxins: field and experimental mycotoxicoses, in: *Mycotoxins and Animal Foods* (J. E. Smith and R. S. Henderson, eds.), CRC Press, Boca Raton, Florida, pp. 455–488.

Tuma, D., Sinha, R. N., Muir, W. E., and Abramson, D., 1989, Odor volatiles associated with mycoflora in damp ventilated and non-ventilated bin-stored bulk wheat, *Int. J. Food Microbiol.* **8:**103–19.

Ueno, Y., 1991, Biochemical mode of action of mycotoxins, in: *Mycotoxins and Animal Foods* (J. E. Smith and R. S. Henderson, eds.), CRC Press, Boca Raton, Florida, pp. 437–453.

van Griensven, L. J. L. D., van Loon, P. C. C., Cox, A. L., Folgering, H. T. M., Wuisman, O. P. J. M., and van den Bogart, H. G. G., 1991, Mushroom worker's lung as an occupational disease in the cultivation of edible fungi, in: *Science and Cultivation of Edible Fungi* (Maher, ed.), Balkerna, Rotterdam, pp. 317–321.

Vincken, W., and Roels, P., 1984, Hypersensitivity pneumonitis due to *Aspergillus fumigatus* in compost, *Thorax* **39:**74–75.

Warnock, D. W., 1991, Introduction to the management of fungal infection in the compromised patient, in: *Fungal Infection in the Compromised Patient,* 2nd ed. (D. W. Warnock and M. D. Richardson, eds.), John Wiley & Sons, Chichester, England, pp. 23–53.

White, W. C., 1990, An overview of the role of microorganisms in "Building-Related Illnesses" *Dev. Indust. Microbiol.* **31:**227–229.

Wicklow, D. T., and Shotwell, O. L., 1983, Intrafungal distribution of aflatoxins among conidia and sclerotia of *Aspergillus flavus* and *Aspergillus parasiticus, Can. J. Microbiol.* **29:**1–5.

Wilkinson, A. P., Denning, D. W., and Morgan, M. R. A., 1989, Immunoassay of aflatoxin in food and human tissue, *Toxin Rev.* **8**:69–79.

Yokotsuka, T., 1985, Fermented protein foods in the Orient, with emphasis on shoyu and miso in Japan, in: *Microbiology of Fermented Foods*, Vol. 1 (B. J. B. Wood, ed.), Elsevier, London and New York, pp. 197–247.

Young, R. C., Bennett, J. E., Vogel, C. L., Carbonne, P. P., and DeVita, V. T., 1970, Aspergillosis: the spectrum of the disease in 98 patients, *Medicine* **49**:147–173.

Yourtee, D. M., and Kirk-Yourtee, C. L., 1989, Aflatoxin detoxification in humans: laboratory, field and clinical perspectives, *Toxin Rev.* **8**:3–18.

Yourtee, D. M., Raj, H. G., Prasanna, H. R., and Autrup, H. (eds.), 1989, *Proceedings of the International Symposium on Agricultural and Biological Aspects of Aflatoxin Related Health Hazards, Toxin Rev.* **8**(1-2), Marcel Dekker, New York.

Zennie, T. M., 1984, Identification of aflatoxin B_1 in grain elevator ducts in central Illinois, *J. Toxicol. Environ. Health* **13**:589–594.

Species Index

Subject Index